PRIVATE, 1742.

GRENADIER, 1751.

GRENADIER. 1768.

OFFICER. 1792.

PRIVATE, 1832.

The 19th Foot waiting to fall in for the Funeral of the Duke of Wellington.

BANDSMAN AND DRUMMER. 1852.

OFFICER AND PRIVATE OF GRENADIERS, 1854.

COLOURS PRESENTED BY QUEEN ALEXANDRA
IN 1875, WITH RUSSIAN DRUMS
TAKEN AT THE ALMA.

Officer and Regimental Colour, 1875.

Photograph by W & D. Downey.

Alexandra

A

History of the Services

OF THE

19th Regiment,

NOW

ALEXANDRA, PRINCESS OF WALES'S OWN

(YORKSHIRE REGIMENT),

FROM ITS FORMATION IN 1688 TO 1911.

BY

MAJOR M. L. FERRAR,

LATE 19TH FOOT.

Author of "With the Green Howards in South Africa."

London:

EDEN FISHER & COMPANY, LTD.

Printed and bound by Antony Rowe Ltd, Eastbourne

PREFACE.

WITH the exception of Cannon's little sketch of the services of the Nineteenth Foot, no history of the regiment has ever been published, and I have therefore written this book in order to amplify and correct Cannon's account, as well as to fill up the gap extending from the year 1848 to the present time.

I may explain here that in April, 1893, the regiment started their monthly journal, "The Green Howards' Gazette," one of its objects being to collect material for the publication of this history, and much that is herein recorded is taken from its pages.

I regret to say that the regimental Digest of Service is most incomplete, and though events of varying interest are detailed since 1796, they are passed over very briefly, and appear to have been written up many years after this date. It has been said that the regimental records were lost at Monk's Corner, when the baggage of the Nineteenth was captured by the Americans in 1781, and it has also been affirmed that they were lost in the wreck of the "Arniston" off the Cape of Good Hope in 1815, when they may have been sent home for safe keeping. The fact, however, remains that I have had little or no orderly room records to work on for the first 180 years of the regiment's existence, and have had to glean my information from various other sources.

My thanks are due in the first place to Colonel J. W. R. Parker for kindly placing at my disposal his valuable notes on the earlier history of the regiment, and to the late Major G. Lidwill, who gave me much information as to the Crimean War. I am also indebted to Colonel G. E. Langford for his account of the Hazara Campaign, and to Lieutenant-General Franklyn for that of the Tirah Expedition, both of which have already appeared in our Gazette.

Mr. Charles Dalton has been most kind in permitting me to make full use of his notes in his Army Lists and Commission Registers, which contain much that is of interest.

The coloured illustrations were painted twelve years ago by Mr. R. Percival Reynolds, under instructions as to details of uniform from the late Mr. S. M. Milne, who was the first authority in England on such matters.

I am greatly indebted to the officers of both battalions of the Green Howards for their ready support in the production of this book. Its preparation has given me a great deal of pleasure, which I hope will be shared by those who glance over its pages.

Lastly, I owe my gratitude to the publishers, Messrs. Eden Fisher & Co., for their unfailing courtesy at all times. No firm could have been more obliging, and they have taken the greatest interest in the work of publication.

M. L. FERRAR.

BELFAST,

28th October, 1911.

CONTENTS.

ILLUSTRATIONS.

CHAPTER I.

The Revolution of 1688—Colonel Francis Luttrell raises the Regiment—Thomas Erle appointed Colonel, 1690—Battles of Steenkirke and Landen—Precedence of Regiments—Siege of Namur—Dress of the Regiment—List of Officers, 1689.

THE arbitrary proceedings of King James II. during the first three years of his reign had filled England with alarm. The rebellion of the Duke of Monmouth in May, 1685, had furnished him with a pretext for augmenting the standing army, which he had maintained at a high establishment in times of peace and increased without the consent of Parliament. Indeed, finding both Houses reluctant to further his design of subverting the reformed religion, he had dispensed altogether with Parliament and aspired to absolute autocracy. All who declined to fall in with his views were removed from office, and their places filled—in defiance of the rigid provisions of the Penal Laws and Test Act—by Roman Catholics.

These high-handed measures alienated the affections of his subjects; many of the most influential and patriotic of whom—seeing the grave danger to which the Constitution was exposed—made overtures to the King's nephew and son-in-law, William Prince of Orange, a staunch Protestant, soliciting his intervention to save the country.

In response to this invitation a fleet of 600 sail was equipped by the States of Holland—then masters of the Channel—and a descent was made—where least expected—on the Devon Coast. On the 5th November, 1688, with a force of over 14,000 men, including six British regiments then in the Dutch service, William of Orange landed at Torbay.

On November 9th he entered Exeter with his army, and a Declaration of the objects of his expedition was read in the Cathedral. The country round flocked to his standard; three regiments of horse deserted King James in a body and joined the Prince, their example being followed by a large number of the King's troops then assembled at Salisbury.

One of the first among the Somerset gentry who rallied round the Prince of Orange was Colonel Francis Luttrell, of Dunster Castle, the head of an ancient and honourable family in that county and Member of Parliament for Minehead, which he had represented since 1679, being at that time twenty years of age. In 1681 he was appointed by the Earl of Winchelsea, Lord-Lieutenant of Somerset, to be Colonel of a regiment of foot in succession to Sir Halswell Tynte, and he was in command of the local forces when the Duke of Monmouth landed at Lynne in 1685. He was, however, obliged to evacuate Taunton on the approach of the Duke, who there assumed the title of King.

In the latter part of the short reign of James II. Francis Luttrell was no longer to be reckoned with as one of his supporters. In 1687 he declined to vote for the repeal of the penal laws, and was one of the first men of importance to join the standard of the Prince of Orange at Exeter in November, 1688.[1]

The high esteem in which Colonel Francis Luttrell was held, and the great reliance placed on him by William of Orange, is evident from the fact that he was at once entrusted by that Prince with a commission to raise a regiment of foot for his service. Nor was the trust misplaced. Somerset folk had not forgotten the "Bloody Assize" after Monmouth's

1 A History of Dunster, by Sir H. Maxwell Lyte.

defeat, when their relations and neighbours had been strung up by scores or sold into slavery by command of King James; and the popularity of the Colonel and the cause which he espoused enabled him in a very few days to place at the disposal of the Prince a strong battalion of enthusiastic supporters from Somerset and Devon.

The establishment was completed by the 19th November, on which day, as a contemporary chronicle relates, "The Prince of Orange left the City of Exeter with his army, and left Mr. Seymor, Governour, with Colonel Luttrell's Regiment to secure it."

The corps thus embodied took precedence as the first regiment raised in England after the landing of William of Orange, and was known in after years as the Nineteenth Foot. Cannon, in his history of the regiment, alleges that it was formed in February, 1689, by the incorporation of several detached companies raised in November, 1688: that such was not the case, and that the establishment was complete on the 19th November, 1688, has been shown above; but all doubt on the matter is ended by the following statement by Colonel Luttrell himself, in a petition to William of Orange after his accession to the throne.[1]

"To the King's Most Excellent Majestie.

"The humble Petition of Coll. Francis Luttrell in "behalfe of himselfe and the rest of the Officers in "his Regiment.

"Sheweth,

"That at your Majestie's happy arrival in this Kingdom "you were graciously pleased to give a Commission to "your Petitioner for raising a Regiment which He "compleated in three days time, and kept the same "fourteen days on his own expence; And that it might "be ready to march where your Majestie's service

1 Extract from the "Green Howards' Gazette," by Major J. W. R. Parker.

"required, your Petitioner clothed the said Regiment at "Exeter, which cost neare £1,500: in which he was so "extreamly abused (the Clothes being already wore out) "that it is absolutely necessary for your Majestie's "Service to cloath a new.

"Your Petitioner therefore most humbly Prayeth "your Majestie That Whereas there still remains due "to the said Clothiers a considerable sume and that its "necessary for the Regiments imediate Clothing; to "direct the Paymaster General to cleer the said Regiment "from the beginning of January last to the first of July "(1689), which will be of great use to your Petitioner "towards Paying for the Old Cloaths, and what remains "then of that sume He is willing to pay out of his own "Estate, so that the Twopences from that time may goe "towards the new Clothing with which and his own "obligaçon (which he is willing to give) He hopes to "have the Regiment in a short time clothed and fitt to "march where your Majestie's Occasions may require.

"And your Petitioner shall ever pray, &c."[1]

In the summer of 1689 the regiment marched to Portsmouth, and was later on stationed in the Isle of Wight. On being relieved by Talmash's[2] Regiment it embarked for Plymouth on the 9th August, and on arrival there took up its quarters in the Citadel with the exception of three companies which were detached to Penryn.

The authority of King William being resisted in Ireland, an army was sent there under Marshal Duke Schomberg, and on the 11th March Luttrell's Regiment received orders to send 520 men to Bideford, there to embark for Belfast, to reinforce the army which had sustained heavy losses in the unhealthy camp at Dundalk. Of this draft 170 men were detailed for Colonel Brewer's[3] Regiment and 350 for

---

1 Treasury Papers.

2 The 5th Foot Northumberland Fusiliers.

3 12th Foot, now the Suffolk Regiment.

COLONEL FRANCIS LUTTRELL.

FROM A PAINTING IN DUNSTER CASTLE.

Colonel Herbert's.[1] Of the 170 there were 74 for Brewer's and the remaining 96 were to be at the disposal of the Duke of Schomberg.[2]

The regiment was unfortunate in losing its Colonel during the summer, as he died at Plymouth on the 25th July, 1690, at the age of thirty-one. His body was brought to Dunster for interment, and the then considerable sum of £300 was spent on his funeral. The hatchment painted on this occasion is still in existence.[3]

The vacancy was not filled for some months till eventually Colonel Thomas Erle was selected for the appointment.[4] This officer had already distinguished himself in the Irish War at the head of his newly raised battalion, and now with both regiments under his command was soon to become still better known.

Colonel Erle's appointment to the command was a great disappointment to the late Colonel's brother, Alexander Luttrell, who held a commission as Captain in the regiment at the time. Naturally he had some claim to succeed his brother, for the latter had not only raised the regiment, but had spent much money on it. When therefore the command was given to another, this Captain Luttrell and several other officers resigned their commissions.[5]

Probably one of these was the Lieutenant-Colonel, William Norcott,[6] for we find that Major Henry Hawley was gazetted to the Lieutenant-Colonelcy on the 15th April, 1691. He

1 23rd Foot, now the Royal Welsh Fusiliers.

2 War Office Marching Order Book.

3 History of Dunster, by Sir H. Maxwell Lyte.

4 Whitehall, 8th January, 1690/91. "Colonel Erle has the command of Luttrell's late Regiment, as well as his own; he has no commission, but only an order to command it." Clark MSS.

5 Papers at Dunster Castle, Somerset.

6 Appointed Colonel of a Regiment of Foot, 16th February, 1694, which was disbanded in 1697.

had joined the Coldstream Guards in 1670 as Ensign, and eight years later had been transferred to the Duke of York's Marines, to be afterwards promoted Major and Captain in Luttrell's new regiment. He was a half-brother to Colonel Erle, and this may have accounted for his advancement.

Whilst the Irish battalion was fighting its first campaign the other remained in England, and no doubt sent reinforcements from time to time to its sister battalion. All this year it was stationed at Plymouth and places adjacent, but on war being declared against France, it received orders to join the Confederate Army commanded by the King of England in person, whose efforts were directed to arrest the progress of aggression pursued by Louis XIV.

Consequently, it embarked at Plymouth on about the 24th February, 1692, but bad weather detained the transports, and it was not for some days that they arrived at Ostend.

On landing there Erle's battalion had its first brush with the enemy, as they were attacked by a French party, who, however, were received very warmly, losing about forty killed and as many prisoners, which caused them to retire in disorder. Mr. Blaythwayte (Secretary for War), in a letter to the Queen, said: "Colonel Erle's Regiment, commanded by Lieutenant-Colonel Freke,[1] landed at Ostend, and going thence to Bruges, the French on notice thereof laid an ambuscade for them, of which the English having advice laid another for them. The French, 4,000 horse, besides foot, had 150 killed, with their chief officer taken, and 50 taken with many horses, etc., which were sold, and the money distributed among the soldiers."[2]

1 Lieutenant-Colonel Robert Freke. He was Captain and Lieutenant-Colonel of Erle's Irish Battalion. It is possible he may have been attached to this battalion for duty, and being senior to Lieutenant-Colonel Hawley, he may have taken command or the latter may have been absent at the time.

2 Luttrell's Diary.

It was not till August that a general advance was made against the French, who were drawn up in a strong position near the village of Steenkirke, twenty miles south-west of Brussels. It was a close and intersected country, where cavalry could only act with great difficulty.

Erle's battalion marched with the main body, most of the men of which, with the exception of those who had fought in the Irish War, had never yet seen a shot fired till their landing at Ostend.

The English Guards and the Danes led the attack in the face of a withering fire from the French, who were posted behind hedgerows on the slope of a hill. There was heavy fighting here, and the French were pushed back by the bull-dog courage of the British troops. The latter were almost in the French camp, and it was imperative that the Duke of Luxembourg, their commander, should make a grand effort to revive the failing spirits of his soldiers and drive the English back ere reinforcements came to their assistance.

The picked regiments of the army were consequently hurried to the front. "The brigades of the French and Swiss Guards came gaily to the charge, led by the Dukes de Bourbon, de Chartres, and the Princes de Conti and Turenne. Not a musket was fired by these regiments. With pikes advanced and swords drawn they came on against the weak line of redcoats, now reduced to half its original strength." The struggle was man to man, but the numbers pitted against them were too much for the Allies, and matters were desperate. Several messages were sent to Count Solmes, who commanded the reserves, asking for support. It is said he refused, saying, "Damn them (the English), they are very fond of fighting; now let them have a bellyful of it."

King William very soon saw that all was over, and made

arrangements for a retreat, which was admirably conducted in the greatest order, and at length the long day closed.

William's errors lay, firstly, in not obtaining more accurate information as to the nature of the ground,[1] which was believed to be much less broken and difficult than it proved, and, secondly, in permitting his advanced guard to get too far to the front. Thirdly, in having a large body of cavalry in front over ground which was quite unsuitable to them, and thereby impeding the action and movements of the infantry.

"The English soldiers were furious with Count Solmes. They declared that his behaviour was too strange to be accounted for, except by the presumption of malice and selfishness. Even the King himself was so disgusted that he could not bear the sight of the man for months afterwards."

The Allies lost nearly 2,000 killed and over 3,000 wounded, and it was, for the numbers engaged, one of the most bloody on record, as the total of the whole force was only about 15,000 men.

The exact loss of Erle's is not known, but the following return will give some idea of the state of the regiment after the battle.

[2]"List of soldiers, sick, deserted, etc., au combat de Steenkirke from 1 June to 20 Sept., 1692:—

BRIGADE DE CHURCHILL, REGMT. DE ERLE.

Present 169, absent 116, deserted 21, morts 5, prisonrs. 1, blessès 7, tuès 20, requéris 56, enrollés 11."

The army now, according to the usual custom of those days, went into winter quarters in Flanders, and in May the following year took the field again, the English getting orders to concentrate near Brussels. King William shortly

1 Colonel Erle, who was a Member of Parliament, said, in the House of Commons, that this was the chief cause of defeat, 21st November, 1692.

2 King William's chest.

afterwards broke up his camp and advanced to Parck, near Louvain, where he entrenched himself so as to command the roads to Brussels and Liege.

"The traveller of to-day who takes his seat at Brussels in the Cologne express rushes as he approaches Liege through two little stations known as Neerwinden and Landen. If by chance he consults his guide-book he will discover that somewhere here in the year 1693 there was fought a great battle between the French and the Allies. That perhaps is too common an announcement in this land of battles to excite his curiosity, but if he enquires further he will learn that this battle was one of the three bloodiest in the whole history of the "cockpit," and remained the most bloody of them all up to the day of Malplaquet."[1]

When the news came on the 14th July that the garrison at Huy had surrendered and that Luxembourg was threatening Liege, the King broke up his camp at Parck and marched in the direction of the latter town, where he was soon confronted by the French, who had swung round past Liege in order to give him battle. The Allies took up a position with their right on the hamlet of Laer, terminating on the left at Romsdorf and Neerlanden, situated respectively on the left and right banks of the River Landen, a little above the village from which that stream takes its name. Romsdorf was held by Brigadier-General Erle's brigade, which consisted of his own battalion, besides Tidcomb's,[2] Collingwood's,[3] and the Earl of Derby's.[4]

Erle had been lying sick of fever at Louvain, when he learned that an action was imminent. He at once rose from

---

1 Landen, by Frederick Dixon.
2 The 14th Foot, the Prince of Wales's Own West Yorkshire Regiment.
3 Disbanded.
4 The 16th Foot the Bedfordshire Regiment.

his bed, and managing to ride to camp, he set a splendid example to his brigade, and was dangerously wounded at the head of it when Romsdorf was attacked.[1]

It was late in the afternoon of the 18th July, when all the troops had got into position and long after the enemy's camp fires had died down, the noise of entrenching tools could be heard along the Allies' front. Thirty men from each regiment were acting as pioneers, and working hard to make the position secure. King William having ridden round his posts a little before midnight, ordered his coach to be brought in rear of Erle's brigade, and there he passed the night.

In the morning the Allies' position was a very strong one. The villages, surrounded after the Flemish custom by moats and fences, and the gardens separated by high mud walls, could only, the King was confident, be carried at terrible cost. It was therefore along the great stretch of open between Neerwinden and Romsdorf that he had concentrated his efforts. The two villages had been linked together by a breastwork strengthened by bastions, the guns in which swept the entire length of the ditch. So complete was its construction that even the most reckless of the French officers, as they scanned it with their field glasses, realised that a day of extraordinary peril was before them.

Luxembourg himself would never have willingly attempted such a position by direct attack, but he decided to pit numbers against entrenchments, and trust to the superior discipline of his army. It was six o'clock when he advanced against the centre, and the full fury of the Allies' batteries was concentrated on the approaching Frenchmen. "For a time the deluge of iron splashed down on the living sea rolling in on the entrenchments, as idly, seemingly, as if it was the ocean.

1 D'Auvergne's History.

Then suddenly the drums and trumpets sounded the 'retire.' The human tide surged backwards out of gunshot, strewing the green strand as it receded with a bloody wrack of mangled men and horses."

Seeing the strength of the centre, Luxembourg now determined to develop the attack on the wings, and massed thirty battalions and forty-eight squadrons between Laer and Neerwinden, where the Allies had only twenty-two. The French, fighting splendidly for a while, carried everything before them, but the Allies being reinforced by the cavalry of the right wing, were rallied, and the English brigade burst back into Laer. "They hurled the French backwards down the narrow streets, chasing them through the gardens, till with a furious rush they flung them over the last fence."

Meanwhile an even more furious struggle had been raging for the possession of Neerwinden. It was held by nine battalions, made up of Hanoverians and English, and was attacked by twenty-four. As the long line swung into the open the Allies could plainly see the uniforms of two battalions of a regiment known as the King's Foot Guards, who had been borne on the Irish Establishment of the Stewarts, and had already won distinction in the Irish War. On the fall of Limerick this regiment had joined Sarsfield, the Irish commander, and after a glorious career on the continent in the service of France, was a century later in the frenzy of the Revolution to lose its identity in becoming the 92nd of the French line.[1]

But the fire they faced was too terrific, and they began to waver and give way. Many of their leaders were killed or wounded, amongst them being the gallant Sarsfield, who died of his wounds at Huy.

---

1 The Battle of Landen, by Frederick Dixon.

About this time Romsdorf was also attacked. This village, as we have before mentioned, was held by Erle's Brigade, and the 14th Foot, from its casualties, appears to have had the hottest work. The brigade was being gradually driven back through the village, which was on the point of being lost, when the King with a little band of officers suddenly appeared, and the whole aspect of the fight changed. "He came into the village with his blue ribbon streaming across his chest, the great Star of the Garter blazing on his breast. When his officers besought him to remove these insignia, he replied, with a laugh, that if they were a good mark for the enemy, they were equally good for his own men to rally on, and rally on them they did. Drawing his sword he headed the rush of pikes across the gardens, and the little phalanx of redcoats led in this way became invincible. The whole mass of French gave way, with the British cheering wildly on their heels."

He was in the act of thanking them when news came of a fresh disaster on the right. By three o'clock no less than fifty battalions, supported by an enormous force of cavalry, had been ranged round Neerwinden, and the attack began as the King left the village.

The scene within Neerwinden was growing momentarily in horror. Four thousand corpses filled the ditches, covered the beds and paths in the gardens, and choked the narrow streets. The little band of defenders was shrinking every moment; the Hanoverians had been crushed, and the Dutch had lost their colours. The troops in and about Neerwinden had at last to give way to force of numbers, and the allied position was forced.

At Laer the final struggle had been fought with true British determination, but by remaining too long the retreat became difficult. The French infantry pushed the British regiments in front, the cavalry galled them in flank and cut them off in rear from the rest of the army.

The King himself remained in front of the foe, making what stand he could with broken ranks in order to get time for the retreat across the river. As he rode away to bring up from the left the English cavalry as a last resource, charge after charge was executed by them, and it was only numbers that enabled the French cavalry to stand against them. The King did all he could, and more than most men would have accomplished. Having seen General Talmash safely off with his left and centre troops, he continued to cover the passage of the river.

The loss was enormous on either side. 140,000 had paraded for action in the morning. Half of these had not been engaged, yet the number in killed alone was certainly not short of 25,000, of whom about 9,000 were Frenchmen.

The news of the victory was received in Paris with acclamation. All the city lined the streets to see the captured banners carried to Nôtre Dame. There were many, however, who declared that a few more of such victories would complete the ruin of the country. The truth was that only the generals at the front knew how dearly the conquerors had paid for their success. Namur was one vast hospital. For days the monks never ceased trampling through the town with the "Host" in their midst; to the very end of summer the open windows along the streets were filled with crowds of convalescents. When the sun of the 20th July rose over the field of the previous day's battle, even those most hardened to the horrors of war shuddered at the sight. All along the hillside throughout the gardens, the hop grounds and the villages the corpses lay like newly mown hay. Neerwinden itself was a city of the dead, wherein five thousand bodies lay waiting burial. Next summer a thick carpet of crimson poppies spread itself in a vivid blaze along the historic slope, and the curious visitor learned from the peasants of the roadside that the flowers had been watered by the blood of twenty-five thousand men.[1]

The casualties amongst the rank and file are not known by regiments in this battle. The returns only state officers. Brigadier-General Erle is mentioned as having been wounded and Captain Charles Barrington is amongst the killed.[2]

After the campaign of 1693 Erle's battalion passed the winter in quarters at Malines.

King William left England in May to take command of the Confederate Army, and in the beginning of the month the regiment left its quarters and pitched its tents near the cloister of Terbanck. The King was most anxious to force the French to an engagement, but this they always avoided, and the year passed away with both armies manœuvring and watching one another.

---

1 The Battle of Landen, by Frederick Dixon. Lord Macaulay, quoting from a letter from Lord Perth to his sister, 17th June, 1694, states in his history: "The next summer the soil, fertilised by 20,000 corpses, broke forth into millions of poppies. The traveller who on the road from St. Tron to Tirlemont saw that vast sheet of rich scarlet spreading from Landen to Neerwinden, could hardly help fancying that the figurative prediction of the Hebrew prophet was literally accomplished, that the earth was disclosing her blood and refusing to cover the slain."

2 The Rawdon Papers.

King William ordered a medal to be struck in commemoration of the battle of Landen, of which an illustration appears opposite. When a heron is hard pressed it throws itself upon its back in the air, and receives the falcon upon the point of its beak. King William is the heron who so severely punished the Duke of Luxembourg when he attacked him at Landen that he was unable to prosecute the original objects of the campaign, and remained inactive. Thus, though Luxembourg was the conqueror of the moment, his plans were defeated by the courage and skill of William. The medal is a rare one, and a specimen in silver is in the officers' mess of the 1st Battalion.—M.L.F.

MEDAL STRUCK BY LOUIS XIV.

TO COMMEMORATE THE VICTORY OF STEENKIRK.

MEDAL STRUCK BY WILLIAM III.

TO COMMEMORATE THE BATTLE OF LANDEN.

Many long and tiring marches were performed in Flanders and Brabant, and in the autumn Erle's battalion returned to the pleasant town of Malines,[1] where it passed the winter in garrison with Churchill's[2] and Trelawney's[3] regiments, with whom it had been brigaded since the battle of Landen, under command of Brigadier-General Erle.

A Board of General Officers was assembled by order of William III. at Gemblours camp, Flanders, on the 10th June, 1694, to fix the precedence of the various regiments of the army, a subject which had been the source of many disputes.

On the recommendation of the Board, the English regiments were directed to reckon their seniority according to date of formation, and Scotch and Irish regiments from the dates of being first placed on the English establishment. This gave Erle's Regiment nineteenth place in order of precedence, and on this principle the number was fixed by the warrant issued by King George II. on the 1st July, 1751, when the numerical titles were adopted for all regiments.

Early in the spring of 1695 this battalion of Erle's marched to the vicinity of Ghent, and was encamped near Marykirk until the army took the field.

The Duke of Luxembourg, who had commanded the French army at the battle of Landen, died during the winter, and was succeeded by Marshals Boufflers and Villeroy.

When King William suddenly drew off the main part of his army to invest Namur he left 30,000 men, under the Prince of Vaudemont, to cover the operations of his force, and amongst the regiments so employed was Erle's.[1]

Marshal Boufflers now shut himself up in Namur, and Marshal Villeroy remained to give battle to the Prince of

1 Cannon's History.
2 3rd Foot the Buffs.
3 4th Foot the King's Own Lancaster Regiment.

Vaudemont. But the latter managed his forces with so much skill, covering all the places on which the French had an eye, that even though Villeroy had twice his strength he could never bring the Prince to an engagement or gain any advantage over him. The officers who served under the Prince magnified his conduct highly, and compared it to anything the greatest general of the age had done, and the officers and men of Erle's Regiment no doubt were proud enough of the part they took in such masterly operations. Villeroy's army towards the close of the campaign had greatly increased in numbers, and had been augmented to 100,000 men, with which large force he finally appeared before Namur.

In the meantime the town surrendered, but the Castle, a strong fortress situated on a rock, still held out. The final assault took place on the 20th August, 1695, when the Bavarians and Prussians, in company with 3,000 British troops, under Lord Cutts, attacked the citadel, which surrendered after heavy fighting.

This event terminated the campaign, and Erle's battalion passed the winter at Dendermond.

As various regiments from the covering army reinforced those taking part in the siege from time to time, there is reason to believe that Erle's was one of these, as a return has been recently found at the Record Office giving a list of the regiments employed in the siege with their casualties, in which Erle's is stated to have had eight men killed and eleven wounded.

It will be of interest to mention something as to the dress and equipment of the officers and men at this time.

Each company of infantry (grenadiers excepted) consisted of about fifty men, made up of pikemen and musketeers. The grenadiers were armed with firelocks, and are said to have used cartridges and had sword bayonets with a pouch for grenades. They also carried hatchets, with which after firing they were on the command "fall on" to rush upon the enemy. Each foot soldier carried a sword, each pikeman a pike 16 feet long, and each musketeer a musquet. The barrels of the latter were about 4 feet long, and carried a ball, fourteen of which went to a pound. All infantry officers carried the spontoon, those of the flank companies having fusils. In 1695 the coats and breeches of the sergeants and men were of a grey or red colour; the coats of the drummers purple with grey breeches. The head-dress was a sort of low-crowned helmet with a brim or a skull cap. The former was sometimes bright and sometimes painted black, and as a distinction officers wore feathers.

We shall now take leave of this battalion for the present, and turn our attention to the junior one, which had already seen much service in the Irish War.

Authorities—D'Auvergne's History; Brigadier-General Richard Kane's Campaigns; Clifford Walton's History of the British Army; An Account of the Battle of Landen, by Frederick Dixon; Fortescue's History of the Army; Lord Macaulay's History; Cannon's History; War Office Records; English Army Lists and Commission Registers, by C. Dalton.

OFFICERS OF THE REGIMENT, 1689.

Colonel: Francis Luttrell.

Captains:

Francis Luttrell, Colonel.
William Norcott,[1] Lieutenant-Colonel.
Henry Hawley,[2] Major.
Robert Carey.[3]
Edmund Bowyer.[4]
Walter Vincent.[5]
William Coward.[5]
Charles Barington.[7]
Joseph Prigman.[5]
Thomas Finch.[9]
Baldwin Mallett.[10]
Hopton Windham.[11]
Alexander Luttrell.[12]

Lieutenants:

John Simmonds,[13] Captain-Lieutenant.
Walter Shepcott.
Wm. Lee.
William Webb.
John Redmore.[14]
John Calmady.[15]
William Willoughby.
John West.[16]
Nicholas Summers.
Richard Williams.
John Dodington.[17]
George Prater.[18]

Ensigns:

Joseph Lewis.
Robert Norcott.[19]
Abraham Hancock.[20]
Thomas Risdon.[21]
John Sidenham.[22]
—— Gregor.
Laurence Coward.[23]
Thomas Adams.
Thomas Robinson.[24]
Capel Stocker.[25]
Thomas Freake.[26]
—— Ensate.

Adjutant: John West.

Quarter-Master: Arthur Balson.

Chirurgeon: Theophilus Allen.[22]

NOTE.—Nearly all these annotations are from Army Lists and Commission Registers, by C. Dalton, F.R.G.S.

1 Luttrell, writing under date 17th November, 1688, says in his diary: "Several gentlemen of Devonshire, we hear, are gone to the Prince as Sir Fras. Drake, Sir Wm. Drake, Mr. Wm. Cary, Major Norcott (sic), Capt. Barrington, &c." Was appointed Colonel of a regiment of foot 16th February, 1694, which was disbanded in 1697.

2 Left the regiment about 1708. Appointed Lieutenant-Governor of Kinsale, and died there in 1724. He was brother to Colonel Fras. Hawley, of the Princess Anne's Dragoons, who fell at Steenkirke.

3 Promoted Major 14th May, 1695. Served at Cadiz and the West Indies. Died on service at Guadeloupe in 1703, at which time he had the rank of Lieutenant-Colonel.

4 Major of Colonel Erle's Regiment of Foot 23rd February, 1694. Commanded one of the independent companies incorporated into Luttrell's Regiment in November, 1688.

5 Commanded one of the independent companies incorporated into Luttrell's Regiment in November, 1688.

7 Brigade-Major in Holland 1st July, 1693. Killed at the battle of Landen.

9 Out before 1694.

10 Out before 1694. Eldest son of Sir Jno. *Malet,* by Florence, daughter of Jno. Wyndham, of Orchard Portman, co. Somerset.

11 Out 1st March, 1690. Grandson of Jno. Wyndham, by Catherine, sister and co-heir of Ralph, Lord Hopton.

12 Younger brother of Colonel Fras. Luttrell. Appointed Lieutenant-Colonel of Colonel Wm. Northcote's Regiment of Foot 16th February, 1694. Lieutenant-Colonel of Geo. Villier's Regiment of Marines 12th February, 1702. Colonel of same 6th February, 1703. Died on 22nd September, 1711.

13 Captain in Erle's 29th December, 1692. Probably the Captain *Jno. Symonds,* who was placed on half-pay before 1700, but reappointed Captain in Erle's Regiment 13th February, 1702. Major before 1703. Served in the Cadiz Expedition and in the West Indies.

14 Captain in Erle's 1st January, 1694. Out of the army before 1702.

15 Captain in Erle's 9th February, 1691. Out of the regiment before 1694.

16 Captain-Lieutenant in Erle's 1st March, 1690. Captain 9th February, 1691. Out before 1702. Served in Flanders, 1693-4. Died in October, 1701.

17 Appointed Adjutant 19th April, 1693. Out of the regiment in 1694.

18 Promoted Captain 1st March, 1690. Served in Flanders.

19 Appointed Quarter-Master 1st April, 1690.

20 Promoted Lieutenant 9th February, 1691. Left the regiment before 1st February, 1693.

21 Promoted Lieutenant 9th February, 1691. Out in 1702. Served in Flanders.

22 Serving as Lieutenant in 1702. Out in 1705. Served in the West Indies in 1702.

23 Lieutenant 1st April, 1690. Out before 1702. Served in Flanders, 1694.

24 Lieutenant 4th March, 1693-4. Serving in 1706. Was at Cadiz and the West Indies 1702.

25 Lieutenant 20th June, 1691. Served in the Brest Expedition, and was killed in Camaret Bay. Widow granted a pension of £30.

26 Son of Robert Freke of Upwey. Died 1698.

27 Put out of his commission by Colonel Luttrell in 1690. Presented a petition to Queen Mary for the restitution of money he had spent on medicines, etc.

CHAPTER II.

Colonel Thomas Erle raises a Battalion—It goes to Ireland—Takes part in the Battle of the Boyne, Siege of Athlone, Battle of Aughrim, Sieges of Galway and Limerick—Present at the attack on Brest, 1694—Disbanded 1697—List of Officers.

NOT long after William III. came to the throne he set about increasing the army, so as to take the field in Ireland against the late King, who had succeeded in gathering together a large force in that country, formed from his own adherents as well as from the Irish Roman Catholics.

Accordingly fourteen new regiments were raised early in the year 1689, only three of which exist at the present day, namely, the old 22nd, 23rd and 24th Foot, all the others having been disbanded. One of these latter battalions, however, raised by Colonel Erle on the 8th March, did not altogether lose its identity, for it was merged into the battalion raised by Luttrell, and the officers and men then incorporated belonged to Erle's Regiment during the whole if not the best part of their service.

Colonel Thomas Erle, who was a Deputy-Lieutenant for Dorsetshire, played an important part at the Revolution. It was in his grounds at Charborough, near Dorchester, that several patriotic gentlemen of the neighbourhood concerted the plan for deposing King James and bringing in his son-in-law, William of Orange, to take the English crown.

Most of the officers of the regiment raised by Erle were naturally county men, and he gave the Lieutenant-Colonelcy to his friend and neighbour, Robert Freke, of Upwey, who induced two of his brothers, Harry and George, to join at the same time. Another brother, Thomas, had already,

GENERAL THE RIGHT HONOURABLE THOMAS ERLE.

FROM A PAINTING BY SIR GODFREY KNELLER.

curiously enough, been given an Ensigncy in Luttrell's Regiment shortly after its formation.

Having recruited its ranks and attained a state of efficiency, the battalion was ordered to Ireland with the Duke of Schomberg's expedition, and landed at Bangor, co. Down, with the rest of the army on the 13th August, 1689.

The first march was to Belfast, where a force was detached to besiege Carrickfergus, which surrendered on the 21st August. The army then proceeded to Lisburn and Hillsboro', and finally to Dundalk. Here a permanent camp was formed, unfortunately on low-lying ground, where owing to bad weather, short rations, and the neglect of ordinary sanitary arrangements, the army lost by fever and dysentery more than half its numbers. So weak did some regiments become that it was found necessary to break them, and Lord Roscommon's was broken into Colonel Erle's.

That the regiment was not in a very good condition at this camp may be gathered from the following report, made after a review of the troops held at Dundalk on the 18th to 28th October:—

"The fact of the Colonel having been ill for some time has not conduced to the benefit of this regiment, because the other officers have not his experience in military matters; so that the regiment is not in a very good state. Nor is it much better than Lord Roscommon's Corps. The clothing is somewhat in disorder, but the Colonel has sent into Scotland for 'surtouts,' which are very necessary in this country, and unless the other regiments (Wharton's[1] and Lord Meath's[2] excepted) are supplied with the same kind of clothing, they will not last more than three months."

---

1 12th Foot the Suffolk Regiment.
2 18th Royal Irish Regiment.

The battalion mustered at this date:—13 captains, 14 lieutenants, 10 ensigns (1 absent, 1 sick), 36 serjeants, 39 corporals, 26 drummers, 581 privates (127 sick, 9 absent).[1]

At this time several small expeditions were constantly being made against the Irish, and Schomberg's army lost a good many men in these excursions, chiefly in the vicinity of their camp and at Newry and Charlemont.

After remaining ten weeks at Dundalk it was decided, owing to the continued ill-health of the troops, to disperse them over the north in different garrisons, Erle's being quartered in the vicinity of Six-mile-water, co. Antrim, with 200 men at Belturbet, co. Cavan. The General's headquarters were at Lisburn, and there was an immense hospital at Belfast, where no less than 3,762 men died between the months of November, 1689, and the following May.

On the 14th June, 1690, King William landed at Carrickfergus, and on the 19th he arrived at Hillsboro', co. Down, where he gave orders for the army to take the field. By the 27th June all the different corps had joined at Dundalk, making in all about 30,000 men.

The main body of the Irish had marched to the River Boyne, and were followed there by King William on the 30th June, where he halted his army, the Irish being strongly posted on the far bank. The next day the famous battle began, the first in the history of our standing army.

The main body of the infantry was ordered to attack the Irish centre at Oldbridge, whilst the left wing, under the King himself, composed of cavalry, was to cross the river at Dunmore. In the meantime a division, under Lieutenant-General Douglas, was to move up the river towards Slane, and effect a passage there. With this force was Erle's

1 King William's Chest.

Regiment, which was in Brigadier-General Trelawny's[1] brigade. Early in the morning General Douglas moved off, and finding the ford of Rosnaree, just below Slane, utilized it, as well as the bridge. There was a fierce fight here, and the crossing was gallantly disputed by two regiments of dragoons, but they were beaten back after an hour's engagement, when the whole of Douglas' force crossed, after which he does not appear to have had much more fighting.

The Dutch Guards, two French regiments, two of the Inniskilling Horse, some Danes, and several English regiments passed the river at Oldbridge. They charged the Irish, posted on the far bank, with great bravery, driving them from their hedges and breastworks with severe loss. The King also had heavy fighting with the left wing, and had to charge the enemy several times at the head of his troops near Dunmore.

The Irish fought with desperate courage, and rallied again and again, but they were finally broken, and retreated at all points in great confusion. Owing to the nature of the ground the English could not carry the pursuit very far, and night coming on, they were unable to ensure such a complete victory as could have been wished for. The English regiments did not, it is said, get much opportunity of distinguishing themselves, but they behaved well, and it was the first time most of them had been under fire. The loss of the English was about 400 killed, that of the Irish between 1,000 and 1,500. Amongst the former was Marshal Schomberg, one of the most distinguished soldiers of the time.

King James fled to Waterford, where he embarked for France, and died in exile.

The battle of the Boyne is celebrated every year with

---

1 The other regiments of Trelawny's brigade were the 2nd, 4th, 13th, and 23rd Foot.

great enthusiasm in almost every town and village in the north of Ireland. Those who commemorate it call themselves Orangemen, after William of Orange, and Orange Lodges exist in every part of Ulster. Processions, accompanied by bands of music, parade the various streets. Huge standards borne on double poles are carried aloft by these enthusiasts, many of them bearing on their flaring colours fanciful pictures of King William crossing the famous "Boyne Water."

On the 5th July the army reached Finglass, about two miles outside Dublin, where the King held a grand review of the troops on the 7th and 8th, 30,000 men being on parade. Erle's Regiment was 693 strong, according to the field state.

The army was now divided up, a strong expedition being sent to besiege Athlone, under Lieutenant-General Douglas, whilst the remainder, amongst which was Erle's, marched southwards, securing the towns of Waterford and Wexford.

On the 8th August Lieutenant-General Douglas' force rejoined headquarters, being unsuccessful in its attempt to take Athlone, and the whole army marched towards Limerick. The King was under the impression that the city would be delivered up to him. But he was sorely disappointed, and after several ineffectual assaults, in which the English lost over 500 killed and 1,100 wounded, King William decided to raise the siege, and moved with the army into Tipperary.

Early in September the troops went into winter quarters, Erle's having two companies stationed at Mount Mellick, two at Phillipstown, two at Tirril's Pass, and six at Streamstown.

King William left about this time for England, and Lieutenant-General Ginckel became Commander-in-Chief.

In the meantime Major-General Kirk, with seven battalions, including Erle's, marched towards Birr, which had been besieged by Sarsfield, the Irish commander, who, in the face of such strong reinforcements, resolved to abandon the siege. The English then once more went into winter quarters.

After a series of small expeditions against the rebels, Erle's Regiment found itself in camp at Mullingar in May, 1691, with seven other battalions and seven regiments of cavalry, all under command of Lieutenant-General Ginckel.

On the 6th June this force marched out from Mullingar to Rathcondra, and was joined by nine infantry and three cavalry regiments, under Lieutenant-General Douglas. Ginckel, although his army was still inferior in point of numbers to that of St. Ruth, who commanded the French and Irish, determined to act on the offensive. So being supplied with new clothes and equipment for his troops, he opened the campaign by attacking Ballymore, in West Meath, which was garrisoned by 1,000 men. This he took in one day, and the way to Athlone was now clear.

On the 21st June the army advanced to the town walls, and a breach was made by a battery of ten guns. The stormers attacked fiercely, and drove the Irish across the bridge which connects the town, but the arch next to the Irish side was broken, and they could not pursue further. Consequently, a wooden structure was raised so as to throw planks across, and by the 28th June some were laid over. The enemy seeing this, sent a sergeant and ten men in armour to destroy the work, all of whom were slain. Another party made a similar attempt, and succeeded in throwing the beams and planks into the river, two of them only surviving the operation.

General Ginckel was not, however, to be diverted from his object, and at length placed a close gallery over the broken arch. Orders were issued at night that 43 grenadiers, 83 privates, and a proper proportion of officers and sergeants from each regiment, with fifteen rounds a man, were to be ready by six o'clock the next morning under the walls, and make an attempt to cross at three different places at the same time. Money was given to the men to encourage them in this dangerous attempt. Unfortunately, the news leaked out, and the Irish were all ready for them, and to add to their discomfiture the gallery was set on fire, so that the attack was countermanded, to the great joy of St. Ruth, who gave a dinner party on the occasion.

Ginckel now called a council of war, and was for retreating, but was opposed by the other generals. The men, too, were very eager to have another try, being taunted by the Irish, who kept telling them that they had poorly earned the money given them by their officers.

Accordingly at guard-mounting next morning, the 30th June, another attempt was made. The ford gave room for twenty men to march abreast, but the bottom was rocky, the stream very rapid, and the shallowest part nearly breast high. On the opposite side was a bastion to defend the passage.

The preconcerted signal was the ringing of the church bell, upon hearing which 2,000 men entered the Shannon by twenties, under the heavy fire of the Irish batteries, and finally forced their way into the works. The Irish fled in confusion, and in half an hour the English were in possession of the town.

In this final attack the enemy had 500 casualties, and the English but 46 killed and wounded, amongst the latter being Captain Strangeways, of Erle's.

On the 10th July Ginckel marched out of Athlone, and came upon the Irish army, which had retreated to a strong position at Aughrim, four miles from Ballinasloe. Their camp was on the side of a hill, on which stood two Danish forts, and half a mile below was a bog with a rivulet running through it, extending along their front and intersected by hedges and ditches on the near side. On the 12th July the army was formed up in two lines of battle, Erle's battalion being in the second line, in which formation they advanced.

After a series of skirmishes the battle commenced in earnest about 4.30 p.m., the principal attack being made on the enemy's right. As the Irish drew off several bodies of horse to meet this, four battalions, Brewer's,[1] Erle's, Herbert's[2] and Creighton's,[3] were ordered by Major-General Mackay to pass a difficult bog, ford a rivulet, and drive the Irish from the hedges of the nearest enclosures. Erle's advanced first with the other battalions following, all the men up to their middles in mud and water, and drove the enemy back from hedge to hedge, but the latter being reinforced came on in great numbers. Colonel Erle seeing this, encouraged his men, telling them that now was the time to show their courage and prove what sort of soldiers they were. On hearing which they pressed on bravely, but being taken in front and in flank, and greatly outnumbered, these battalions had to retire with heavy loss. The Irish followed, shouting and plying them with musketry, but supports arriving, under Major-General Talmash, the four regiments faced about, repulsed their pursuers, and by a spirited effort recovered their lost ground. The cavalry passed the bog near the Castle of Aughrim, and by a determined charge completed the overthrow of the Irish, who had lost heart owing to the death of St. Ruth, who was killed by a cannon ball.

---

1 12th Foot, now the Suffolk Regiment.

2 23rd Foot the Royal Welsh Fusiliers.

3 Disbanded.

Major Robert Tempest, writing to Sir Arthur Rawdon, of Moira, co. Down, says:—

"At six in the evening begun the fight, Kirk's[1] and Gustavus Hamilton's[2] regiments then marching out on the right to a ditch before the Castle. Sir Harry Bellasy's[3] and my Lord George Hamilton's[4] against all the lined hedges and ditches, who making first to one hedge, perceiving the broad way where their horse should come down, as is said before, upon our cannon, we crossed and barricaded it with turnpikes; and both regiments joining to one another close without interval, unanimously went together over a plain field, and received the enemy's fire, likewise theirs from the Castle, and took possession of their works, which the enemy perceiving would not stay to charge, but immediately retreated; he poor Jellet was killed. By this time Colonels Erle and Brewer, on our left, advanced through the bog, upon which the enemy fell down again upon them and us, with bodies of foot and squadrons of horse, which caused both Erle and Brewer's to retreat, they not being able to stand their force. Here Colonel Erle, with Captains Bingham and Gooking of his regiment, were taken prisoners, but rescued afterwards by our horse, who passed a defile one by one through a boggy trench."[5]

George Story, chaplain to the Earl of Droheda's Regiment, gives the following account of the part taken by Erle's in the fight:—

"Colonel Earl, Colonel Herbert, Colonel Creighton, and Colonel Brewer's Regiments went over at the narrowest place where the Hedges on the Enemies side run furthest into the

1 4th King's Own Regiment.
2 20th Lancashire Fusiliers.
3 6th Royal Warwickshire Regiment.
4 7th Royal Fusiliers.
5 The Rawdon Papers.

Bogg. These four Regiments were ordered to march to the lowest of the Ditches, adjoining to the side of the Bogg, and there to post themselves till our Horse could come about by Aghrim Castle, and sustain them, and till the other Foot had marched over the Bogg below, where it was broader, and were sustained by Col. Foulks' and Brigadeer Steuart's. Col. Earl advanced with his Regiment, and the rest after him over the Bogg, and a Rivulet that ran through it, being most of them up to their Middles in Mudd and Water. The Irish at their near approach to the Ditches fired upon them, but our Men, contemning all Disadvantages, advanced immediately to the lowest Hedges and beat the Irish from hence. The Enemy however did not retreat far, but posted themselves in the next Ditches before us: which our Men seeing, and disdaining to suffer their Lodging so near us, they would needs beat them from thence also, and so from one Hedge to another, till they were got very nigh the Enemies main Battel. But the Irish had so ordered the matter, as to make an Easie Passage for their Horse, amongst all those Hedges and Ditches, by which means they poured in great numbers both of Horse and Foot upon us: which Colonel Earl seeing, encouraged his men, by advancing before them, and saying, *There was no way to come off but to be Brave.* As great an Example of true Courage and Generosity as any Man this Day living. But being both flanked and fronted, as also exposed to all the Enemies Shot from the adjacent Ditches; our Men were forced to quit their Ground, and betake themselves to the Bogg again, whither they were followed or rather drove down by the strength of Horse and Foot, and a great many killed. Colonel Earl and Colonel Herbert were here taken Prisoners; the former, after twice taking and retaking got free at last, tho' not without being wounded.

* * * *

"Then Major General Talmash, seeing the Disadvantages our Foot laboured under in the Centre, he shewed at once an extream concern for his Countrymens being repulsed, and as much Generosity and Courage, by hastening to Succour those that at that time stood most in need of it; coming up therefore in all haste with some fresh Men, he gave Orders for our broken Regiments to halt, and face about, which they did immediately, and returned the same Measure to the Irish that some of themselves had met withal, the very Minute before that is, they knockt them on the Head; for the Irish followed us towards the Centre of the Bogg, which tho' not two hundred yards from the lowest Ditches, yet before the Enemy could recover those again, our Men had killed above three hundred of them, and then marched boldly up to their old Ground again from whence they had been lately beat: which is only natural to Englishmen; for it's observable that they are commonly fiercer, and bolder, after being repulsed than before; and what blunts the Courage of all other Nations, commonly whets theirs, I mean the Killing of their Fellow Soldiers before their Faces."

The English took nine brass cannon and all the tents and baggage, besides 11 standards and 32 pairs of Colours. The total loss was 73 officers and 600 men killed, and 111 officers and 960 men wounded.

Story gives the following casualties as having occurred in ERLE'S Regiment:—

1 Colonel wounded (Colonel Erle).
1 Major killed (Major Devenish).
2 Captains killed.
1 Captain wounded.
1 Lieutenant killed.
2 Lieutenants wounded.
87 Soldiers killed.
70 Soldiers wounded.

The number of wounded officers does not, however, agree with the following list in the accounts of Colonel Erle at the Record Office, Dublin.

Extract from Accounts of Coll. Thomas Erle:—

Paid to severall officers in consideration of their wounds received at ye Battle of Aughrim, on accot of 14 daies pay per warrt 14 July, 1691.

	£	s.	d.
The Collonell	16	16	0
Capt. Bingham	7	0	0
Capt. Pepper[1]	7	0	0
Capt. Geoghegan	3	5	4
Lt. and Q.-Mr. Wills	6	10	8
Lieut. Dolling	3	5	4
Lieut. Discorissin	3	5	4
Lt. Gookin	3	5	4
Ensigne Penny	2	11	4

The loss of the Irish was computed at 500 officers and 7,000 men killed and 450 prisoners. The strength of their army was 20,000 foot and 5,000 horse; that of the English 17,000 horse and foot.

After a few days' rest the army marched on to Galway. There was no opposition approaching the town, but at the north-west gate the suburbs were burnt, and there was some show of preparing for a vigorous resistance. A trumpeter was sent to demand surrender, but the answer from the enemy was that they would fight it out to the last. Owing to information from an officer who had deserted from the Irish, the General was shown the way to an important fort on the south-east of the town, which was not quite finished. The attack on this fort was successful, and it was an important capture, as it commanded a great part of the wall on that

1 Brigadier-General 1707. Major-General 1st January, 1710. Commanded a cavalry brigade at the battle of Almanara, where he distinguished himself, and gained additional honours at Saragossa. Taken prisoner at Brighuega in 1710. M.P. for Oban 1715. Died at Montpellier 22nd December, 1725.

side of the town. After it was taken the Irish turned their guns on it, killing and wounding several men. They then proceeded to set fire to the suburbs, the houses of which had formerly belonged to the English, and on the 24th July the Governor sent a drummer to the General asking for terms. Articles were then signed, one of which was that the garrison could march out to Limerick, with their arms, colours flying, drums beating, matches lighted, and as much ammunition and rations as each officer and soldier could carry with him.

All Ireland was now in possession of the English, except the old city by the Shannon. Ginckel, taught by the events of the previous year's siege, resolved to act with great caution. He had the passes of the Shannon secured by armed vessels, and a powerful escort with his artillery. By the 25th August the English army was occupying the same position it had held the year before.

Operations proceeded very slowly, partly owing to wet weather, and on the 13th September Colonel Erle was sent to England to acquaint the Queen how things went on, and should the town not be taken that season, it would be necessary to send more frigates, so as to completely block up the Shannon.

About the same time the trenches were opened, and the guns in a few days began a brisk fire on the Irish town, the inhabitants of which were obliged to seek refuge in the English town, or on the King's Island, where a hasty camp was formed. But Ginckel found that nothing could be done unless he cut off intercourse with the County Clare, from whence the Irish received their supplies.

This was carried on by Thomond Bridge, of which it became necessary for Ginckel to get command. A considerable body of troops were accordingly sent to the mainland of Clare on the 22nd September, and after a sharp contest, forced their way to the works that protected Thomond Bridge. Orders for storming being given, the grenadiers of the different regiments, supported by two battalions, rushed boldly forward, through a tremendous fire of guns and small arms, forced the Irish from their position, and finally routed them completely.

At this juncture a French major, who was in command at the bridge, fearing that the English would enter with the fugitives, raised the drawbridge, and before the fury of the English could be stayed, 600 of the enemy were actually slain on the bridge and 150 drowned in the Shannon.

The next day the garrison beat a parley, when a truce of three days was granted. On the third day terms of capitulation were offered to Ginckel which he would not grant, but after further negotiations the articles were signed.

This capitulation putting an end to the war, Ginckel and Sarsfield, the Irish commander, endeavoured to secure the Irish regiments for their respective masters. Sarsfield and the clergy urged the soldiers in favour of France, adding that they might soon be able to renew the conflict under more favourable circumstances.

On the 6th October the Irish army, 14,000 strong, assembled on the Clare side of the river. About 3,000, including the Ulster Irish, withdrew to their homes or joined the English army. By far the larger number volunteered for foreign service, and in accordance with the terms of the treaty, were conveyed to France, where they formed the nucleus of the famous Irish Brigade, whose valour was conspicuous on almost every battlefield in Europe.

Erle's battalion now returned to its Irish quarters, and left for England early in 1692, where it joined its sister battalion at Plymouth. The latter remained in the citadel whilst the Irish battalion was quartered at Plympton and places adjacent. A muster taken on the 31st December shows its strength to have been as follows:—

Colonel absent, 11 captains, 14 lieutenants, 12 ensigns, 39 serjeants, 39 corporals, 26 drummers, and 662 privates.

Colonel Erle now received orders to send a battalion to Flanders, to embark on February 1st, 1693, the Irish one being detailed for the duty. This order was subsequently amended, and Erle was directed to embark "thirteen companies of the Two Battall," under his command. From subsequent evidence it appears that the thirteen companies selected comprised ten of Luttrell's and three of the Irish battalion.

The battalion that remained at home had one more tour of active service, when in 1694 an expedition was fitted out for the purpose of harrying the French coast, but without any good results.

Since its arrival from Ireland, in 1692, the battalion had been quartered near Plymouth, and in February, 1694, it moved to Portsmouth, where it was encamped on the Portsdown Hills till it embarked on the 24th May with the above-mentioned expedition. This consisted of a mixed battalion of Guards, besides Babington's,[1] Stewart's,[2], Hastings',[3] and Venner's[4] regiments, in addition to other corps since disbanded. General Talmash was in chief command, and it was resolved to attempt a landing at Camaret Bay, at the mouth of Brest harbour.

On the 8th June the attempt was made, the attack being

1 The 6th Foot the Royal Warwickshire Regiment.
2 The 9th Foot the Norfolk Regiment.
3 The 13th Foot Prince Albert's Somersetshire L.I.
4 The 24th Foot the South Wales Borderers.

led by 600 grenadiers, under Lord Cutts. The enemy's works were very strong and well manned, and there was a large body of cavalry manœuvring in the open. The British troops, which numbered 8,000, set out in a confused manner, and the fire from the ships was ill-directed and badly sustained, whilst that from the forts was heavy and continuous. The plan of attack was not adhered to, and only 2,000 men actually landed. Of these 700 were killed and wounded, and it was with some difficulty the remainder were re-embarked. General Talmash was mortally wounded, and the affair was a complete failure.[1]

On the return of the expedition the battalion landed in the Isle of Wight, and proceeded to Portsmouth at the end of June.

Early in October it was split up into detachments, which were quartered at Lymington, Weymouth, Melcombe Regis, Abbotsbury, Winchester, Salisbury, and Christchurch, and later on in the year it moved further west to Taunton, Ilchester, Yeovil, Wellington, Tiverton, Beminster, Ceon Abbas, Sturminster, and other villages.

These companies marched into Plymouth on the 7th February, 1695, detachments being sent to Pendennis and Scilly.

On the 27th January, 1696, a draft of six men per company was ordered to be prepared for Flanders to reinforce the battalion on service there.

The peace of Ryswick was signed in September this year, and King William saw his efforts for the preservation of liberty and the balance of power in Europe attended with success.

---

1 The casualties in Erle's are not known, but Lieutenant Capell Stocker was amongst the officers killed, his widow being granted a pension of £30.

In October the out companies at Pendennis and Scilly were ordered to rejoin headquarters at Plymouth, and the regiment marched to Exeter a month later, throwing out detachments at Topsham, Sidmouth, Crediton, Chudleigh, Newton Bushell, Moreton and Chagford.

Not long afterwards, consequent on the reduction of the army, this battalion, as well as several others, was disbanded, when the following letter was received by Major-General Erle:—

Sir,

His Majesty having been pleased to order that a company be formed out of the officers of the regiments that have lately been broke, which are to march at the head of the 1st Regt. of Foot Guards, if any of the officers are willing to enter into this service, you will send them to Colonel Shrimpton, Major of the said Regiment, as soon as may be convenient.

I am, etc.,

GEORGE CLARK,

Secretary at War in the absence of

Mr. Blathwayt.

Major-General Erle.

Similar letters to all other regiments disbanded.

Many of the officers and men were, of course, incorporated with the Flanders battalion, but it is not known whether any of the former availed themselves of the above offer.

COLONEL THOMAS ERLE'S REGIMENT.

(Many of the annotations are from Army Lists and Commission Registers by Charles Dalton, F.R.G.S.)

CAPTAINS.	LIEUTENANTS.	ENSIGNS.	Effective Men.	Sick Men.	Dead Men.
Coll. Erle ...	Joanes [10] ...	Williams [15] ...	33	7	20
Lt. Coll. Freek [1] ...	Freek ...	Trimloe ...	31	9	20
Mar. Davenish [2] ...	Goarden ...	Enclin ...	22	9	29
Harwood [3] ...	Ramsay [11] ...	Izard	29	15	16
Sorton [4] ...	Dobb ...	Tredenham ...	27	12	21
Gookin [5] ...	Willis [12] ...	Gookin [16] ...	32	11	17
Abington [6] ...	Floyer ...	Henning [17] ...	27	8	25
Davenish ...	Bourne [13] ...	Lee [18]	36	2	22
Gras. Strangeway [7]	Mering and Dolling [14]	... ...	32	10	18
Phillips [8] ...	Hanserd ...	Phitzherbert [19]	27	7	16
Winn ...	Winn ...	Montgomery	60	—	—
Bingham [9] ...	St. Mullen	Cadogan [20] ...	60	—	—
Warring ...	De la Couer	... ...	60	—	—
			486	90	204

The above list is from a bundle marked "1692 undated" in King William's Chest, but evidently anterior to the death of Captain Strangeways, who was killed at the siege of Athlone 30th June, 1691.

---

1 *Robert Freke.* Born at Upwey, 1655. Son of Robert Freke of Upwey. Brevet-Colonel 16th February, 1694. Deputy-Governor of Plymouth in 1696. Died in 1709. Buried at Upwey.

2 *Francis Devenish.* Major 8th March, 1689. Killed at the battle of Aughrim.

3 *Oxenbridge Horwood.* In ranks of Royal Horse Guards. Ensign Prince George of Denmark's Regiment of Marines (disbanded 28th February, 1689). Captain Erle's Regiment 8th March, 1689, when first formed. Served in the ranks of the Royal Horse Guards at the battle of Sedgemoor. Received a bounty of £30 8s. 4d. for wounds.

4 *Samuel Sourton.* Captain 8th March, 1689.

5 Vincent Gookin. Captain 8th March, 1689. Ric. Coote's Regiment 21st March, 1692. Son of Vincent Gookin, Surveyor-General of Ireland, and grandson of Sir Vincent Gookin, Knt., who settled in Ireland.

6 Andrew Abington. Captain 8th March, 1689.

7 *William Strangeways.* Captain Grenadier Company, 1691. Killed at the siege of Athlone.

8 Thomas Phillips. Lieutenant 28th June, 1689. Captain before July, 1691. Gustavus Hamilton's Regiment 31st January, 1692. Colonel Farrington's Regiment of Foot 18th February, 1696. Half-pay in 1698.

9 Charles Bingham. Captain Lord Roscommon's Regiment, 1689. Transferred to Erle's when the former was broken, as Captain. Killed at the battle of Aughrim, leaving a widow and three children.

10 *Edward Jones.* Ensign 8th March, 1689. Captain-Lieutenant 2nd August, 1689. Captain before 1697.

11 John Ramsay. Lieutenant 8th March, 1689.

12 *Charles Wills.* Born at St. Germain's, Cornwall, 1666, son of Anthony Wills. Captain in Erle's 1st July, 1691. Major in Colonel Thomas Saunderson's Regiment 6th November, 1694. Afterwards the Right Hon. Sir Charles Wills, K.B., M.P., a most distinguished general. Died in London 25th December, 1741.

13 John Bourn. Ensign 8th March, 1689.

14 Joseph Dolling. Lieutenant 1st November, 1689. Captain-Lieutenant 2nd April, 1695. Half-pay in 1698. Captain Earl of Huntingdon's Regiment 10th March, 1702. Paid 14 days' pay on account of wounds received at the battle of Aughrim, £3 5s. 4d.

15 George Williams. Ensign 2nd August, 1689. Lieutenant in 1690.

16 Charles Gookin. Lieutenant 27th May, 1691. Half-pay in 1698. Lieutenant in Erle's 1st March, 1702; Quarter-Master 21st January, 1702/3. Captain before 1706. Appointed Deputy-Governor of Pennsylvania in 1708. Paid 14 days' pay on account of wounds received at the battle of Aughrim, £3 5s. 4d. Served in the West Indies 1702.

17 Richard Henning, of Plympton, Devon. Born 1664. Fourth son of Robert Henning of Henning's Crookson, Dorset. Lieutenant Erle's Regiment 26th December, 1690. Captain-Lieutenant 1st May, 1694. Captain 22nd April, 1695. To Colonel John Gibson's Regiment of Foot, 10th March, 1702.

18 *William Leigh.* Captain 1706. Captain and Lieutenant-Colonel Coldstream Guards 22nd July, 1715. In the Malplaquet Roll. Died 12th January, 1732/3.

19 *Fitzherbert.* Ensign 2nd August, 1689.

20 William Cadogan. Born at Liscartnan 1672. Eldest son of Henry Cadogan, of Dublin, Counsellor at Law. Captain Erle's Regiment 4th March, 1693/4. Major Inniskilling Dragoons 1st August, 1698. Brevet-Colonel 1st June, 1701. Quarter-Master-General 25th August, 1701. Colonel 5th Dragoon Guards 2nd March, 1703. Major-General 1st January, 1707. Lieutenant-General 1st January, 1709. General 12th July, 1717. Colonel Coldstream Guards 11th October, 1714; 1st Foot Guards 18th June, 1722.

Served through all the wars in Flanders in the reigns of William III. and Queen Anne. Wounded at the siege of Mons. Taken prisoner at the siege of Menin in 1706, but soon afterwards exchanged. Second in command to Duke of York at close of 1715 rebellion.

The one man whom Marlborough delighted to honour. Created Baron Cadogan 30th June, 1716, and advanced to an Earldom 8th May, 1718. D.S.P. 17th July, 1726, in the 57th year of his age. Buried in Westminster Abbey.

CHAPTER III.

Return of Battalion (late Luttrell's) to England—Death of King William III.—Expedition to Cadiz and the West Indies—Embarks for Flanders—Battle of Malplaquet—The Malplaquet Roll—Sieges of Douay, Bethune, and Bouchain—Expedition to Vigo, 1719—Ireland and Scotland—Dress—Regimental Marches.

FINDING the progress of his arms arrested and the fortune of war in favour of the Confederates, the French King contemplated detaching England from the alliance against his interests by replacing King James on the throne, for which purpose preparations were made for invading England, and a conspiracy was organised in London for assassinating King William.

In consequence of this, Erle's battalion and a number of other regiments were ordered to return to England. The battalion accordingly sailed from Sas-van-Ghent in March, 1696, and eight companies disembarked at Tilbury on the 18th of the month. They were quartered at various towns and villages in Essex, and moved into Cambridgeshire a week later.[1]

The remaining five companies were captured by French privateers flying English colours, and were taken to Dunkirk on the 17th March, as well as five companies of Bellassis'[2] Regiment and a battalion of Scots Guards. These were all redeemable according to the cartel with a month's pay.[3]

The battalion remained in England till the summer of 1697, when eleven companies were ordered to embark on the 1st July "for the first ports in Zealand they can make."[3] It

1 Marching Order Book.
2 The 22nd Foot.
3 Luttrell's Diary.

joined the army encamped near Brussels on the 14th July, and two days later it was reviewed by King William. On peace being concluded the regiment returned to England in November. Its strength at this time was 13 companies, consisting of 41 commissioned and 68 non-commissioned officers, 466 privates.

In March, 1699, Erle's proceeded to Ireland,[1] where it remained until 1702, when the accession of the Duke of Anjou, grandson of Louis XIV., to the throne of Spain had produced another war. England might have abstained from open hostilities with France had it not been for the following circumstance.

The death of James II. occurred at St. Germain on the 6th September, 1701, and his son, the titular Prince of Wales, was immediately proclaimed by order of Louis XIV. as King of England, Scotland and Ireland by the title of James III. This indignity to the British Sovereign and nation, added to the contemplated union of the crowns of France and Spain, made war inevitable, and King William with the Emperor of Austria and the States General concluded the "Grand Alliance," the principal objects of which were to procure the Spanish Netherlands as a barrier for the Dutch and to prevent France and Spain becoming eventually under the sway of the same prince.

War was thus on the eve of being proclaimed, when King William met with the accident which terminated his life on the 8th March, 1702. The accession of Queen Anne, however, caused no alteration in the policy of her predecessor, and war was declared on the 4th May following. Additional forces were sent to Flanders, and the Earl of Marlborough was appointed to command the British, Dutch and auxiliary

1 Luttrell's Diary.

troops with the rank of Captain-General. The contest which ensued is known as the "War of the Spanish Succession."

Six regiments had been added to the regular army in the year 1702 as Marine Corps, and six others, amongst which was Erle's, had been detailed for sea service.

Early in the year an expedition against the port and city of Cadiz was resolved upon by the British Government, Erle's Regiment being one of those selected for this enterprise.

Accordingly on the 15th June the battalion proceeded to the Isle of Wight, where it embarked on board the fleet, which was commanded by Admiral Sir George Rooke.

Lieutenant-Colonel Hawley, as well as several other officers had the Queen's leave to be absent on this occasion,[1] and Major Robert Carey was in command. The marching out strength was 29 officers, 24 sergeants and 584 rank and file.

---

1 The following officers had the Queen's leave to be absent:—Major-General Thomas Erle, Lieutenant-Colonel Henry Hawley, Captain Leonard Craddock, Captain George Freke, Captain-Lieutenant John Pratt (acting as regimental agent in Ireland), Lieutenant John Penney.

* NOTE.—AN ACCT. OF THE STRENGTH OF MAJR.-GENLL. ERLE'S REGIMENT, JUNE 15TH, 1702.

	Sargts.	Corplls.	Drus.	Privat Men.	
Granadeers	2	3	2	44	Bedford
Majr.-Genll.'s Comp. ..	2	3	2	44	Expedition
Colln. Freke	2	3	2	43	
Lt-Colln. Hawly	2	3	2	43	Barford
Majr. Cary	2	3	2	45	Kent
Capn. Norman	2	3	2	44	Kent
Capn. Dej. Cophan	2	3	2	45	Barford
Capn. Morgan	2	3	2	43	Bedford
Capn. Edgeworth	2	3	2	44	Eagle
Capn. Cradock	2	3	2	44	
Capn. Symons	2	3	2	41	Plymouth
Capn. Prater	2	3	2	44	
	24	36	24	524	

(Endorsed) The strength of Majr.-Genll. Erle's Regiment, June 15th, 1702. From Harleian MS.

Officers abscent—
Lieut.-Colln. Hawly
Capt. St. John Pratt
Lieut. John Penney

The land forces were under the orders of the Duke of Ormonde, and consisted of about 12,000 men.

Towards the end of July the expedition sailed from Spithead, its immediate object being the reduction of Cadiz. By the middle of August the fleet arrived off that port, but the town was found so well protected by the strong fort of Matagorda, which stands at the mouth of the harbour, that the 4,000 English and Dutch troops sent to attack the fort were obliged to retire, after many lives had been lost, as the Spanish batteries within the harbour poured such a destructive fire upon the troops who were busy raising batteries on the low marshy ground destitute of cover. To atone for this repulse, another body of troops took and occupied the adjacent towns of Rota and Port Saint Mary. The latter was a wealthy town, and full of loot. "Scarce a man but got something," says a contemporary writer, "for Port St. Mary was so thoroughly plundered that the very iron about the houses was brought away." Looting and marauding had been strictly prohibited by the Commander-in-Chief, as it was hoped that many Spaniards in Cadiz favourable to the cause of the Austrian Archduke Charles, whose claims to the Spanish crown the Grand Alliance had backed up, would come over to the side of the Allies. But disgusted with the excesses of the British and Dutch troops, who even despoiled and robbed churches, none of the Carlist sympathisers joined Ormonde's army.

Cadiz being found too strong, the troops re-embarked. Following out his instructions, the British Admiral despatched Captain Hovenden Walker with a dozen ships to the West Indies, having on board Erle's, Hamilton's,[1] Donegal's,[2] and

1 20th Foot the Lancashire Fusiliers.
2 35th Foot the Royal Sussex Regiment.

Charlemont's[1] regiments. The remainder of the expedition sailed for Vigo Bay to attack the Spanish treasure ships there.

Captain Walker's squadron arrived at the West Indies at an opportune moment. Admiral Benbow, who had engaged single-handed the French squadron of seven ships, under Du Casse, in August, had just died of his wounds at Port Royal, Jamaica. Cheated out of victory by the base desertion of five of his captains, who kept their ships astern during the fight, which continued several days, the gallant Benbow's death was inexpressibly sad. Colonel Christopher Codrington, Captain-General of the British Leeward Islands, organised an expedition against the Island of Guadeloupe in the winter, and on the arrival of the British squadron at Antigua, Codrington joined Walker with Colonel Whetham's Regiment and a small body of Colonial troops.[2] The expedition then sailed for Guadeloupe, arriving there early in March, and anchored off Basse Terre, where the "Boyne," flagship of Commodore Walker drew the fire of the French batteries, and some men were killed. A council of war, attended by the captains of the various ships and the commanding officers of the different regiments, with General Codrington as president, assembled on board the "Boyne" on the 11th March, and it was determined to attempt a landing in the morning. The force told off for this purpose consisted of two subalterns and 220 men from Erle's, under command of Major William Morgan, and a similar number from the other four regiments as a first detachment. Two other parties of soldiers and marines were in support, and were ordered to land at four o'clock in the morning.

After a sharp fight the landing was effected, and the

1 36th Foot the Worcestershire Regiment.

2 Army Lists and Commission Registers, by Charles Dalton, F.R.G.S.

French retired to the fort and Castle of Basse Terre. Major Morgan was mortally wounded, and sent on board the "Boyne," where he died a few days later.[1] The other casualties in this morning's fight are not known.

Codrington now applied to the Commodore for some ships' guns with which to bombard the fort, as he had no artillery, and as he could not get what he wanted there appears to have arisen some friction between the two officers. On the 28th March Codrington wants more guns, but Walker refuses, as he has only retained thirty rounds per gun. He may have to fight the French fleet, and declines to accept the responsibility of so weakening his ships without the sanction of a council of war. This he held on board the "Boyne" on April 1st, and it was resolved to send a further supply of ammunition ashore for the troops, retaining twenty rounds per gun for each ship.

A few days later the squadron which had been lying in the "Petits Habitants" stood out to Basse Terre to attack the castle and forts, and on the 4th April the French blew up the castle and retired across the river Galleon.

All this time the men had been undergoing much privation. They had no tents, and their meals of porridge and peas were sent ashore from the ships ready cooked, so that they were always eaten cold. By the end of April the bread ran short, and the men had to be put on half rations. Consequently, there was much sickness amongst them, and many deaths took place daily. Notwithstanding this, the French were driven to the woods, and two of the chief towns, La Bayliffe and Basse Terre, were stormed and captured.

Early in May the command devolved on Colonel Charles Wills, then commanding Lord Charlemont's Regiment, as

---

1 Log of the "Boyne," flagship of Sir Hovenden Walker.

the two senior officers, Colonels Codrington and Whetham, were obliged by illness to leave Guadeloupe.

This distinguished soldier had been appointed a Captain in Erle's Regiment on 1st July, 1691, and was promoted Major in Colonel Thomas Saunderson's Regiment on the 6th November, 1694. After a brilliant career he rose to the rank of General. He died in London in 1741, and is buried in Westminster Abbey.

After destroying the French settlements and fortifications along the coast, Wills embarked his troops on board the squadron during the first week in May, bringing away all the captured French guns. The retreat was well timed, as the French had sent a relieving force from Martinique, which arrived just too late.

The actual casualties in Erle's at Guadeloupe between the 12th March and the 3rd May were 2 officers killed (Major William Morgan and Lieutenant Lucy Craven), 1 officer wounded (Lieutenant Sankey Godfrey, afterwards died), 27 men killed, 31 wounded, besides 1 officer, Lieutenant-Colonel Robert Carey and 12 men dead from sickness.

The actual strength of the regiment on embarkation at Guadeloupe was only 11 officers and 309 of other ranks, so that from the 15th June, when it sailed from the Isle of Wight, the mortality, including the losses at Guadeloupe, was 18 officers and 299 men, a terrible death-roll in such a short period.

While Guadeloupe was being devastated by the British, a squadron, having on board Columbine's Regiment, was on its way to the West Indies, under Vice-Admiral Graydon. This Admiral had orders to collect all the forces that were scattered throughout the plantations, and then attempt the capture of Placentia, so as to drive the French out of the Newfoundland trade.

Having collected the shattered British regiments in the West Indies, Graydon sailed for Placentia. On arriving off the south-eastern cost of Newfoundland a terrible fog set in, and navigation became impossible. "This fog," says Secretary Burchett, the naval chronicler of this expedition, "was a fog to admiration, lasting thirty days." After the sweltering heat of the "Indies," the sudden change to damp and cold had a baneful effect on the soldiers. "They were," continued Burchett, "benumbed in their limbs, and subject to fluxes and scurvies." The five regiments on board the fleet were reduced to a total strength of 1,038. When the fog at last lifted it was decided by a council of war that an attack on Placentia was impracticable. Nothing remained then, in Admiral Graydon's opinion, but to return to England.[1]

The regiment was stationed in Ireland in 1704, and the following year embarked for England, and landed near Chester in October. It remained on garrison duty for the next three years, and once more embarked for Flanders in 1707, where it remained till the end of the war.

On the retirement of Lieutenant-Colonel Henry Hawley in March, 1709, Major George Freke was selected to fill the vacancy. He had been appointed a Captain in Erle's Irish battalion in 1689, of which his brother Robert was the first Lieutenant-Colonel. He had seen much active service, had been through the Irish War, and had served in Flanders with Luttrell's battalion. He had also taken part in the expedition to Cadiz and the West Indies in 1702.

"After the successful battles of Blenheim, Ramillies and Oudenarde, King Louis had sued for peace, but the terms

1 English Army Lists and Commission Registers, by Charles Dalton, F.R.G.S.

offered by the exulting Allies were too hard of acceptance, and the old King appealed to his starving people for one last effort."

Exhausted as France was, she had plenty of men and arms, and the best General of France, Marshal Villars, was placed at the head of 100,000 men for the 1709 campaign.

Villars had entrenched his troops near Douai, and Marlborough finding the camp too strong to storm, moved off to invest Tournay. After a murderous siege of two months it surrendered, and Marlborough now turned his attention to the great frontier fortress of Mons. Villars, however, pushed forward to save it, and the allied generals moved to meet him.

South of Mons, near the village of Malplaquet, lies a plateau, where the French took up their position, their right on the wood of Lanière and their left in the wood of Taisnière. Between the two lay a little plain, across which a line of entrenchments was drawn.

Villars meant to fight a defensive battle, and Malplaquet was to be one of the sanguinary battles of the American Civil War, an attack on earthworks through a forest. The allied position extended from Aulnois on the left to Frameries on the right, and the men lay down to sleep, anxiously waiting the dawn of the eventful morrow.

At 3 a.m. on the 11th September divine service was performed at the head of every regiment with the utmost decorum. The greatest regularity pervaded the ranks, as with slow but steady step the troops marched from their bivouacs to the posts assigned to them in the field. A thick fog hung over everything, under cover of which the troops marched, but at half past seven it brightened up, and the morning mists cleared away. The action commenced by a vigorous fire of the artillery on both sides.

The Dutch, who were on the left, agreeably to the orders they had received, which were to threaten and not make a real attack for half an hour, halted when within range of grape, and a violent cannonade was merely exchanged on both sides. The centre continued to press on, regardless of the fire by which it was assailed, and inclining to the right, attacked the right of the wood of Taisnière. At the same time forty battalions on the right of the centre advanced against the wood in front, while Lord Orkney, with fifteen battalions, menaced the entrenchments in the opening. Meanwhile Marlborough in person led on the cavalry in support, and charging at their head he entered the entrenchments, which were at the same time turned by the centre battalions.

While this bloody conflict was raging in and around the wood of Taisnière, the half-hour during which the Prince of Orange had been directed to suspend his attack had elapsed. He now advanced with great impetuosity, and carried the first and second entrenchments on the French right. But with no reserve, and having only forty battalions against the French seventy, he was taken in close order and driven back with dreadful slaughter.

Hearing that matters were in this precarious state on the left, Marlborough galloped over, and with his reserves the combat was in some degree revived in that quarter.

Order was hardly restored when intelligence arrived from the right that the enemy were assuming the initiative in the wood of Taisnière, for Villars had drawn considerable reinforcements from his centre and had sent them to the threatened quarter. This gave Marlborough his chance. He directed Lord Orkney to advance, supported by a powerful body of horse on each flank, and if possible force the enemy's entrenchments in the centre, now stripped of their principal defenders. He charged through the half-manned lines, carried the entrenchments at that side, and poured his men into the open plain. Marlborough, upon seeing this advantage, instantly gave orders for the grand battery of forty guns in the allied centre to advance. They opened a tremendous fire of canister and grape on the dense masses of the French cavalry. These splendid troops, however, bore up gallantly against the storm, and even charged the allied horse before they had time to form within the lines; but they were unable to make any impression, and retired from the attack sorely shattered by the allied artillery.

The battle was won. In vain did Marshal Boufflers, now in command, Villars having been badly wounded, charge to recover the central works, and it only remained for him to make good his retreat, which he did in perfect order, and the French, in deep dejection, reunited near Quesnoy and Valenciennes, about twelve miles from the field of battle.

The Allies were too exhausted to follow up. The French did not lose 500 prisoners and very few guns; of standards and colours they had taken as many from the Allies as the Allies had taken from them, and as to their killed and wounded, they did not exceed 14,000, whilst the Allies lost 20,000.[1] This was, of course, due to the French occupying a defensive position and fighting under cover till the close of the action.

The battle of Malplaquet was a desperate duel between France and England, in which the whole strength of each nation was put forth, and the successful result was rather

1 Stanhope's History.

owing to the superior talent of the English General, and the unconquerable resolution he had communicated to his followers, than to any superiority either of military skill or national resources enjoyed by the victorious party.

The battles of Landen, Malplaquet and Waterloo have been classed as the three most bloody in the whole history of the "cockpit," and Marlborough himself, in a letter to his Duchess from the field, wrote: "I have fought a very murdering battle."

No list of Erle's Foot is given in the Flanders Army List for 1709, and the only writer who includes the regiment in the list of British infantry at Malplaquet is Boyer. The claim of the regiment to commemorate Malplaquet on its colours was *inter alia* considered by a War Office Committee that sat in 1882 to determine the claims of regiments to add certain unrecorded victories to their battle honours. Erle's was probably in the reserve at Malplaquet.[1]

The roll of officers present at the battle was compiled by Charles Dalton, F.R.G.S., from an MS. Army List at the Public Record Office, corrected to May, 1709, which gives all the regiments that served at Malplaquet, with the exception of a battalion of the Coldstream Guards and Erle's. The staff, artillery train and the aforesaid two regiments have been added by Mr. Dalton to the roll.

At Waterloo there were 69,686 men in Wellington's army, and the loss was 22,469, or one in three nearly; at Malplaquet it was one in five; at Talavera one in four—5,000 being killed and wounded out of 19,800 engaged.[2]

It is interesting to compare the losses at Malplaquet with the slaughter at the battle of Vionville, fought on the 16th

1 English Army Lists and Commission Registers, by Charles Dalton.

2 Siborne's Waterloo.

August, 1870, where the French lost 16,000 men in ten hours, and the Germans fared quite as badly.

According to Boyer's Annals of Queen Anne, the following infantry regiments were present at Malplaquet:—

2nd Batt. Guards.		Lalo's	... 21 F.
Lord Orkney's 1st Batt.	1 F.	Sabine's ...	... 23 F.
" " 2nd "	1 F.	Primrose's ...	... 24 F.
Argyle's ...	... 3 F.	Preston's ...	... 26 F.
Webb's	... 8 F.	Orrery's ...	Disbanded
North's	... 10 F.	Temple's ...	"
How's	... 15 F.	Evan's	"
Godfrey's ...	... 16 F.	Meredith's ...	... 37 F.
Ingoldsby's ...	... 18 F.	Prendergast's ...	Disbanded
ERLE'S... ...	... 19 F.		

THE MALPLAQUET ROLL.

From Army Lists and Commission Registers by Charles Dalton, F.R.C.S.

LIEUT.-GENERAL ERLE'S REGT. OF FOOT.

Captains.	*Lieutenants.*	*Ensigns.*
	Saml. Norman,[12] Captain Lieutenant.	Edwd. Rawson,[25] 5th April, 1708.
Geo. Freake,[1] Lieutenant-Colonel, Comg., 23rd March, 1709.	Jno. Whiting, W.[13] 7th May, 1708.	
Leonard Cradock,[2] Major, 23rd March, 1709.	Mons. Simonds.[14]	Gabriel Simonds,[26] 1st August, 1707.
Ric. Hawley,[3] 21st January, 1703.	Wm. Court,[15] 5th April, 1708.	Tho. Inwood,[27] 7th May, 1708.
Wm. Leigh.[4]	Ric. Tovey.[16]	Wm. Butts,[28] 24th June, 1708.
Harry Freake,[5] 7th May, 1708.	Sir. Warren Crosbie, Bt.[17] 24th April, 1707.	
Ric. Stone,[6] 7th March, 1706.	Robt. Eaglesfield.[18]	Wm. Davenport,[29] 24th June, 1707.
Ric. Wills.[7] Gr.	Wm. Martin,[19] 1st Lieutenant. } Tho. Piercy,[20] 2nd Lieutenant. 7th May, 1708. }	
Tho. Woodhouse,[8] 24th April, 1707.	Tho. Holland,[21] 21st January, 1703.	Michl. Legge,[30]
Geo. Paxton,[9] 23rd March, 1709.	Wm. Mercer,[22] 18th October, 1703.	Mat. Waller,[31] 24th April, 1707.
Tho. New,[10] 23rd December, 1708	Jas. Philips[23]	
Wm. Taylor,[11] 1st May, 1709.	Tho. Pratt.[24]	

Chaplain: Saml. Dunster,[32] 7th May, 1708.

Adjutant: Tho. Pratt, 6th December, 1706.

Quartermaster: Erle Simonds,[33] 23rd March, 1709.

---

1 See text.

2 Appointed Ensign 25th August, 1693. Captain 18th May, 1699. Had the Queen's leave to be absent from his regiment when it was sent to the West Indies in 1702. Brevet-Major 7th March, 1706. Out of the regiment 10th June, 1710.

3 Younger son of Henry Hawley of Brentford. Appointed Ensign 7th March, 1692. Lieutenant 13th August, 1695. Major 10th June, 1710. Serving in the same regiment in 1740.

4 A certain "Wm. Lee" was appointed Lieutenant in the regiment in 1689. Serving in Ireland as Captain in 1706. Brevet-Lieutenant-Colonel 1st January, 1712. Transferred to the Coldstream Guards as Captain and Lieutenant-Colonel 22nd July, 1707. Services untraced after 1709.

5 Brother to Lieutenant-Colonel Geo. Freake. Appointed Captain-Lieutenant 6th December, 1706. Services untraced after 1709.

6 Died as Captain in 1716.

7 Appointed Lieutenant 4th March, 1694. Untraced after 1709.

8 Commission renewed in January, 1715. Half-pay in 1735.

9 Not in any subsequent list.

10 Appointed Ensign 20th August, 1695. Lieutenant before 1706. Captain-Lieutenant 7th May, 1708. Untraced after 1709.

11 A certain Wm. Taylor was appointed Colonel of a regiment of foot to be raised for H.M. service in the West Indies 1st April, 1710.

12 Appointed Ensign 31st August, 1693. Half-pay in 1698. Reappointed Ensign in 1702. Lieutenant before 1706. Captain-Lieutenant in 1709. Commission renewed 11th January, 1714/15.

13 The name of Lieutenant Jno. Whiting is erroneously given among the officers of Colonel Godfrey's Regiment of Foot, wounded at Malplaquet in Records 16th Foot. Whiting's name appears as Lieutenant in Sutton's Regiment in the MS. Army List for 1715, and was promoted Captain-Lieutenant in the 16th Foot same year.

14 Captain 23rd December, 1726. Commission renewed in 1727. Serving in 1730. Out in 1733. Served in the Expeditions to Cadiz and the West Indies.

15 Tried by General Court Martial held at Warde, Flanders, on 6th May, 1711. Charge: Making a complaint of a paper delivered to Colonel Freake signed by several officers of the regiment, reflecting on his honour and reputation. Captain James Phillips and several other officers gave evidence. Acquitted, the Court being unanimously of opinion that the accusation was grounded on errors. Further services untraced.

16 Out of the regiment in 1715.

17 Of Crosbie Park, co. Wicklow. Retired from the regiment 17th June, 1746. Died 30th January, 1759.

18 Ensign 25th November, 1706. Serving in Colonel Stanwix's Regiment of Foot as Lieutenant before 1714. Lieutenant of a company of invalids at Carlisle 3rd April, 1719. Serving in 1740.

19 Untraced after 1709.

20 Commission renewed 11th January, 1714/15.

21 Ensign in Erle's 4th March, 1694. Lieutenant 21st January, 1702/3. Captain 6th December, 1716. Half-pay in 1718.

22 Promoted Captain 31st March, 1737. Retired 4th June, 1744.

23 Promoted Captain 10th January, 1710/11. Major 27th January, 1740/41. Lieutenant-Colonel 24th September, 1744. Out of the regiment 6th February, 1746/7.

24 Appointed Quarter-master 21st September, 1705. Captain of Grenadiers in January, 1715.

25 Commission renewed 11th January, 1714/15.

26 Promoted Lieutenant 16th November, 1713. Half-pay in 1713.

27 Lieutenant and Captain 1st Foot Guards 11th January, 1715. Captain and Lieutenant-Colonel 18th July, 1718. Died in 1747 on the full-pay list while holding the appointment of Deputy-Ranger of Enfield Chase.

28-29 Services untraced after 1709.

30 Lieutenant 23rd January, 1710/11. Commission renewed 11th January, 1714/15. Captain 27th January, 1740/41.

31 Serving as Lieutenant in 1715.

32 Out of the regiment 2nd November, 1713.

33 Placed on half-pay in 1713.

After wintering in Flanders the regiment again took the field in the spring, and advanced with the army towards Tournay, afterwards taking part in the movements by which the French lines were forced at Pont à Vendin.

When the siege of Douai was undertaken Erle's was one of the corps selected to take part in this service. To ensure success it was found necessary to pass the strong and well fortified lines of La Bassée, which had been greatly strengthened by new works, and were defended by forty battalions and thirty squadrons. It was anticipated that the passage could only be effected with considerable loss, but the defence was abandoned, and the Allies passed without the loss of a man at Pont à Vendin and Carrières.

Douai was then invested on the 23rd April, and trenches opened on the 4th May. But though Marshal Villars assembled his army and professed his intention of risking an engagement for its relief, he was only "very bountiful in threats," and after several advances, ultimately retired to a distance, and the place surrendered, but not without obstinate fighting, after fifty days of open trenches. In carrying on the attacks, and storming the outworks, the regiment had terrible losses, and probably in all its career it was never engaged in such a bloody business. The garrison at last finding they could get no help from Villars, beat a *chammade* on the 25th June, and articles of capitulation were signed two days afterwards. On the 29th June the garrison was permitted to march out with military honours.

At this siege the regiment had 3 sergeants and 91 rank and file killed, and 1 major, 2 captains, 8 subalterns, 10 sergeants, and 197 rank and file wounded.[1]

---

1 Cannon's History.

MEDAL TO COMMEMORATE THE BATTLE OF MALPLAQUET.

MEDAL TO COMMEMORATE THE SIEGE OF DOUAY.

The following return shows the different regiments present at the siege of Douai, with their killed and wounded[1]:—

Regiments	Officers K.	Officers W.	Men K.	Men W.
Erle's, 19th Foot	—	11	94	207
Mordaunt's, 21st Foot	1	7	49	182
Sabine's, 23rd Foot	2	9	54	147
Primrose's, 24th Foot	1	9	35	148
Preston's, 26th Foot..	1	6	50	186
Hans Hamilton's, 34th Foot ..	1	5	81	125
Sutton's, Disbanded	5	8	110	113
Honeywood's, Disbanded	2	6	86	170
Total ..	13	61	559	1278

The success of the Allies was dearly bought, and they lost altogether 8,700 killed and wounded, in numbers exceeding the strength of the garrison.

Erle's now formed part of the covering army during the siege of Bethune, which fortress surrendered on the 29th August. Aire and St. Venant were afterwards invested and taken. Marshal Villars was a spectator of these conquests, with an army superior to that of the Allies, but he did not venture from the strong position that he occupied. When he wished to frighten the Allies he represented his army as 160,000 strong, but when apologising for his inactivity, he stated his force at 40,000. " No doubt the lustre of Blenheim, Ramillies and Malplaquet, which played round the bayonets of Marlborough," choked him off somewhat; any way he retired without fighting.

After these surrenders Erle's regiment marched to Ghent, where it passed the winter.[2]

Advancing up country in the spring of 1711, the regiment encamped for a short time at Warde, where it was joined by a fine body of recruits from England.[2] It took part in the operations by which the enemy's fortified lines were passed

1 Battle Honours of the Army, by C. B. Norman.

2 Cannon's History.

at Arleux on the 5th August. This campaign was conducted by the Duke of Marlborough in person. He resolved to invest Bouchain, but to do so it was necessary to pass the lines which Marshal Villars had constructed in the direction of Arras and Cambray with such strength that he deemed them impregnable, and called them the *non plus ultra* of Marlborough.

By a succession of skilful movements and manœuvres Villars was utterly deceived, and Marlborough passed the lines without the loss of a man. He then invested Bouchain in the face of a superior army of the French. This proved a difficult service; but every obstacle was overcome by the skill and perseverance of the generals and engineers, and the innate bravery of the soldiers, who on more than one occasion fought up to their waists in water. The garrison surrendered on the 13th September.[1] The losses in Erle's are not known.

In the spring of 1712 the regiment quitted its winter quarters. Before the army was assembled General Erle was succeeded in the Colonelcy by Brigadier-General George Freke, who had held the Lieutenant-Colonelcy since 1709. The former consequently severed his connection with the regiment with which he had been associated for twenty-three years. His last active service with it had been when employed with the covering army during the siege of Namur. His promotion was then rapid, as on his return home he was appointed Governor of Plymouth and promoted Major-General in 1697. Further advancement came in 1703, and Erle was made a Lieutenant-General of the Ordnance on the recommendation of Marlborough. At the battle of Almanza he commanded the centre, and a year later was appointed Commander-in-Chief of a combined expedition

1 Cannon's History.

to the coast of France. This was Erle's last active service, and after he sold his Colonelcy to Brigadier-General Freke he took up several commands at home, being appointed respectively Commander-in-Chief in South Britain and Governor of Portsmouth.

Owing to his political views, he was removed from his position on the Ordnance, and except in 1715, when he was sent down to put Portsmouth in a state of defence, General Erle was not employed again.

His death took place at Charborough, Dorset, on the 23rd July, 1720, and he was buried in the vault of the Parish Church beside his wife Elizabeth, second daughter of Sir William Wyndham, Bart., of Orchard Wyndham, Somersetshire, who died before him.

Brigadier-General Freke only retained the Colonelcy for a short time, as he sold it to Brigadier-General Richard Sutton in April, 1712, and then retired from the army. He died in 1730, and was buried at Upwey, Dorset.

The following sketch will show what a distinguished officer Richard Sutton was at this point in his career:—

In 1690 he had been appointed an ensign in Viscount Castleton's Regiment, and was with it in the Castle of Namur when Fort William surrendered to the French in June, 1692. The following year he took part with his regiment in the attack on the French fortified lines between the rivers Scheldt and Leys, and it is recorded that Castleton's, which led the right attack on Beau Verd redoubt, gained great honour in this business and lost many lives.

At the famous siege and capture of Namur in 1695, Richard Sutton earned the goodwill and confidence of the fighting Lord Cutts—that heaven-born leader of forlorn hopes and storming parties, who loved to fight in the deadly breach with any weapon that came to hand, and whose immunity under the hottest fire had earned for him in the British Army the name of "Salamander."

It happened that at midnight on the 18th August the French made a sortie from the Castle of Namur with 200 mounted dragoons and 500 grenadiers. Half this force made an attack on the advanced guard of the British right, while the remainder tried to rush the advanced guard of the British left. Count de Rivers and Lord Cutts, who had just posted the advanced guards to cover the workmen in the trenches, beat back the attacking party on the right. While they were doing this 100 dragoons charged the advanced guard to the left, which consisted of about forty soldiers, under Captain Sutton. This officer told his men to keep their ground and not to fire till he gave the order. When the dragoons were within a few yards of Sutton he gave the order to fire, and during the confusion that followed the discharge he marched his men quietly towards the British main body. But the dragoons pressing again upon him, he directed those of his men who had reserved their fire to give the enemy another volley. "The French," says the historian who relates the foregoing incident, "received the volley undauntedly, pressed on, and if their grenadiers had charged at the same time, could doubtless have caused a great confusion in the trenches; but the Spanish and Bavarian horse, who were near, fell upon the enemy with so much vigour that they drove them to the very palisades of the Castle, killing some and taking others prisoners."

After the Peace of Ryswick his regiment was disbanded, and Sutton was placed on half-pay till 1701, when he was appointed Major in Princess Anne of Denmark's Regiment, the present King's (Liverpool) Regiment. Under the Duke of Marlborough he took part in the siege of Fort St. Michael, the taking of Ruremond, and the storming of the citadel of Liege.

In 1702 he was promoted Lieutenant-Colonel, and led his regiment to victory in the glorious battles of Schellenberg and Blenheim, gaining the brevet of Colonel for his services. Sutton was also in command at the battle of Ramillies as well as at the sieges of Menin and Aeth. At the battle of Oudenarde his corps led the attack on the village of Eyne, which was held by seven battalions of Swiss. After a brave defence Eyne was carried sword in hand, and three entire Swiss battalions were taken prisoners, when Sutton received their colours.

In 1709 Colonel Sutton was given a foot regiment, which, however, was disbanded at the Peace of Utrecht, and on the 1st January, 1710, was appointed a Brigadier-General in Marlborough's army. He commanded a brigade at the forcing of the French lines at Arleux in August, 1711, and had the honour of taking home the despatch announcing this important victory. He commanded a brigade under the Duke of Ormonde in 1712, and in 1714 brought home the British troops from Flanders.[1]

Naturally, coming to his new regiment with such a splendid record, the officers and men were proud to have him at their head.

There was now again a vacancy for a Lieutenant-Colonel in the regiment, which Brigadier Sutton conferred on Major George Grove in 1713. This officer had been appointed an Ensign in Colonel Edward Fitzpatrick's Regiment in 1692, and in August, 1693, was transferred to the Royal Fusiliers.

1 From a Memoir by Charles Dalton, Esq., in "The Green Howards' Gazette."

Nearly two years later he was promoted Captain into Brigadier-General William Evans' Foot. He had served in Flanders with the Royal Fusiliers, and had been present at the battle of Malplaquet.

The army in Flanders was this year commanded by the Duke of Ormonde, and advanced to the frontiers of France; but negotiations for a treaty of peace having commenced, a suspension of hostilities was proclaimed, and the British troops retired to the vicinity of Ghent, Sutton's and Orrery's[1] regiments being quartered in Bruges, the garrison there being under the command of Brigadier Sutton.

The death of Queen Anne took place on the 1st of August, 1714, when the English troops were ordered to return to England. The regiment landed at Dover in August, and was quartered, three companies at Sheerness, two at Tilbury Fort, and seven at Gravesend, Northfleet, Southfleet, Grays and Horndon. Two companies afterwards proceeded to Languard Fort, "and such part of these two Comp^{s} for whom there shall not be room in the Fort are to Quarter in the neighbouring villages of Walton, Felixstow and Trumless, subject nevertheless to the orders of the Govr of Landg^{d} Fort."[2]

Two companies later on marched to Hull and two to Tynmouth, with orders to "Rest the Sundays & every third or fourth day on their mar: as the Offi^{r} in Ch^{r} shall see cause."[2]

These were followed in September, 1714, by two more companies from Grays and Horndon to Tynmouth, and three more to Hull at the same time.

In March, 1715, the regiment proceeded to Ireland, and was employed on home service till it embarked for Vigo.

1 The Earl of Orrery's, later 21st Fusiliers.

2 War Office Marching Order Book.

Another change in the Colonelcy occurred in August, when it was conferred on the Lieutenant-Colonel, George Grove, in succession to Major-General Sutton.

The new Lieutenant-Colonel was Thomas Handasyde from Brigadier-General Thomas Handasyde's Regiment, his commission being dated the 5th August, 1715.

England had declared war against Spain in 1718, and the following year an expedition was fitted out under command of Lord Cobham, and sailed on the 21st September, Vigo being the objective, where a quantity of military stores had been accumulated for the invasion of Britain in favour of the Pretender. Colonel Grove's Regiment formed part of this force.

The expedition entered the harbour of Vigo on the 29th September, seized seven ships, three of which were fitted up as privateers, whilst the grenadiers landed three miles from the town. A sharp fire of musketry was opened on them from the hills, but at too great a distance to do much harm. The battalion companies then gained the shore, and bivouacked.

On the 1st October they approached the town. Abundance of wine was found in the houses deserted by the inhabitants, and the soldiers drank of it freely, so that a strict discipline was necessary. The enemy abandoned the town of Vigo and the Fort of St. Sebastian, but 1,000 Spaniards retired to the citadel only to surrender after losing 300 men. Ten thousand barrels of powder, 8,000 muskets, 15 brass guns were found in the castle.

Whilst the siege of the citadel was in progress 500 men were detached against Rodonella, who took and burnt the town. General Wade, with 1,000 men, embarked on the 12th October, and proceeded to the upper end of Vigo Bay. Here they landed, and marched thirty miles to Pont a Vedia, when thirteen companies of Spaniards fled in a panic, and the inhabitants hurried from their homes. The town of Pont a Vedia, with its arsenal, barracks for 2,000 men, 13 pieces of brass and 86 of iron ordnance, 5,000 stand of small arms, 300 barrels of powder, and abundance of stores, fell into the hands of the English.

The King of Spain being now oppressed on all sides and his projects defeated, made overtures for peace, and the troops returned to England about the end of October, and landed at Portsmouth.

As regards the dress of the officers and men, all chivalric costume had ended in the reign of Queen Anne. The hat ornamented with feathers, with a broad brim, two sides of which were turned up, was the head-dress; a full flowing wig, square-cut coat and long flapped waistcoat, with large pockets to both. Breeches tied below the knee, with stockings drawn over up to the middle of the thigh; sword slung over the right shoulder, sash worn round the waist and knotted on the right side. This was the dress of the officer.

Armour for the infantry had been completely set aside, as well as pikes, and every soldier was armed with a musket, bayonet, and sword. The grenadiers also ceased to carry grenades. The men wore an easy scarlet coat with facings, a cocked hat, breeches and stockings, with a strap below the knee to keep them up.

After the return of the regiment from the expedition to Vigo its movements at home are not recorded. Cannon states that it marched to Scotland in 1723, and Stocqueler in his history says that it formed part of the force under Brigadier-General George Wade which assisted in disarming the clans in 1724. Brigadier-General Henry Grove's[1] Regiment is constantly referred to in connection with these operations in Scotland, but there is absolutely no mention of Colonel George Grove's. We are led to believe that the regiment was in Ireland at this period, as in the year 1725 it was inspected by Major-General Thomas Pearce when quartered in the County Cork. The headquarters and four companies were at Kinsale, with detachments of two companies at Ross Castle and Macroom, and one company at both Ross Carbery and Bantry. The strength of the battalion was 11 field officers and captains, 10 lieutenants, 10 ensigns, 4 staff, 19 sergeants, 20 corporals and 334 privates. The inspecting officer reports as follows.:—

"In obedience to your Excellency's Command I have viewed the above Regiment. The Bodies of the men are very good. The Regt is very well disciplined, performing their Exercises and firing very well. The Cloathing is good, delivered to the men at different times from ye 18th to ye 28 July. The arms and Accoutrements of this regiment are bad. I found the men were regularly heard and no Complaints. This Regiment have tents.

"(Sd) THOS. PEARCE."

On returning to England in January, 1727, the regiment landed at Bristol, and was quartered at Wells, sending out detachments to Frome, Shepton Mallet, and in June one company to Winchester.[2] It was now augmented from 10 companies of 34 men, by 1 sergeant, 1 corporal, 1 drummer, and 26 privates each, and 2 additional companies.

During the year the regiment was held in readiness to assist the Dutch in the expected war with the Emperor of Germany, but no embarkation took place.

---

1 10th Foot the Lincolnshire Regiment.
2 War Office Marching Order Book.

Colonel George Grove died on the 13th October, 1729, from the effects of a fall from his horse, and King George II. restored Richard Sutton to the Colonelcy of his old regiment. He had been promoted Major-General in 1727.

Orders now came for Ireland, and the command changed hands on the 12th June, 1731, when Lord Sempill succeeded Lieutenant-Colonel Thomas Handasyde. He had been appointed Adjutant of Colonel George Preston's[1] Regiment in 1708, and with it had served in Spain and Flanders, taking part in the battle of Malplaquet. As a Captain he was transferred to Brigadier-General Grant's Regiment in 1715, and was promoted Major three years later.

In June, 1732, the regiment marched from Dublin to Waterford, where it took up its quarters.

Major-General Sutton was promoted Lieutenant-General in 1735, and on his death in 1738 the Colonelcy was conferred on the Honourable Charles Howard, of the Coldstream Guards, his commission bearing date the 1st November.

This officer was the second son of Charles, third Earl of Carlisle, and had joined the Coldstream Guards in 1715. Two years later he was promoted into the 16th Foot, from which he exchanged to the 9th Dragoons almost at once. Returning to the Guards as Captain and Lieutenant-Colonel in 1719, he remained with them till 1734, when he was appointed A.D.C. to the King and promoted Colonel.

He greatly distinguished himself in the War of the Austrian Succession, having command of a brigade at Fontenoy, where he was wounded in four places, and also took part in the battles of Roucoux and Lauffeldt, in command of the infantry.

1 26th Foot the Cameronians.

GENERAL THE HONOURABLE CHARLES HOWARD, K. B.

FROM A PAINTING IN NAWORTH CASTLE.

General Howard acted as Deputy-Governor of Carlisle in 1725, and in 1752 was appointed Governor of Fort George and Fort Augustus. He was one of the Grooms of the Bedchamber to King George II., and for many years member of Parliament for Carlisle. His death occurred on the 11th May, 1765.

In 1739, when war was declared against Spain, the regiment was withdrawn from Ireland and embarked for Scotland, where it arrived in July, and was quartered in Linlithgow, pending the departure of Barril's[1] Regiment from Edinburgh. This took place on the 23rd July, when Colonel Charles Howard's marched in and took up garrison duty in the ancient capital.

On the 8th August Colonel Howard inspected his regiment on parade, and afterwards gave a splendid entertainment to the officers, at which the Duke of Buccleugh and several other persons of distinction were present.[2]

Recruiting went on briskly during this time, and the companies shortly after their arrival were over fifty strong.

On the 23rd September Colonel Howard again reviewed his regiment in the Canongate.

"Edinburgh, 13th November, 1739.

"His Excellency General Clayton reviewed Colonel Howard's Regiment in St. Anne's yards, and although it was an extreme frost, they performed their exercises well; the veterans, as well as the recruits, being all new cloathed, made a fine appearance."[2]

On the 7th July, 1740, four companies of the regiment marched from the Canongate for Berwick to relieve General

1 4th Foot the King's Own Royal Lancaster Regiment.

2 Extract from "The Scots Courant," 1739.

Howards' Regiment,[1] which had gone to Newcastle to quell some riots there.

General Clayton reviewed, near Leith on the 29th July, the six remaining companies of the regiment. Writing on the same date to his brother, Lord Carlisle, Colonel Howard says: "General Clayton saw my six companies this morning; they did their business very well, and gave the General satisfaction and me much pleasure. They march next Wednesday for their quarters in Scotland, so I have nothing further to do here, but shall go the latter end of the week to Berwick."[2]

On the 31st July these six companies marched from Edinburgh *en route* to Glasgow, being relieved by Colonel Cornwallis' Regiment.

On the promotion of Lieutenant-Colonel Lord Sempill[3] to the 42nd Foot he was succeeded on the 22nd January, 1740/41, by Lieutenant-Colonel Anthony Rotolph Ladaveze[4] from Sir Harry Goring's[5] Regiment. He had been appointed Second-Lieutenant in Colonel Joshua Churchill's[5] Foot in April, 1707, of which he was Adjutant in 1712 and promoted Major in 1732.

The regiment was quartered in Dundee in April, 1741, the last company arriving there on the 15th of the month. Here it was again inspected by Colonel Howard, who writing to Lord Carlisle from Edinburgh on the 6th June, says: "I came here yesterday, having been round my regiment in their quarters, and I have received an order from the Lords

1 3rd Foot the Buffs, commanded by Lieutenant-General Thomas Howard.

2 MSS. at Castle Howard.

3 Lord Sempill was transferred to the 25th Foot on the 9th April, 1745. He commanded the left wing of the army at Culloden. Died in Nov., 1746.

4 Born at the Hague in 1689. Second son of the Minister there. Received the freedom of Carrickfergus in 1724, and died 25th November, 1769.

5 The 31st Foot.

Justices for them to go to Berwick upon General Howard's and Cornwallis' being ordered from the North for this encampment in Essex."[1]

Consequently the regiment forthwith embarked at Berwick for the Tower of London, where it arrived during the last week in June. Orders then came for it to join the camp at Colchester in company with six other regiments. At the close of the training the troops were reviewed there by H.R.H. the Duke of Cumberland on the 2nd of September, who was attended by the Duke of Marlborough, the Earl of Albemarle, and Generals Wade, Honeywood and Howard.

On the breaking up of this camp in September the regiment was ordered to Glasgow, where it remained till August, 1742, taking the place of Colonel Desgrangue's Regiment at Edinburgh during the month.

The accompanying illustration will give some idea of the dress worn at this period. The voluminous redcoat is similar in many respects to that worn in 1715, excepting that the skirts are hooked back and show the colour of the regimental facing. The edging of the coat, the lapels, cuffs, pocket flaps, and the red waistcoat are trimmed with the distinctive regimental lace, white with two stripes, red and green. A white neckcloth was worn, and the hat was three-cornered, trimmed with white lace. The breeches were red, and white gaiters were worn high above the knees, fastened with dark-coloured garters. The ammunition pouch was supported by a broad leather belt over the left shoulder.

On the 4th November, 1742, Colonel the Hon. Charles Howard's Regiment marched out of Edinburgh to Berwick, which it made its headquarters, sending out a detachment of two companies to Newcastle. The latter were joined by the

1 MSS. at Castle Howard.

rest of the battalion in January, 1742/3, when it was relieved by Colonel Price's Foot.

It was at this period that Colonel Howard was sent on a diplomatic mission to the Court of Vienna, and regimental tradition has it that he was presented by the Empress Maria Theresa with three marches for his regiment, a quick, grand or slow, and a funeral march. The two latter are still in use, but the quick-step has been lost for many years, certainly since 1848. A copy of these two marches will be found in the Appendix.

CHAPTER IV.

"The Green Howards"—The Battle of Fontenoy—Roll of Officers, 1745—The Rebellion in Scotland—Battles of Roucoux and Lauffeldt—Introduction of Fifes—The Nineteenth Foot—Siege of Belle Isle—The Belle Isle Roll.

A BRITISH Army had proceeded to Flanders in 1743 to support the interests of the house of Austria against the designs of France and Bavaria, but Brigadier-General Howard's Regiment did not leave England till the spring of 1744, when it proceeded to join the army, and served in the campaign of that year under Field-Marshal Wade. The regiment was encamped for some time on the banks of the Scheldt, and afterwards advanced into the territory subject to France as far as Lisle; but no serious fighting took place, and it passed the winter in quarters at Ghent.

Up to this time, and in fact up to 1751, it was customary for regiments to be called by their Colonel's names, but in this campaign of 1744 a difficulty presented itself, for there was serving in the same command another "Colonel Howard's Regiment," viz., that of Lieutenant-General Thomas Howard. The confusion that must result in a campaign in which two regiments of the same name are serving side by side was soon apparent; but though alike in name, there was in this case a marked difference in the appearance of the two corps, for the facings of Colonel Charles Howard's Regiment were green, whilst those of the other were buff. Small wonder then that the former was quickly dubbed "The Green Howards," a name that has been retained ever since by the regiment, and which is at the present day most popular with all ranks, and, we might add, its best known designation.

In September of this year Lieutenant Colonel Ladaveze was succeeded in the command by Major James Phillips, who had exactly forty-two years' service, all of it spent in the regiment to which he had been gazetted in September, 1702.

The Green Howards were called from their winter quarters in Ghent in April, 1745, and advanced with the army commanded by the Duke of Cumberland to the relief of Tournay, which fortress was besieged by a numerous French force, under Marshal Saxe.

Cumberland, after leaving Soignies on the 5th May, arrived on the 9th at Bruffœil, within sight of Saxe's army, which was drawn up on the slope before the village of Fontenoy, the centre of his position. In his front were a lot of copses and woods, filled with irregular troops, fighting on the French side. Cumberland did not advance further this day, but fixed his headquarters at Maubrai, a village within musket shot of the enemy's advanced posts. Bruffœil in the rear served as his headquarters and a base for hospitals and baggage. On the 10th May the French advanced posts were pushed out of the copses, and the Duke went forward to reconnoitre the position.

Saxe's army occupied the crest of the slope running from Antoing on the Scheldt to Fontenoy, facing south, and from Fontenoy to the forest of Barri, facing east. Three redoubts ran between Antoing and Fontenoy, the latter village being fortified with works and cannon. Up to the forest ran a double line of entrenchments, and at its edge were two more redoubts.

By 5 a.m. on the 11th May the battle had begun. Cumberland's plan was simple enough. A brigade, consisting of three English regiments and a Hanoverian battalion, were to attack the French left, whilst the remainder of the infantry were simply to march across the thousand yards of open ground and sweep the enemy out of the entrenchments. This order seems to have been misunderstood, for this first attack was never made, and Cumberland, getting impatient, placed himself at the head of the British line. This was composed of three battalions of Guards and fourteen line regiments, together with some Hanoverian troops. They advanced in two lines, the Green Howards being on the extreme left of the first line.

"Forward tramped the ranks of scarlet, silent and stately as if on parade. Full half a mile of ground had to be traversed before they could close with the invisible enemy that awaited them in the entrenchments over the crest of the slope, and the way was marked clearly by the red flashes and puffs of white smoke that leaped from Fontenoy and the redoubts on either flank. The shot plunged fiercely and more fiercely into the serried lines as they advanced into that murderous cross-fire, but the gaping ranks were quickly closed, the perfect order was never lost, the stately step never hurried. Only the Hanoverians, finding they were cramped for want of space, dropped back quietly and decorously, and marched in a third line behind the British. Silent and inexorable the scarlet lines strode on. They came abreast of village and redoubt, and the shot, which had hitherto swept away files, now swept away ranks. Then the first line passed beyond village and redoubt, and the French cannon took it in reverse. The gaps grew wider and more frequent, the front grew narrower as the men closed up, but still the proud battalions advanced, strewing the sward behind them with scarlet, like some mass of red blossoms that floats down a lazy stream, and sheds its petals as it goes."[1]

1 Fortescue's History of the Army.

The English advance is also described in the well-known lines by the poet Thomas Davis.

> Six thousand English veterans in stately column tread;
> Their cannon blaze in front and flank, Lord Hay is at their head;
> Steady they step adown the slope—steady they climb the hill;
> Steady they load—steady they fire, moving right onward still,
> Betwixt the wood and Fontenoy, as through a furnace blast,
> Through rampart, trench, and palisade, and bullets showering fast;
> And on the open plain above they rose and kept their course,
> With ready fire and grim resolve, that mocked at hostile force:

The first French line was utterly shattered and broken, and the redcoats continued their triumphant advance for full three hundred yards into the heart of the French camp, and Saxe thought that the battle was lost. But the Dutch and the Austrians, who were co-operating with the British, had been defeated in their attack on Fontenoy, and the British being unsupported had to fall back. The Allies made a second attempt on the village, and Cumberland, relying on this, again advanced at the head of the British. Saxe, however, had brought forward his reserves, and amongst them the famous Irish brigade, to meet him; the French troops had rallied, and the task of the British was desperate.

Yet even in their second attack they carried their advance as far as their first, and though their losses had been enormous, their splendid discipline enabled them to beat back the Irish brigade for a time. Being closed in by the French on both flanks, there was nothing for it but a retreat, and the army fell back on Aeth, where it encamped under the guns of that fortress.

The regiment had 2 officers (Lieutenant Le Grand and Ensign Gibson) and 17 men killed; 7 officers (Major Petitot, Captains Cochrane and Douglas, Lieutenant Coote, Ensigns Cheape, Martin and Potterfield), 1 serjeant and 69 rank and file wounded; 13 men were missing.

Major-General the Hon. Charles Howard, who commanded a brigade in the action, was wounded in four places.

The British casualties amounted to 4,041 men, the rest of the Allies lost 3,000, and the French about 5,000.

In this battle the British infantry won undying fame. Duncan, in his history of the Royal Artillery, says: "Fontenoy was a defeat, but hardly one which can be said to have tarnished in the slightest the British arms."

Their firmness and discipline exhorted admiration even from the foe.

"The British behaved well, and were excelled in ardour by none of our officers. . . I cannot say as much for the other auxiliaries, some of whom seemed as if they had no great concern in the matter whichever way it went. In short, we gained the victory, but may I never see such another."[1]

Skrine, in his admirable account of the battle, says: "Every infantry regiment did its duty nobly. Fontenoy should be borne on the colours of all who shared in the glory of that day."

---

1 An Account of Fontenoy. Paris. 1745.

ROLL OF OFFICERS—1745.

RANK.	NAME.	REMARKS.
Colonel	Hon. Charles Howard	See text.
Lieut.-Col.	James Phillips... ...	In the Malplaquet Roll. Out of the regiment 6th February, 1746/7.
Major	William Petitot ...	Lieut.-Col. 2nd July, 1747. Major-Gen. 2nd Mar., 1761. Wounded at Fontenoy and Lauffeldt. Died at Northallerton 26th July, 1764.
Captain	Lord George Bentinck	Capt. and Lieut.-Col. 1st Foot Guards 27th May, 1745. Present at Dettingen. Died at Bath 2nd March, 1759.
,,	Sir Warren Crosbie, Bt.	Retired 17th June, 1746. In the Malplaquet Roll. Died 30th Jan., 1759.
,,	Richard Legge ...	In the Malplaquet Roll.
,,	Thomas Leake ...	Taken prisoner at battle of Roucoux, 1746.
,,	Robert Douglas ...	Lieut.-Col. 10th April, 1758. Wounded at Fontenoy. See Belle Isle Roll.
,,	James Halliburton ...	Out of the regiment 2nd Feb., 1747/8.
,,	Charles Lumsden ...	See Belle Isle Roll.
,,	Patrick Cochrane ...	Wounded at Fontenoy. Died 23rd May, 1746.
,,	Robert Farmar ...	See Belle Isle Roll.
,,	William Masters ...	Wounded at Lauffeldt, 1747. Captain 36th Foot, 1755.
,,	William Cholmley ...	Retired 24th April, 1755.
Capt.-Lieut.	Alexander Moor ...	Died at Caldwell 3rd March, 1790.
Lieutenant	George Coote	Wounded at Fontenoy. 15th Foot, 1756. Died at Ipswich, 1785.
,,	Henry Goddard ...	Wounded at Lauffeldt, 1747. 24th Foot 1754.
,,	George Sempill ...	Captain 53rd Foot 31st Jan., 1756.
,,	Thomas Mainwaring...	
,,	James Campbell ...	Taken prisoner at Roucoux, 1746.
,,	Daniel Le Grand ...	Killed at Fontenoy, 1745.
,,	Thomas Cuthbert ...	Wounded at Roucoux, 1746. See Belle Isle Roll.
,,	William Ogilvie ...	
,,	Robert Aston... ...	
Ensign	Hugh Sempill... ...	Capt.-Lieut. 25th August, 1756. Died January, 1764.
,,	Alexander Gibson ...	Killed at Fontenoy, 1745.
,,	Robert Skeene ...	Captain 59th Foot 14th April, 1756. Colonel 48th Foot 31st March, 1783. Lieut.-General 10th Nov., 1782.
,,	Alexander Cheape ...	Wounded at Fontenoy, 1745.
,,	George Brown ...	Wounded at Lauffeldt, 1747.
,,	Boyd Potterfield ...	Wounded at Fontenoy.

RANK.	NAME.	REMARKS.
Ensign	William Murray ...	66th Foot April, 1758.
,,	Edward Paunceforte...	
,,	John Collett Mawhood	Captain 3rd Foot 3rd February, 1746.
,,	William Farquhar ...	
,,	William Martin ...	Wounded at Fontenoy and Lauffeldt.
,,	James McFarlane ...	Wounded at Roucoux, 1746. Lieut. 55th Foot 30th December, 1755.
Lieut. and Adjutant	Nicholas Ford ...	Captain 2nd July, 1747.
Surgeon	Walter Hamilton ...	

The operations of the Allies subsequent to the battle of Fontenoy were of a defensive character, and the army was so much inferior in numbers to the enemy that it was unable to prevent the capture of several fortified towns in the Austrian Netherlands.

Owing to the rebellion which broke out in Scotland when Charles Edward, eldest son of the Pretender, gained some advantages at the head of the Highland clans, several regiments were ordered to return to England. Cannon, in his little history of the regiment, says it was one of those which remained in Flanders, but this is evidently a mistake, as will be seen from the following facts.

Lord Harrington wrote on September 15th to the Duke of Cumberland as follows:—

"The rebels in Scotland have found means to steal by at a distance from the King's troops, which were moving on through the Highlands to give them battle, and were, when the last letter came away, in full march for Edinburgh, having left Sir John Cope by that motion two or three days behind them. It is, therefore, his Majesty's pleasure that your Royal Highness should . . . give orders for the marching of ten of the best English battalions directly for Willemstadt, to be commanded by Sir John Ligonier, who is to take them to Gravesend with the utmost expedition. ."

Cumberland replied to this September 20th:—

"I must do both the Marshal and Prince Waldeck the justice to say that they were equally concerned for the occasion, and the diminution of our present small force. The ten battalions which I have named for this service are the three of Guards, Sowle's,[1] Pulteney's,[2] Charles Howard's, Bragg's,[3] Douglas's,[4] Johnson's,[5] and Cholmondeley's.[6] I can assure his Majesty that last Fryday I had the satisfaction to see the whole army under arms, and can with the greatest truth say that the battalions were equally fine and in good order; but if there were any prefference to be given it was to these ten, which I have pick'd out for that very reason."

The review alluded to was held on September 17th, on an extensive plain near Vilvorden. General orders prove that it was an unqualified success:—

"September 18th. It is H.R.Hs. pleasure to declare that he is so thoroughly pleas'd and satisfied with the clean appearance and good order of the troops he saw yesterday, that he orders it to be signified to all ye officers that he is extremely obliedged to them for their care."

All the above regiments on arrival in England were hurried up to the Midland Counties, but only two of them, Pulteney's and Cholmondeley's, took part in the battle of Culloden, fought on April 16th, 1746.

Other reinforcements from the army that followed on after the first contingent had better luck, and most of the regiments took part in the Scottish campaign.

---

1 11th Foot The Devonshire Regiment.
2 13th Foot Prince Albert's Somersetshire Light Infantry.
3 28th Foot The Gloucestershire Regiment.
4 32nd Foot The Duke of Cornwall's Light Infantry.
5 33rd Foot The Duke of Wellington's West Riding Regiment.
6 34th Foot The Border Regiment.

The war against France was still going on, and no sooner had the rebellion in Scotland been extinguished than troops were again poured into Flanders. Towards the end of June Sir John Ligonier brought over four cavalry and six infantry regiments, amongst the latter being the Green Howards. He had also under his command 6,000 Hessian mercenaries and 13,000 Hanoverians, but the English force only comprised the above ten regiments. In addition to these troops there were 49,000 Austrians and 24,000 Dutch.

Crossing the river Jaar on the 7th October, the Allies found themselves in the beautiful plain of Liège, with the river Meuse in their rear, and an army of twice their strength, under Marshal Saxe, in their front on the Jaar.

The Marshal crossed the Jaar on the 10th October with the intention of attacking. The Allies' position faced due west, their right on the Jaar, and their left on the village of Ans. In the centre of the position, near the village of Liers, were posted four British battalions, besides some Hanoverians, the villages of Roucoux and Varoux being on their left front.

On the 11th October the garrison of these two villages was withdrawn to cover the left, and their defence was entrusted to the four British regiments in addition to four battalions of Hanoverian and Hessian troops. The Green Howards were posted in Roucoux, together with Sowle's Regiment, and were both under command of Brigadier-General Douglas.

The battle was opened by a furious assault on the left flank, and at the same time fifty-five battalions in three columns were launched upon Liers, Varoux and Roucoux. The disparity of numbers was about one to six, yet the British, Dutch and Germans, outmatched though they were, stood their ground gallantly, and beat back the leading brigades of the enemy. But no sooner were two French brigades repulsed in each village, than a third brigade ran in, and the eight battalions, though they still held Liers, were forced to withdraw from Roucoux and Varoux. About this time Sir John Ligonier came galloping up and ordered Howard's and Sowle's[1] to take up a position in a hollow-way, where they defended themselves till a retreat was ordered, when the army crossed the Meuse, and encamped near Maestricht.

The regiment had 1 serjeant, 1 drummer and 33 rank and file killed on this occasion; Lieutenant Cuthbert, Ensign McFarlane, 3 sergeants and 19 rank and file wounded; Captain Leake, Lieutenant Campbell and several men prisoners.

The army after this battle, according to its usual custom, went into winter quarters in Holland.

If the cause of Queen Maria Theresa was to be saved, it was evident that great efforts were imperative in the coming campaign of 1747. The contribution of the British to the allied army was four cavalry and fourteen infantry regiments, and the whole army was to be under the command of the Duke of Cumberland.

The French were encamped at Malines, and towards the end of May the Allies marched straight against them, but Saxe, entrenched to the teeth, remained immovable for three weeks, and Cumberland despaired of bringing him to action. However, hearing that a detached force of 30,000 French was near Tongres, Cumberland marched to the Demer in order to cut them off, but Marshal Saxe was on his guard, and on the 29th June prepared to concentrate his whole army at Tongres. An Austrian corps was thereupon ordered to

1 11th Foot Devonshire Regiment.

occupy Bilsen, and the Allies' main body was marching there when Saxe suddenly confronted them on the 2nd July, having made a forced march.

The Duke now prepared for battle, and took up a position extending from some rising ground known as the Commanderie, a little to the south-west of Bilsen, along a chain of villages to the river Jaar, south of Maestricht.

The Commanderie was held by the Austrians, with the right thrown back to Bilsen. Then came the Dutch in the centre of the line, whilst the Hanoverians and British held the villages of Val or Vlytingen, Lauffeldt and Kesselt on the extreme left.

The enemy began by attacking Val and Lauffeldt, the former being occupied by three English and one Hanoverian regiment, who repulsed the attacks of their numerous opponents for some time, but were eventually forced to give way. They were reinforced by Onslow's,[1] the Green Howards and Conway's,[2] and returning to the charge recovered the village. "It was just such another fight as that of Landen in 1693. From hedge to hedge, and from wall to wall, the French, notwithstanding all their superiority in numbers, were driven back headlong from the village with terrible loss. Saxe's first attack had been brilliantly repulsed."

He at once replaced the beaten troops by two fresh brigades, but these met with no better success. Few moments elapsed before a fresh body of assailants came rushing forward, but the British were again victorious, and a fourth attack on the position was also repulsed, the Irish brigade losing no fewer than sixty officers in the struggle. But numbers finally told, and a retreat had at last to be ordered.

1 8th Foot the King's Liverpool Regiment.
2 48th Foot the Northamptonshire Regiment.

This was successfully effected in good order, with little further loss, and the army arrived at Maestricht the same evening.

The total casualties of the English cavalry and infantry were close on 2,000 men.

Colonel the Hon. Charles Howard was in command of the infantry both at this battle and also at Roucoux. In the Duke of Cumberland's words, "they behaved so well one and all, that he could not commend any one regiment without doing injustice to the rest."

The casualties in the Green Howards were:—Killed: Lieutenant-Colonel James Williams, 2 serjeants, 1 drummer and 32 rank and file. Wounded: Major Petitot, Captain Masters, Lieutenants Goddard, Brown, Martin and Phillips, Ensigns Dobson and Fuller, 3 serjeants and 103 rank and file. Missing: 15 rank and file.

Lieutenant-Colonel Williams[1] had only taken over the command from Lieutenant-Colonel Phillips in February this year. He was succeeded by Major William Petitot, whose commission as Ensign dated from 1721. He had served with the regiment in Flanders, and had been wounded both at Fontenoy and Lauffeldt.

In 1747 fifes were introduced into the army generally. The Guards had adopted them in 1743, and the Green Howards were the first regiment of the line to make use of them. Our authority for this statement is Francis Grose, the antiquarian, who, in his history of the British Army, states:—

"The fife was not adopted in the marching regiments till the year 1747. The first regiment that had it was the

1 No commissions for Lieutenant-Colonel Williams are forthcoming at the Record Office, so that his previous service cannot be traced.—M.L.F.

19th, then called the Green Howards, in which corps I had the honour to serve; and I well remember a Hanoverian youth, an excellent fifer, being given by his Colonel to Lieutenant-Colonel Williams, then commanding that regiment at Bois-le-duc, in Dutch Flanders."

The regiment was again in the field in the summer of 1748, but hostilities were at length terminated by the treaty of Aix-la-Chapelle, which was signed on the 7th October. By it all the great treaties, from that of Westphalia in 1648, which first recognised the principle of a balance of power in Europe, to that of Vienna in 1738, were renewed and confirmed. Prussia retained Silesia, and the Empress Maria Theresa was guaranteed the possession of her hereditary dominions according to the Pragmatic Sanction. France surrendered her conquests in Flanders and the English those in the West Indies, so that Great Britain gained nothing by the war, save the glory of having supported it.

On the 15th March, 1748, Major-General the Hon. Charles Howard was removed to the Colonelcy of the 3rd Dragoon Guards, and was succeeded by Colonel Lord George Beauclerk from the 8th Marines. He was the sixth son of Charles, first Duke of St. Albans, and was born on the 26th December, 1704. Entering the army as Ensign in the Grenadier Guards, then the 1st Foot Guards, in 1723, he was transferred in 1728 as Captain to the Royal Horse Guards, in which regiment he served till 1736, and was then retransferred to his former regiment with the rank of Captain and Lieutenant-Colonel.

He continued to serve in the Foot Guards till 1745, when he was appointed Aide-de-Camp to King George II. with the rank of Colonel, and in the year 1747 the Colonelcy of the 8th Marine Regiment was conferred on him. This regiment was disbanded in 1748, and Lord George then became head of the Green Howards.

He was made Governor of Landguard Fort in 1753, and two years later was promoted Major-General, and commanded for a time the troops in North Britain. In 1758 he was promoted Lieutenant-General.

The regiment returned to England in November, 1748, and was stationed at Winchester, with detachments at Southampton, Petersfield and Romsey. The headquarters moved to Salisbury in March, 1749, with small detachments at Fisherton, Milford, Harnham, Andover, Ambresbury and Stockbridge.

Orders came again for foreign service in June, 1749, when the Green Howards embarked at Gosport for their first tour at Gibraltar. On their arrival there they took up their quarters in the South Barracks.

On the 1st July, 1751, a Royal Warrant was issued for establishing uniformity in the clothing and colours of the various regiments, by which the facings of Lord George Beauclerk's were ordered to be green. The first or the King's Colour was directed to be the Great Union; the second or Regimental Colour to be the colour of the facings, with the Union in the upper canton; in the centre of the colour, the number signifying the precedence of the regiment in gold Roman characters, within a wreath of roses and thistles on the same stalk, surmounted by the Crown.

Up to this time the regiment had been called by the name of its Colonel, but henceforward the numbers came into general use in the army, and though they were given up officially under the territorial system in 1881, most of them have died very hard.

The illustration shows a grenadier of the regiment at this period. The motto over the white horse on the little flap of the cap was ordered to be "*nec aspera terrent.*" The front of it was the same colour as the facings, with the King's cypher and crown embroidered over it. The number was on the middle part behind.

On the 24th May, 1752, the Green Howards embarked for England, and on arrival at Portsmouth marched to Salisbury, where they were quartered, throwing out detachments at Fisherton, Milford and Harnham. In July the same year they marched to Berwick viâ Reading, leaving on the way a detachment of three companies at Newcastle and Gateshead. The route showing the different halts may prove of interest.

ROUTE[1] FOR LORD GEORGE BEAUCLERK'S REGIMENT OF FOOT FROM READING TO BERWICK.

Henley
Chesham and Amersham
Dunstable
Newport Pagnal
Northampton
Harborough

(To halt the Sundays and Thursdays in every week on their march.)

Leicester, where each Division is to halt five days, and the Divisions are to proceed from thence in the manner they marched into Leicester to
Loughborough
Nottingham
Mansfield
Bligh and Bantree and Harworth
Doncaster
Sherborn
Tadcaster
Boroughbridge
Northallerton, where each Division is to halt five days, and the Divisions are to proceed from thence in the manner they marched into Northallerton to
Darlington
Durham

1 War Office Marching Order Book,

Newcastle and Gateshead, where the last Division is to relieve the Companies of Guise's, and the two first Divisions to continue their march to

Morpeth
Alnwick
Belford
Berwick, where the seven Comp^ys are to relieve the five Comp^ys of Guise's in the Duty of that Garrison.

H. Fox.

After being quartered at Edinburgh the regiment returned to England in 1756, and was stationed at York.

Owing to a dispute respecting the extent of the British dominions in North America having occasioned another war with France, the Nineteenth was augmented to two battalions, the new one being raised at Morpeth on the 25th August, 1756.

In April, 1758, the new battalion was formed into the 66th Foot, now the 2nd Battalion Princess Charlotte of Wales's Royal Berkshire Regiment. Its first Lieutenant-Colonel was Rowland Phillips, promoted from Major in the Nineteenth, and all the other officers, with the exception of Major Charles Beauclerk, who came from the 18th Foot, were posted from the Nineteenth. Exactly a hundred years were to elapse before the Green Howards were again to have two battalions.

The regiment marched to Newcastle in April, 1758, sending out two companies to Hull, and later on in the year one to Stockton and two to Durham. The detachment at Hull was employed in guarding the French prisoners of war there.

The regiment was reviewed on the 21st June, 1758, at Newcastle by Colonel John Parslow, and as it is the first inspection of which we have any complete account, we give a full detail of it. There were present on parade 1 lieutenant-colonel, 1 major, 1 captain, 11 lieutenants, 7 ensigns, 1 adjutant, 1 quartermaster, 1 surgeon, 2 mates, 36 serjeants, 20 drummers, and 856 rank and file.

REMARKS OF THE INSPECTING OFFICER.

OFFICERS. Properly armed, ready in their exercise. Salute well. Uniform good and conformable to regulation.

MEN. Well limb'd and well siz'd; a few old soldiers of service. The Grenadier Company of a large size. Of a soldier-like air; silent and steady under arms; accoutrements well fixed, and hats well cock'd.

EXERCISE. The Manual and Evolutions in good time and well; exact in marching and wheeling. The Serjeants know their duty. The Drummers and Fifers are good.

FIRINGS. Twenty-four rounds close and well—by Platoons standing. Advancing and Retreating twice repeated by Sub-Divisions in the same manner—and likewise by Grand Divisions—By Firings—By Platoons in the Square, formed by Files—By Faces in the Square, formed by Grand Divisions.

The Street Fire and a General Discharge.

ARMS. Clean and Good; Serjeants and Drummers have swords—swords are bespoke and shortly expected for Rank and File.

ACCOUTREMENTS. Good and clean.

CLOATHING. Good—The men have white and marching gaters.

RECRUITS. Of a low stature, but young and in General Strong and well made.

ACCOUNTS. Kept regularly.

COMPLAINTS. None.

UNIFORM. Red Lapelled Faced and lined with Yellowish Green. Bound and Looped with a Binding Striped.

A Regiment properly appointed good and well diciplined and fit for service.

(Sd.) JOHN PARSLOW,
Col. of the 70th Regiment of Foot.

Edinburgh, 4th July, 1758.

In July, 1759, the Nineteenth moved south, five companies being stationed at Hertford and Ware, the remainder at Hodson, Barnet, Hicksend and Hatfield. During the summer it went into camp at Brentwood, in Essex, after which it was quartered at Dover and Canterbury. On the 27th June, 1760, it was again encamped, this time at Barham Downs, near Canterbury, in company with the 14th and 21st Foot, under command of Lieutenant-General Campbell. The camp was broken up on the 14th October, and the Green Howards marched for Winchester, but at Sandheath on their way there the order was countermanded, and they were ordered instead to Hilsea, where they arrived on the 25th October.

The war with France, known as the "Seven Years' War," had now lasted five years, and was evidently drawing to a close, and the English Government determined on the seizure of Belle Isle, a little island off the coast of Brittany, in order that it might have another valuable asset in its hands for exchange when the peace negotiations came to be entered upon.

The island, which is about the same size as Jersey, lies abreast of Quiberon Bay, and some nine miles from the French coast. It is encircled by cliffs about two hundred feet high, interspersed with small sandy bays, most of which were guarded by forts, so that the little left undone by nature in rocks and crags had been supplied by art.

The troops destined for this expedition consisted of eleven infantry regiments, in addition to artillery and marines, amounting in all to over 8,000 men, under the command of Major-General Studholme Hodgson. The Nineteenth, which formed part of this force, was now commanded by Lieutenant-

Colonel Robert Douglas,[1] who had succeeded Lieutenant-Colonel Petitot on the 10th April, 1758, on the promotion of the latter to the 71st Foot. He had been appointed Ensign in the regiment in 1739, Captain in 1743, and Major in 1747, had served in the campaign of 1745 in Flanders, when he was wounded at Fontenoy.

The transports were collected at Spithead about the beginning of March, 1761, where they were joined by the "Sandwich," 90 guns; the "Valiant," "Torbay," "Temeraire," "Dragon," and "Swiftsure," each of 74 guns; the "Prince of Orange," 70 guns; the "Achilles," 60 guns; and several frigates, the whole convoy being under the command of Commodore Keppel.

The garrison of Belle Isle was commanded by the Chevalier de St. Croix, and was composed of the Régiment de Nice (two battalions), the Régiment de Bigorre, the Battaillon de Dinan, which, with the Artillery, Engineers and Coast Guard, formed a force of 150 officers and 2,500 men.

Every advantage had been taken of the rocky and precipitous nature of the coast to place the island in a perfect state of defence, and the works of Le Palais had been but lately strengthened. The three harbours of Palais, Sauzon and Goulphar were of little use to the fleet, being all more or less exposed, shallow or dangerous at the entrance.[2]

On the 7th April the transports bearing the expedition passed the south end of Belle Isle close along the shore, and at one of the bays by Point Locmaria, called Port de Andro, the General and Commodore Keppel thought that a landing might be tried. Orders were thereupon given for the

1 On the 4th May Lieutenant-Colonel Douglas was appointed a Brigadier, the command probably falling to Captain R. Farmer as Major Lumsden had been wounded.—M.L.F.

2 British Minor Expeditions.

attacking force to be in the flat-bottom boats by 4 o'clock the next morning. It was composed of seven companies of grenadiers, including those of the regiment, and four battalions, the Nineteenth, 67th, 85th and 90th, the whole being under the command of Major-General Crawfurd.

On the 8th this force rendezvoused under the stern of the "Achilles," which, together with three other ships, immediately began to bombard the fort of Port de Andro, where the landing was to be attempted. In about an hour the fort was silenced, and 300 men were at once landed. The French, who were prepared for their reception, charged them with fixed bayonets, and took a great many prisoners, upon which the boats that had not yet disembarked the remainder of the troops returned to their ships. At the same time another attempt was made to land at Sauzon, on the northern part of the island, by two battalions and 500 marines, but these also were repulsed, the total loss being 37 killed, 76 wounded and 315 prisoners, total 428; the share of the Nineteenth being nearly half, viz., Lieutenant Dougal Stuart, 1 drummer and 20 rank and file killed; Major Charles Lumsden, 1 serjeant, 1 drummer and 38 rank and file wounded; Captain-Lieutenant John Scrymsour, Lieutenants Forbes and Nugent, 4 serjeants, and 115 rank and file prisoners.

The regiment, mustering 800 men, evinced great gallantry on this occasion in the assault on the cliffs; the officers and soldiers rushed to the attack with great bravery, but were unable to ascend the steep acclivity under a heavy fire. They attempted to help one another up, but failed, and the men were ordered to re-embark.

A second attempt was made on the 22nd April. At 5 p.m. all the forts and works were simultaneously attacked, and at the same time feints were made for landing at several other points. Whilst the enemy's attention was thus distracted, Brigadier Lambert, with the Nineteenth, made for a part of the coast which the enemy considered unassailable, and had consequently neglected.

The grenadiers of the regiment, under the command of Captain James Paterson, followed by Captain Murray and some marines, effected a landing on the rocks near Locmaria, and gained the summit before the enemy saw what was intended, and just in time to repulse the attack of 300 Frenchmen sent against them. The grenadier company, though they lost heavily, maintained their ground till the remainder of General Lambert's troops got up. Captain Paterson lost an arm, and there were thirty men killed and wounded.

The whole of the troops now landed, and drove the enemy into Le Palais, the capital of the island, where the French Commander, the Chevalier de St. Croix, prepared for a stubborn resistance, expecting very soon to receive help from Brest. His forces were joined by the local militia, raising the strength of the garrison to 4,000 men.

The English were reinforced during the siege by the 3rd Buffs and the 36th Foot, in addition to the 16th Light Dragoons, and other troops.

On the 3rd May the French attacked the trenches with great determination, and notwithstanding the efforts of Brigadier-General Crawfurd, who was commanding on the left, all the works at that point were destroyed. Many men were killed, and General Crawfurd and his two aides-de-camp were taken prisoners. The French then retired, satisfied with their success, and in less than twenty-four hours the works were repaired and redoubts constructed to guard against a second surprise. The various regiments at this time were brigaded as follows:—

1st Brigade—		
Whitmore's, 9th	743	
Erskine's, 67th	766	
Rufane's, 2nd Battalion, 76th ...	460	
Crawfurd's, 85th	117	
		2,086
2nd Brigade—		
Beauclerk's, 19th	632	
Loudon's, 30th	780	
Morgan's, 98th	496	
Crawfurd's, 85th	176	
		2,084
3rd Brigade—		
Panmure's, 21st	750	
Rufane's, 2nd Battalion, 76th ...	576	
Colville's, 69th	666	
Crawfurd's, 85th	152	
		2,084
		6,254

The siege was actively carried on, and by the middle of May the French had retired to the citadel of Le Palais, the town being entirely abandoned. The heavy batteries kept up a fire upon this stronghold, and it was obvious that, as our fleet commanded the passage between the island and the mainland, the place must eventually surrender, but it was not till the 26th May that the fire of the stubborn Frenchmen began to slacken to any extent. On the 7th June the place was summoned prior to the final assault.

For two months the Governor had held out against great odds, and with the conviction that he could obtain no support. He therefore wisely determined to prevent unnecessary bloodshed by capitulation, and was allowed to march

MEDAL TO COMMEMORATE THE TAKING OF BELLE ISLE.

CITADEL.

BELLE ISLE, 1761.

through the breach with the honours of war, drums beating, colours flying, lighted matches, and three pieces of cannon with twelve rounds each.

Upon this taking place the Port de l'Avancée, or principal gate into the citadel, was taken possession of by a Captain and 50 grenadiers of the Nineteenth Foot.

Belle Isle remained in the hands of the English until the Peace of Paris, 1763, when it was restored to France.

Return of killed and wounded from 8th April to the surrender of the citadel of Palais on the 7th June:—

REGIMENTS.	KILLED.				WOUNDED.			
	Officers.	Sergeants.	Drummers.	Rank and File.	Officers.	Sergeants.	Drummers.	Rank and File.
Burgoyne's 16th Light Dragoons	1	—	—	1	—	—	—	1
Artillery	1	1	—	8	4	—	—	2
Buffs 3rd	—	—	—	1	—	—	—	3
Whitmore's 9th ..	—	2	—	22	2	1	1	30
* Beauclerck's 19th ..	1	—	1	54	4	1	2	91
Panmure's 21st	1	2	—	17	—	1	—	35
Loudon's 30th	—	—	—	22	1	1	—	40
Manner's 36th	—	—	—	2	—	—	—	2
Erskine's 67th	2	1	—	26	—	1	1	30
Colvill's 69th, 1st Batt.	—	—	—	11	4	5	—	28
Rufane's 76th, 1st ,,	1	1	—	16	—	2	—	8
,, ,, 2nd ,,	1	—	—	9	—	1	1	21
Crawfurd's 85th ..	1	—	—	18	—	2	—	21
Morgan's 90th	1	—	—	8	—	1	—	20
Stuart's 97th	—	—	—	8	1	—	—	20
Grey's 98th	1	—	—	4	—	—	—	12
Marines	—	2	1	34	8	—	1	46
Total	11	9	2	261	24	16	6	410

* Killed—Lieutenant Dugald Stuart.

Wounded — Major C. Lumsden, Captain J. Paterson, Lieutenants J. Majoribanks and H. Hutchinson.

THE BELLE ISLE ROLL—1761.

From a MSS. Return in the Record Office dated Bangor, Belle Isle, 31st May, 1761.

RANK.	NAME.	REMARKS.
Lieut.-Col.	Robert Douglas ...	Retired 17th June, 1767. Died at Strathendry, Scotland, 13th December, 1803.
Major	Charles Lumsden ...	Wounded in the attack on Fort de Andro 8th April, 1761. Retired 24th November, 1761.
Captain	Robert Farmar ...	To 34th Foot 15th July, 1761.
"	James Paterson ...	Wounded 22nd April, 1761. Lieut.-Col. 63rd Foot 15th June, 1763. Major-General 20th November, 1782. Colonel 28th Foot 13th July, 1787.
"	Thomas Cuthbert ...	Retired 19th July, 1778.
"	Thomas Pemberton ...	Retired 2nd April, 1762.
"	Samuel Townsend ...	Lieut.-Col. 34th Foot 5th May, 1769. Lieut.-Gen. 12th Oct., 1793. Died in London 17th May, 1794.
Capt.-Lieut.	John Bird	Lieut.-Col. 15th Foot 30th Jan., 1776. Killed at Germantown, near Philadelphia, 4th October, 1777.
Lieutenant	John Scrymsour ...	Taken prisoner 7th April, 1761. Resigned 8th January, 1768.
"	Alexander Gordon ...	Half-pay 27th Foot 27th February, 1764.
"	William Hatsell ...	Retired 17th November, 1780.
"	Andrew Forbes ...	Taken prisoner 7th April, 1761. Half-pay 19th Foot, 1768.
"	John Majoribanks ...	Wounded 22nd April, 1761. See American War Roll, 1781.
"	James Nugent ...	Taken prisoner 7th April, 1761. Lieut. 70th Foot 24th June, 1768.
"	John Leslie	Retired 27th April, 1772.
"	William Gordon ...	Retired 12th November, 1778. Afterwards Sir W. Gordon, Bart. Died 7th January, 1804.
"	Henry Hutchinson ...	Wounded 22nd April, 1761. To 110th Foot 18th April, 1762.
"	John McGill	See American War Roll, 1781.
"	Robert Menzies ...	Major 71st Foot 14th May, 1776. Killed outside Boston July, 1776.
"	Archibald Bogle ...	Half-pay 19th Foot, 1763.
"	Dougal Stuart... ...	Killed at Port de Andro 7th April, 1761.
"	John Stuart	Half-pay 1763. Capt. 37th Foot 25th December, 1770.
Ensign	Marshall Wright ...	108th Foot 22nd October, 1761. Died 12th August, 1824, as Fort-Major, Dartmouth.
"	William Wood ...	Half-pay 19th Foot, 1768.

RANK.	NAME.	REMARKS.
Ensign	Robert Hutchinson ...	Half-pay 83rd Foot 8th Dec. 1769.
,,	Richard Foley... ...	Half-pay 19th Foot, 1763.
,,	Robert McGill ...	Resigned as Captain 3rd Sept., 1773.
,,	Nicholas Colthurst ...	Capt.-Lieut. 106th Foot 17th Oct., 1761.
,,	Bacon Wastell ...	Resigned 1st October, 1761.
Lieut. and Adjutant	Robert Harper ...	Captain 100th Foot 2nd April, 1762.
Lieut. and Qr.-Master	James Hebden ...	Wounded at the battle of Lauffeldt when serving as a volunteer. Retired 15th June, 1763.
Chaplain	Edward Taylor ...	Resigned 11th January, 1764.
Surgeon	William Hamilton ...	To Hospital Service 20th July, 1761.

CHAPTER V.

The American War of Independence, 1781—The Campaign in North Carolina—Relief of Fort Ninety-Six, Monk's Corner, and the Battle of Eutaw Springs—Roll of Officers, 1781—The North Riding Regiment—Lord Moira's Expedition to France, 1794—Actions at Tuyl and Geldermalsen—Retreat of the Army through Holland and Germany—Roll of Officers, 1794-95.

ON its return from Belle Isle in December, 1761, the Nineteenth was stationed at Guildford, with detachments at Godalming, Haslemere, Farnham, Odiham and Alton. The war with France had terminated in 1762, with the peace of Fontainebleau.

In February the regiment marched to Hilsea, and in May to Plymouth. Before the end of the year it had gone to Falmouth, with a detachment at Penryn, in relief of the 2nd Battalion of Devonshire Militia.

Service abroad was not long again in coming, and on the 1st September, 1763, the Green Howards landed at Gibraltar, where they were quartered for the following eight years.

Major Charles Mawhood, of the 3rd Buffs, was brought in to command the regiment on the 17th June, 1767, in place of Lieutenant-Colonel Douglas, who retired. He had been appointed a Cornet in the Royal Dragoons in August, 1752, and was promoted Captain-Lieutenant in the 15th Light Dragoons, on the formation of that regiment in March, 1759. Four years later he was transferred as Major to the Buffs.

The next year saw a change in the Colonelcy by the death of Lord George Beauclerk, on the 11th May, when it was conferred on Major-General David Graeme.

This officer was the eldest son of James Graeme, of Braco, co. Perth, and during the Seven Years' War had raised a corps of Highlanders, which was honoured with the title of the 105th, or Queen's Own Royal Regiment, of which he was appointed Colonel in 1761. This regiment was disbanded in 1763, and the following year Colonel Graeme was nominated to the Colonelcy of the 49th Foot. He was promoted Major-General in 1762, Lieutenant-General ten years later, and General in 1783.

General Graeme assisted to escort Princess Charlotte of Mecklenburg Strelitz to England for her wedding, and it is said was responsible for the selection of that Princess as the Queen of George III. He was appointed Secretary to the Queen elect on the 5th September, 1761, and in the commission given him a month later to raise and command a regiment of two battalions, under the title of "The Queen's Own Royal Regiment of Highlanders," may be seen a further mark of royal favour.

It was probably due to his influence that many Scottish officers joined the Green Howards during the next thirty years.

On the 19th December, 1768, a warrant was issued for regulating the standards, colours, clothing, etc., of the cavalry and infantry, by which it was directed that the facings of the Nineteenth should be deep green. The style and cut of the uniform underwent a considerable change. The grenadiers' mitre-cloth cap was replaced by one of bearskin, similar in shape, on the front of which was a badge of the King's crest with the motto "*Nec aspera terrent*," the whole in white metal on a black metal ground (see illustration). Instead of the white neckerchief, a turned-down collar and black tie was now worn, and the coat was scantier and cut away to show the figure.

The waistcoat was shorter, and was changed from red to white. The officers' coats were lapelled to the waist with the facing of the regiment, and had cross-pockets and sleeves with round cuffs. The lapels and cuffs were the same breadth as the men's. The regimental lace was white with two stripes, red and green, and the number of the regiment was for the first time put on the officers' and men's buttons. The gilt gorget worn by the officers had engraved on it the Royal cypher, and was suspended from the first button on the lapel on either side by a coloured ribbon and rosettes. The sword belt of white buff was suspended over the right shoulder with an oval silver breast-plate in the centre, on which appeared the regimental number.

In 1771 a light company was added to the Nineteenth, as well as to all other infantry regiments. The officers of the grenadiers and light companies wore two epaulettes with a grenade in case of the grenadier company and a bugle in that of the light company, embroidered on each. The officers of the battalion companies only wore one epaulette, on the right shoulder.

In May, 1771, the Nineteenth was relieved at Gibraltar, and on disembarking at Portsmouth it was ordered to march to Richmond, Surrey, "by the shortest and most convenient route, taking care not to interfere with the quarters of the 30th Regiment at Farnham and Alton."[1]

It had barely arrived at Richmond when it was ordered north, four companies going to Newcastle, five to Tynmouth, and one to Clifford's Fort, in relief of the 37th Foot, which had gone further north.

In April, 1772, the regiment marched to Berwick, and was stationed in Scotland till 1775, when it proceeded to Ireland, its establishment at the time being 35 officers, 20

1 War Office Marching Order Book.

serjeants, 30 corporals, 10 drummers, 2 fifers and 380 private men.

The Nineteenth remained in Ireland for upwards of five years, and was quartered in the south at Cork, Limerick, Kinsale and Clonmel.

Lieutenant-Colonel James Coates had taken over command of the battalion on the 26th October, 1775, on Lieutenant-Colonel Mawhood being transferred to the 17th Foot. He had originally joined the regiment as Ensign in December, 1755, but had been promoted into the 66th Foot on the formation of that regiment in April, 1758.

Lieutenant-Colonel Mawhood was appointed Colonel of the 72nd Foot, or Manchester Volunteers, in December, 1777. He served with the 17th Foot in the American War of Independence, and Beatson, in his Naval and Military Memoirs, states that "the bravery and abilities of Colonel Mawhood deservedly gained him the highest applause in the fighting near Princetown, 4th January, 1777." He died at Gibraltar during the last siege on the 29th August, 1780, when in command of the 72nd.

The American War of Independence had commenced in 1775, and during the time the regiment was in Ireland had continued with varying success.

In 1781, on further reinforcements being required, the Nineteenth, in company with the 3rd Buffs and 30th Regiment, embarked from Cork on the 17th of March, and arrived at Charlestown on the 3rd of June.

At this time the enemy was besieging Fort Ninety-Six, a village so called from being that number of miles from Kecowee, in the Cherokee country.

Lord Rawdon, afterwards the famous Marquis of Hastings, was at this date commanding the Charlestown district, with his headquarters at Monk's Corner. Now that reinforcements had arrived he thought it would be a good opportunity to advance to the relief of the fort, so he set out on the 7th June with the flank companies of the three regiments and some of his own troops, in all about 150 horse and 1,800 foot.

After fourteen days' marching they reached the fort, only to find that the Americans, under Greene, had been repulsed in an attempt to storm the place, and had beaten a retreat. He was followed by Lord Rawdon as far as the banks of the Enoree, when the latter, finding further pursuit hopeless, returned to Ninety-Six. Arrangements were now made to abandon the fort, and leaving 800 of his force to bring off the garrison, Lord Rawdon marched for the Congaree with the remainder, intending to establish a post there at Friday's Ferry.

General Greene, hearing that the British force was divided up, ordered Lee, who was to be joined by Sumpter and Marion, to repair to Friday's Ferry, on the opposite side to the British, and keep them in check till he himself should overtake them.

Lord Rawdon, determining not to risk the approach of Greene, forced the passage of the river, and drove Lee before him, who, fortunately, had not met either Marion or Sumpter, and advanced to Orangeburgh, where he halted, being joined next day by a convoy under escort of the Buffs.

General Greene now came up and carefully examined the British position on the 12th July, but thought it was too strong to attack, so drew off and retired to the hills of Santee. Before effecting this, however, he ordered the light troops, under Marion, Lee and Sumpter, to make a raid towards Charlestown, and to unite at Monk's Corner for the purpose of dislodging the Nineteenth Foot, which was holding this post. Sumpter hastened towards Monk's Corner, and, joined by Lee, arrived before the place on the 16th July.

Lieutenant-Colonel Coates had three lines of retreat to Charlestown, and thought it prudent to decamp in the night, setting fire to the church which had been used as a magazine and destroying stores which he did not wish to leave to the enemy. He was pursued by the Americans, and overtaken eighteen miles from Monk's Corner. The rear-guard, numbering about 100 infantry, who were escorting all the regimental baggage, were charged with great impetuosity by the Americans and completely broken up, all the baggage being captured and most of the men taken prisoners.

The enemy now followed up the main body of the regiment, but Lieutenant-Colonel Coates had occupied a strong position in rear of a bridge on which he had posted a howitzer. Here he was vigorously attacked, and some hard fighting took place, the Lieutenant-Colonel and other officers being actually engaged in hand to hand combats, but the enemy's horse were unable to pass the bridge, and they finally retreated, leaving the Nineteenth free to proceed on their way to Charlestown.

On Lord Rawdon being invalided home, Colonel Stewart, of the Buffs, took over the command. Finding that Greene had retired to the hills, he again advanced to the Congaree, a movement that at once brought on the Americans. This caused Stewart to fall back on Eutaw Springs, forty miles from the Congaree and sixty from Charlestown.

The English were at this time very hard up for provisions. They had no bread, and when some 400 men went out in the fields foraging for potatoes and vegetables they were surprised by the enemy, and many were killed, wounded and taken prisoners.

On the 8th September Stewart, being in close touch with the Americans, drew up his force in the following order on a small hill which ran at right angles to the road from the Springs.

On the right was a provisional battalion composed of the flank companies of the Nineteenth and 30th, under the command of Major John Majoribanks (19th). In the centre across the road were two guns with an escort. On the left lay the 63rd and 64th Regiments, and beyond them the Buffs. The reserve, composed of provincials, was in rear of the guns.

A furious onslaught was made on the left by Greene, and at the same time the American cavalry charged down upon the right flank. The Nineteenth and 30th companies met this charge with such steadiness that the cavalry retired in confusion, leaving their Colonel and many prisoners behind them. While both flanks were thus being attacked, Greene sent forward his main body right up to the guns, and seemed on the point of gaining the day when Major Majoribanks, wheeling up his battalion, took him in flank, and the Americans were finally driven from the field.

The flank companies lost 1 officer[1] and 12 men killed, and 2 officers, 5 sergeants, 2 drummers and 81 men wounded; total loss, 85 killed, 351 wounded, 257 missing.

Extract from Colonel Stewart's despatch to General Earl Cornwallis:—

"Eutaw,

"9th September, 1781.

"My particular thanks are due, etc. . . . .

"But to Major Majoribanks and the flank Battalion under his command I think the honour of the day is greatly due." . . . . . .

---

1 Casualties of Officers, Nineteenth Foot—Lieutenant Robert Hickman killed; Lieutenant Lord Edward Fitzgerald wounded.

The action at Eutaw Springs was the last of the war. Six weeks later Lord Cornwallis, with 7,000 men, the bulk of the British army in America, surrendered to General Washington at Yorktown. The negotiations for peace were very protracted, and it was not till November, 1782, that they were brought to a successful termination.

"Thus ended," says Stedman, the historian of the war, "the most extensive, difficult and burdensome war in which Great Britain had till then been engaged, which cost her £115,000,000 and the lives of 43,600 men, whilst the Americans during the same period lost not less than 100,000. Although the issue of that war was unfortunate, our national character was not impaired, nor the contest, while it was maintained on the whole inglorious; neither martial ardour was wanting, nor military enterprise, nor patriotic zeal.

"In that rank and those circumstances of life which are at once a temptation and an apology for dissipation, and a love of pleasure, the military spirit of Britain shone forth with undiminished lustre, and the noblest families exhibited bright examples of true courage, exalted genius, and consummate wisdom."

ROLL OF OFFICERS
Who served in the War in America, 1781.

RANK.	NAME.	REMARKS.
Lieut.-Col.	James Coates, Bt.-Colonel	General 29th April, 1802. Colonel 2nd Queen's 20th December, 1794. Died at Heslington, near York, 22nd July, 1822.
Major	John Majoribanks ...	Was in command of the flank companies at Eutaw Springs. Died 23rd October, 1781.
Captain	John McGill	Retired as Bt.-Lieutenant-Colonel 3rd June, 1794.
"	John Smith	See Holland Campaign Roll.
"	John Skerrett ...	Major 48th Foot 7th December, 1791. Afterwards Lieut.-General. Died 18th August, 1813.
"	Ralph Hanson ...	Retired 12th April, 1782.
"	William Sleigh ...	Present at Eutaw Springs. Lieutenant-Colonel 83rd Foot 15th June, 1794. Died at Stockton 13th Feb., 1825.
"	David John Bell ...	See Holland Campaign Roll.
"	Colin Campbell ...	Taken prisoner 24th December, 1781. Retired as Lieut.-Col. 24th Foot 23rd April, 1799. Died in 1811.
Capt.-Lieut.	William Giles ...	Retired as Captain 20th March, 1793.
Lieutenant	Richard Gem ...	Retired as Captain 25th April, 1793.
"	George Taggart ...	Captain 55th Foot 30th April, 1792. Died 27th January, 1795.
"	Marcus Lowther Crofton	To 71st Foot, 1787.
"	William Wray ...	Retired as Lieutenant 1st June, 1789.
"	Matthew Scott ...	See Holland Campaign Roll.
"	Conway Blennerhasset	Died 17th July, 1786.
"	William Vincent ...	See Holland Campaign Roll.
"	Robert Hickman ...	Killed at Eutaw Springs 8th Sept., 1781.
"	William Beamish ...	Retired 25th August, 1783. Died 17th April, 1828.
"	Lord Edward Fitzgerald	Wounded at Eutaw Springs. Afterwards Major in 90th and 54th Regiments. Removed from the Army on political grounds, 1792. Died in prison in Dublin 4th June, 1798.
Ensign	John Bloomfield ...	Retired in 1784.
"	Charles McDonald ...	Lieut.-Colonel McDonnel's Regiment (New Corps) 27th November, 1794.
"	John Barrington Perryn	Retired as Lieutenant 9th June, 1786.
"	William Caulfeild ...	Retired as Lieutenant 30th Jan., 1790.

RANK.	NAME.		REMARKS.
Ensign	Robert Bloomfield	...	Retired as Lieutenant 23rd Aug., 1785.
,,	William Grant	...	See Holland Campaign Roll.
,,	Arthur Cavenagh	...	Resigned 4th October, 1786.
Qr.-Master	William McLarin	...	Retired 2nd January, 1794.
Surgeon	John Denholme	...	
Mate	Edward Bishop	...	35th Foot 14th November, 1782.

The garrison of Charlestown, to which the English troops had been withdrawn, prepared to evacuate that town in December, 1782.

The Nineteenth embarked for Jamaica on the 13th December, and the ravages made by a year in garrison in a town utterly destitute of sanitary arrangements may be judged when it is stated that the three regiments which only eighteen months before had landed from England, viz., the Buffs, 19th and 30th could only muster, with details of the 17th, 23rd and 33rd (regiments which had surrendered with Lord Cornwallis), 75 officers and 882 men.[1]

Before the regiment left America the following War Office circular was received, granting for the first time the county designation:—

"London,

"31st August, 1782.

"His Majesty having been pleased to order that the 19th Regiment of Foot which you Command should take the County name of the 19th or 1st York North Riding Regiment and be looked upon as attached to that Division of the County, I am to acquaint you that it is His Majesty's further Pleasure that you should in all things conform to that Idea and endeavour by all means in your power to cultivate and improve that connection so as to create a mutual attachment between the County and the Regiment which may at all Times be useful towards recruiting the Regiment.

1 30th Foot Regimental Record.

"But as the completing of the several Regiments now generally so deficient is in the present Crisis of the most important national Concern, you will on this Occasion use the utmost possible Exertion for the purpose, by prescribing the greatest diligence to your officers and recruiting parties, and by every suitable attention to the Gentlemen and considerable Inhabitants; and as nothing can so much tend to conciliate their Affections as an Orderly and polite behaviour towards them and an Observance of the strictest discipline in all your Quarters, you will give the most positive Orders on that Head; and you will immediately make such a disposition of your Recruiting parties as may best Answer that end.

"I have the honour to be,

"Sir,

"Your most obedient and most humble Servant,

"(Sd.) H. R. CONWAY."

The Nineteenth remained in Jamaica till the 8th April, 1791, when it embarked for England, and landed at Portsmouth after a voyage of two months.

In December, 1792, the establishment was fixed at 1 colonel, 1 lieutenant-colonel, 1 major, 7 captains, 1 captain-lieutenant, 11 lieutenants, 8 ensigns, 1 chaplain, 1 adjutant, 1 quartermaster, 1 surgeon, 1 mate, 22 serjeants, 30 corporals, 10 drummers, 2 fifers and 400 privates.

The French revolution had commenced in 1788, and in 1793 Louis XVI. was beheaded, and a Republican Government established, which was followed by war between England and France.

On the 3rd September, 1793, the Nineteenth embarked at Portsmouth for Flanders, and landed at Blend on the 14th of the same month. Circumstances necessitating its recall to England, the regiment re-embarked at Blend on the 6th

November. In this short campaign three rank and file were killed.

The same year the French Royalists of La Vendée and La Loire took up arms against the regicide Government, and solicited aid from England. An expedition was accordingly fitted out under Major-General the Earl of Moira, to aid these Royalists, the Green Howards being one of the regiments composing it.

The expedition was held in suspense waiting for the Vendeans to gain a sufficient portion of the coast for the English troops to land, and the fleet was driven about the Channel in stormy weather in the winter of 1793-4.

No opportunity of landing in France and of joining the Royalists having occurred, the regiment disembarked on the coast of Devonshire in January, 1794, and went into quarters, the health of the men having suffered from being so long detained on board the fleet.

After a few months' repose in garrison the Nineteenth again embarked, and sailed with the troops under the Earl of Moira for Ostend, which fortress was menaced by a numerous French force. They disembarked there on the 26th June.

Lord Moira's situation was a most unpleasant one. France during the preceding winter had been making great preparations, anticipating a war in Flanders, and had mustered an army of 200,000 men. The partial disaffection of Prussia, after having accepted the British subsidies, increased the difficulties of the situation, more particularly as regards the troops under the command of the Duke of York at Malines.

The original destination of the force under Lord Moira, which was the West Indies, was therefore changed to the arena of Flanders, which has been aptly termed the "cockpit of Europe."

Lord Moira resolved not to limit his services to the defence of Ostend, but to attempt to effect a junction with the army commanded by the Duke of York. He therefore gave orders to his troops, who were encamped on the sand-hills near the town, to stand to their arms, and without anyone knowing his intention, marched through the town and halted two miles outside.

The officers left all their clothing and camp equipment behind, thinking they would be encamped where they halted, but they were off at 2 o'clock the next morning, not knowing where they were going. After a halt, ten miles from Ostend, the march was resumed at midnight, and Alost was reached on July 3rd. Whilst there they had a skirmish with a patrol of the enemy about 400 strong, who were mistaken by the picquets for Hessian cavalry, and who reached the market place of Alost before they were discovered, when they were driven out by the mounted and infantry picquets.

On the 9th July the Duke of York passed through Malines, and there met Lord Moira's column. The march from Ostend had been very fatiguing; neither officers or men had taken off their clothes, nor had they any tents or houses to enter from the time of their landing. There were constant skirmishes with the enemy up to the 20th July, when Lord Moira was ordered home, and was succeeded in his command by Lieutenant-General Abercromby.

Various unimportant movements took place up to the 6th November, when the army crossed the Vaal at Nymeguen, and in this position there were several smart engagements. On the 20th November the French made a general attack on the British advanced posts. The enemy were repulsed and driven back with great gallantry, but the 37th Foot suffered severely through mistaking a detachment of the enemy for Rhoan Hussars, and allowing them to approach too closely. The address of the French was surprising; they formed corps and dressed them in the exact uniforms of the emigrants (Royalists) in the British service, and the English were in consequence easily deceived, as a part of the Rhoan Hussars had been acting with the 37th Regiment during the day.[1]

The enemy a few days later renewed their attempts on our outposts, and many men were lost on each side, General Abercromby being amongst the wounded. Finally the outposts were driven in, and the enemy established themselves in front of Nymeguen, and began to erect batteries preparatory to a siege. To destroy these a sortie, consisting of 2,500 men, English, Swiss and Hanoverian troops, was made, and though it was entirely successful at the time, the French losing heavily, it did not hinder the eventual repair and completion of their batteries, so that it was found necessary to evacuate the town.

After this evacuation, which took place on the 7th November, the troops were cantoned along the banks of the Vaal. The weather at this time was very severe; there was a great deal of frost, and when not freezing it was continually raining. The English, in consequence, suffered greatly, especially as their new clothing had not been received, and they were badly in want of shoes and socks. Luckily in the middle of December a large quantity of donation clothing from Arnheim was given out, and saved many lives. The ice on the river was now so strong that the heaviest cannon could cross it, and the French were making every preparation to do so.

1 Historical Journal of the British Campaign on the Continent in 1794, by Captain L. T. Jones.

On the 27th they crossed near Tuyl, with 600 men, taking possession of the post of Tiel, and soon after brought over a large force. Orders were at once given to dislodge them, and 8,000 men, mostly British, set out on this service.

At Geldermalsen, Lord Cathcart with his brigade, consisting of the Nineteenth, 33rd, 42nd, 78th and 80th Regiments, with some Hussars, struck off from the British column so as to get in rear of Tuyl, and at daylight on the 30th attacked the position with such resolution that the French were driven across the river with great loss, that of the British being 24 killed and wounded, the share of the Nineteenth being one man killed and four wounded.

The army was now very much on the alert, and it might have been considered as a line of picquets between the Vaal and the Leck rivers. The enemy were not content, for again on the 4th January they crossed the river and retook Tuyl. The British retired upon the village of Geldermalsen, where the Nineteenth and 42nd, together with some other regiments, had halted to cover the retreat through the village. The French cavalry, however, cut through the retreating picquets, and attacking the regiments holding Geldermalsen, met with a severe repulse. As the French retired they seized two guns which had been posted in front of the village and abandoned by the picquets, and were dragging them off when the 42nd, under Major Dalrymple,[1] charged with great impetuosity, retook the guns and brought them safely into the village. For their gallantry on this occasion the 42nd were rewarded with a distinctive badge, "The red heckle or vulture plume."[2]

The severity of the weather—it was one of the coldest winters known in Holland for years—and the heavy and

1 Appointed Lieutenant-Colonel of the 19th Foot, 31st December, 1794.
2 History of the 42nd Highlanders.

prolonged duties which pressed on the British troops in consequence of the superior numbers of the enemy, were such as few constitutions could endure for any length of time. It was therefore determined to take up a more defensive position behind the Leck. In this movement Lord Cathcart's brigade was constantly assailed by the French, sometimes being repulsed, and sometimes driving back the enemy. But the latter steadily pressed the British troops, now reduced by disease and exposure, and superior numbers obliged them to retire at all points. The sufferings of the army during this retreat are beyond the power of words to describe. The cold was intense, the water collected in the eyes of the men congealed as it fell, and hung in icicles from their eyelashes; the breath froze and lodged in incrustations of ice about the face, and on the blankets and coats wrapped round the unfortunate soldiers.

"There have been few situations where the courage, constancy and temper of British soldiers have been more severely tested than in the concluding period of this unfortunate campaign; pursued by an enemy of more than three times their numbers, and through regions so hostile that every Dutchman was only too eager to refuse the meanest shelter to the harassed soldiers. Exhausted by an accumulation of obstacles and hardships, which at that time had not been experienced by any modern army, and has since only been exceeded by the retreat from Moscow, the army at the beginning of April reached Bremen."[1]

Here the British for the first time experienced kind treatment, and by the 14th April were in a condition to embark for England. They had been almost decimated by their toils and the severity of the season. As many as two

1 Fortescue's History of the Army.

hundred in each regiment were victimised by exposure, want of attention and the harassing pursuers.[1]

During this campaign Colonel Coates had command of a brigade, with Captain Herbert Beaver as his Brigade-Major, the command of the regiment consequently devolving on Major William Houston.

The regiment embarked at Bremerleche in the "Convention" and "Concord" transports, and sailed on the 24th April. It arrived in England early in May after a tedious voyage, and was encamped together with six other regiments on the Nursling Heights, Southampton.

---

1 Stocqueler's History.

ROLL OF OFFICERS

Who served in the Campaign in Holland, 1794-95.

RANK.	NAME.	REMARKS.
Lieut.-Col.	James Coates, Major-General	Commanded a brigade during the war. See text.
Major	William Houston ...	Promoted to 84th Foot 21st March, 1795. Afterwards Sir Wm. Houston, Bart., G.C.B., G.C.H. Colonel 20th Foot 5th April, 1815. Died 8th April, 1842, at Bromley, Kent.
,,	Wroth Palmer Acland	To Coldstream Guards 10th May, 1800. Afterwards Lieut.-General Sir W. P. Acland, K.C.B., Col.-Commandant 60th Rifles 9th August, 1815. Died at Bath 8th March, 1816.
Captain	John Smith	Retired as Bt.-Lieut.-Colonel 28th May, 1795. Died in Edinburgh, 31st October, 1822.
,,	David John Bell, Bt.-Major	Died 11th May, 1795.
,,	Samuel Graham, Bt.-Major	Lieut.-Col. 1st West India Regiment 20th May, 1795. Afterwards Lieut.-General and Deputy-Governor of Stirling Castle. Died in Edinburgh January, 1831.
,,	Matthew Scott ...	Lieut.-Col. 28th Foot 11th Nov., 1795. Died 12th July, 1796.
,,	John Wauchope ...	Retired as Major and Bt.-Lieut.-Col. 17th July, 1801.
,,	William Vincent ...	See Ceylon Roll, 1802.
,,	Dunbar James Hunter	See Ceylon Roll, 1802.
,,	Herbert Beaver ...	Acted as A.D.C. to Br.-Gen. Coates. See Ceylon Roll.
,,	Robert Honner ...	See Ceylon Roll, 1802.
Capt.-Lieut.	William Grant ...	See Ceylon Roll, 1802.
Lieutenant	John Creighton ...	Retired 13th March, 1795.
,,	Thomas Alexander Kennedy ... Adjt.	Commanded five companies with an Embassy to Kandy in March, 1800. Died at Colombo 15th April, 1800.
,,	P. Groves	Captain 3rd Foot, 18th March, 1795.
,,	Alexander Black Vilant	See Ceylon Roll, 1802.
,,	James McNab ...	See Ceylon Roll, 1802.
,,	Edward Henry Madge	See Ceylon Roll, 1802.
,,	Chas. Albert Vigoreux	Captain Corsican Corps 4th April, 1795. Lieut.-Col. 45th Foot 20th December, 1826. Served in the Peninsula and at Waterloo with 30th Foot. Died in London 24th December, 1841.
,,	John Jewell	See Ceylon Roll, 1802.

RANK.	NAME.	REMARKS.
Lieutenant	Chas. Morgan Clayfield	Transferred to 85th Foot 24th April, 1801. Died at Downend, Gloucestershire 20th September, 1828.
"	Fenwick Hutchinson	Half-pay 82nd Foot 8th July, 1795. 90th Foot 18th September, 1804. Died 26th April, 1808.
Ensign	James Christie ...	Resigned 20th September, 1799.
"	George Hewson ...	Lieut. 44th Foot 8th January, 1795.
"	John Nairn	Died 8th August, 1799.
"	Thomas Wetherall Ottley	See Ceylon Roll, 1802.
"	Sir James Colquhoun, Bart.	Died at Errod 24th April, 1799, on the march to Seringapatam.
Qr.-Master	James Le Hey ...	See Ceylon Roll, 1802.
Surgeon	Charles Lind ...	Appointed 28th March, 1787. Surgeon to the Forces 20th June, 1795. Deputy-Inspector of Hospitals 1804. Out of the Army List 1816.

CHAPTER VI.

The 19th sails for Ceylon—A detachment proceeds to Seringapatam, 1799—An Embassy to Kandy, 1800—The Campaign of 1803—The 19th marches to Kandy—Massacre of the detachment left at Kandy—Captain Johnston's Expedition—List of Officers, 1802.

COLONEL COATES had been nominally if not actually in command of the Nineteenth for nearly twenty years, when he was gazetted to the Colonelcy of the 2nd Foot in December, 1794, the vacant command being given to Major George Dalrymple, of the 42nd Highlanders, who had been appointed an Ensign in that regiment in 1773, and had served with it in the American War of Independence and in the campaign in Holland. We have already referred to his distinguished conduct at Geldermalsen in the last chapter.

In October, 1795, the Nineteenth embarked on board the "Britannia," destined for the West Indies, but after experiencing a very severe gale the ship was obliged to put back and the men disembarked.

In February, 1796, the regiment marched to Ealing, where drafts amounting to 427 men were received from disbanded corps.

On the 28th April, 1796, the Nineteenth again embarked for service abroad, under the command of Lieutenant-Colonel Dalrymple. On its way to India it landed at Capetown in August, and remained there a few weeks, when the transport again put to sea, and finally disembarked the regiment at Madras on the 15th November. Before leaving Capetown a very flattering order was published by Major-General Craig as to the conduct of the men whilst quartered there.

Madras proved only a temporary halting place, as orders were received directing the Nineteenth to proceed to Colombo, where it arrived on the 8th December.

A short time previous to the arrival of the regiment in India the beautiful and mountainous island of Ceylon had been captured from the Dutch, who had become united with France in hostility to Great Britain, and in this island the Green Howards were fated to remain for a period of twenty-four years, with the exception of a few short tours in India.

General David Graeme died in Edinburgh in January, 1797, and the Colonelcy was conferred on Major-General Samuel Hulse, from the 56th Foot.

This distinguished officer was the second son of Sir Edward Hulse, Bart., and entered the army in 1761 as Ensign in the 1st Foot Guards, to which he was promoted Captain and Lieutenant-Colonel in 1776. At the commencement of the French Revolutionary War he went on active service, and commanded his battalion in Flanders in 1793, being present at the siege of Valenciennes and distinguishing himself at the action of Lincelles, for which he was thanked in orders by H.R.H. the Duke of York. He was later engaged in the operations before Dunkirk, and in the subsequent movements until October, when he was promoted Major-General and returned to England.

In May, 1794, Hulse again proceeded to Flanders, and commanded a brigade before Tournay, where several partial actions occurred. After the retreat through Holland he returned to England early in 1795, when he was appointed Colonel of the 56th Foot and placed on the home staff, where he continued for three years.

On promotion to Lieutenant-General he was given a command in Holland under H.R.H. the Duke of York in 1799, and was in several engagements from the 19th September to the 6th October.

FIELD MARSHAL SIR SAMUEL HULSE, G.C.H.

On returning to England General Hulse took over the command of the southern district, and in 1810 was removed to the 62nd Foot. He was appointed Governor of Chelsea Hospital in 1820, Ranger of Windsor Park, a Privy Councillor, a Knight Grand Cross of the Royal Hanoverian Guelphic Order, and on the accession of King William IV. was promoted to the rank of Field-Marshal. He died at Chelsea Hospital on New Year's Day, 1837, aged 90 years.

About this period the officers and men of the regiment subscribed one day's pay, amounting to £84, for the relief of the widows and children of the men killed in the victory obtained by Lord Nelson over the French fleet at the battle of the Nile on the 1st August, 1798. The men also subscribed 2,500 pagodas (8s. 9d. each) towards carrying on the war in Europe.

On the 16th February, 1799, five companies embarked for India on board the "Joyce," under the command of Captain William Vincent, for the purpose of joining the army formed for the capture of Seringapatam, the capital of Mysore. The Sultan of this territory, Tippoo Sahib, had sided with the French and shown marked hostility to the British, and it was therefore decided to send this expedition against him.

The Nineteenth detachment marched from Trichinopoly on the 29th March in company with a force of about 2,500 men, and was employed in settling the country in the neighbourhood, while at the same time it escorted a large convoy for revictualling General Harris' army. In company with other detachments it arrived before Seringapatam on the 13th May, just nine days after the fall of the fortress, which must have been a great disappointment to the officers and men composing the reinforcements.

Later on, in August, 1799, a force consisting of these five companies of the Nineteenth and thirteen companies of native infantry, under command of Major John Bannerman, was sent to reduce the fort of Panjamcorchy, which was from twenty-five to thirty miles north-east of Tinnevelly and of Palamcottah, with further orders to capture the chief and disarm the whole of the Southern Polygars. At the beginning of September, Bannerman advanced from Palamcottah, and arriving before the fort on the 5th, attempted to storm it immediately with his native troops, only the Nineteenth having not yet come up, he was repulsed owing to the misbehaviour of his men, with the loss of four European officers killed and two wounded, and 93 Sepoys killed and wounded. However, on the arrival of the Nineteenth on the following day, the enemy evacuated the fort, and within six weeks the Polygar was caught and executed, forty-four forts were destroyed, several chiefs were imprisoned at Palamcottah, and Bannerman's mission was declared to have been accomplished.[1]

The five companies rejoined headquarters at Colombo early in 1800.

On the 10th March, 1800, the light company and four battalion companies, under command of Captain Thomas Kennedy, in addition to some native troops, four guns and two howitzers, attended General McDowall on an embassy to the King of Kandy. The object of the mission was to keep up friendly intercourse with the King, as well as with a view to political objects of importance.

The General, with his staff and escort, started from Colombo, and after an interesting though tedious march reached Kandy on the 10th April. The road after leaving Sittivacca became exceedingly difficult, and the men of the

1 Fortescue's History of the Army.

regiment were frequently obliged to assist in dragging the guns through the ravines and defiles. Finally it became so bad that they had to be left five miles in rear with an escort of Sepoys, whilst the Nineteenth and the rest of the native troops marched to the King's royal gardens near Ruanelle, where they encamped. The men were infested in this march by leeches to a most alarming degree, and most of them had their legs and different parts of their bodies streaming with blood. They swarmed amongst the bushes and in the grass, and could not be avoided. The Dutch always said that these leeches were one of the worst enemies they had to encounter.

The General stayed at Ruanelle a week, and whilst there several sick men of the Nineteenth, as well as Captain Kennedy, were sent down to Colombo by boat. Unfortunately, the latter died a fortnight after his arrival there.

Finding the road to Kandy a very difficult one and almost impracticable for the guns, the weather besides being so unfavourable and storms of frequent occurrence, General McDowall determined to leave the Nineteenth detachment and the greater part of the native troops at this encampment and proceed himself to Kandy. Permission was granted to any officer of the regiment who could be spared to accompany the General.

During the six weeks the five companies of the regiment remained in this camp very few days passed without heavy rain. Thick, damp fogs were the rule at night, succeeded by excessive heat in the early part of the day, and the result was lots of sickness amongst the men, chiefly dysentery and liver complaint.

The General, after paying his visit to the King, returned to Ruanelle on the 6th May. The whole force left four days later for Sittivacca, where they encamped the same night, and reached Colombo on the 14th.[1]

The five companies, seven days after their return, were ordered to embark with the remainder of the regiment for the coast of Coromandel for service against the Polygars. They landed at Tatucorreen and Kilkerrie, from whence they marched to Palamcottah, Ramnah, Mattura, Dindigul and Trichinoply, and continued serving in the southern parts of India for eight months, when they returned by way of Nagapatam to Trincomalee in April, 1801. This town is situated on the north-east side of the island, and detachments were sent by the regiment to various posts in the interior. The inland provinces were under the rule of the King of Kandy, and were difficult of access to Europeans owing to the unhealthy climate of this part of the island.

At this time the regiment was under orders to join Sir David Baird's force for service with Sir Ralph Abercromby's expedition to Egypt. This reinforcement proceeded from Bengal and rendezvoused at Trincomalee in December, 1800, but owing probably to the absence of the regiment in India, the order was countermanded.

Early in 1803 it was found necessary to commence hostilities with the Kandians, with whom attempts had been made to establish a treaty of friendship and commerce but without success. The expedition which was now undertaken was the outcome of the following incidents.

In the month of April, 1802, some inhabitants of Putlam who were British subjects were on a trading journey in the Kandian country when they were forcibly despoiled of a quantity of areka nuts, together with their cattle. It was satisfactorily proved that one of the confidential agents of

1 Captain William Macpherson, 12th Foot, Brigade-Major to Major-General Hay McDowall, kept a diary of events during this expedition.—M.L.F.

the Adikar (Prime Minister) conveyed the areka nuts to Ruanelle, and sold them there to some traders from Colombo. Reparation was repeatedly demanded on the part of the local government from the court of Kandy, and no redress being obtained the Governor determined to march a force to Kandy to enforce his demands.

The troops destined for this service were formed into two divisions, the 1st or Colombo Division being under the command of Major-General Hay McDowall, and consisted of the 51st Foot (625 strong), two companies of the Nineteenth, two incomplete companies Bengal Artillery, the 2nd Ceylon Regiment (1,000 strong), and one company of Malay Pioneers. The 2nd or Trincomalee Division was under the command of Lieutenant-Colonel Barbut, 73rd Foot, and was formed as follows:—Five companies Nineteenth Foot, a Malay regiment, and one company Madras Artillery,

The 1st Division marched on the 31st January, 1803, and the streets of Colombo were lined with spectators to see the troops march through the town. "The music of the 51st band animated the march, and the faces of the soldiers were full of cheerfulness at the prospect of active service."[1] The troops halted on the 1st February, and were inspected by the Governor, the Hon. Frederick North. At daybreak the next morning the force crossed the Kallaniganga river in boats and marched to Jaellé, half-way to Negombo, about nine miles.

Next day they crossed the river Dandegame and marched eight miles to Karrenagame, and on the 4th February went on ten miles to Halpy. They reached Ketany the same day, and on the 5th camped at Allagooly, a pleasant spot on the banks of the Kaymelle river. On the 6th they arrived at Katadenia, which was fixed as a depôt for forwarding supplies, and was garrisoned by a party of Ceylon infantry. Here the

1 History of Ceylon, by the Rev. W. Cordiner.

division remained four days, and a redoubt was constructed, being named Fort Frederick, in honour of the Governor.

Katadenia, situated in an open plain, is traversed by a great river, called the Maha Oia, and possessed all the advantages requisite for a depôt, having a communication with the canal and lake of Negombo, so that stores of all kinds could be forwarded to it by boat. On account of the difficulty in procuring transport, most of the baggage and tents were left at this depôt.

The roads in many parts of the interior of Ceylon were, one hundred years ago, so bad and so hilly that wheeled transport was quite impracticable, and the whole of the baggage had to be carried by natives.

On the 11th February the division reached Dambadenia. Part of the route lay through paddy fields, where the men had to wade knee deep in water and mud. The most correct discipline was observed, property was protected, and no license allowed. Satisfactory accounts were here received of the progress of the troops under Lieutenant-Colonel Barbut, who had hitherto advanced without opposition.

On the 16th February the force left Dambadenia, reaching Kadroo Ellie the same day. This was the most fatiguing march the men had as yet experienced. The guns and baggage were only got over the bad road with the greatest difficulty. The banks of the Kadro Koomra were reached on the 17th, and over fifteen miles further was recorded the next day, when they surprised some Malays and Kandians, taking one of the latter prisoner.

On the 19th February the advanced guard forced the two strong and important posts of Galle Gegedra and Gerria Gamme. At the former no resistance was made, and the enemy fled precipitately, but the latter fortress was in a very strong position on the summit of a rocky mountain. Had it been defended by skilled troops it would have been impregnable. The fire from the battery was maintained so long that it was thought that the advanced guard had found more than the expected opposition in carrrying the post. The whole force was therefore ordered forward. Access to the fort was by a kind of natural staircase, winding up the side of the mountain, intersected by a succession of perpendicular rocks, all of which were within range of the Kandian battery. The fort was finally charged by the grenadiers of the Nineteenth, under Captain Honner, their loss being one sergeant and one private severely wounded. The sergeant was shot through the lungs and recovered, but the private had his thigh broken and died soon afterwards.

Although these were the only casualties, many of the men soon fell victims to the toil and exertion of this day's work, and the greater part of the officers and men felt its effect long after the campaign was over. On the march being resumed Lieutenant Nixon (19th Foot) was left with a party to garrison the fort.

On the 19th February, the division from Trincomalee, which had started on the 4th, stormed the stronghold of Canavetty, about twenty miles north of Kandy. This force crossed the Mahaville Ganga on rafts on the 20th February, and halted at the village of Wattapoloa, within a mile and a half of Kandy. The same day General McDowall's troops reached Kallagostolle, a few miles from the town. Three guns were fired as a signal which had been agreed upon, and were answered by Lieutenant-Colonel Barbut. Notwithstanding all the difficulties of the road and want of information of each other's movements, the two forces arrived at the gates of Kandy almost at the same moment.

The advanced guard, under Brevet-Major Evans (19th Foot), now entered the town, which was found deserted except by a few pariah dogs, and on the 24th and 25th of the month the Nineteenth and 51st Regiments marched into Kandy.

The following tables set forth the marches of the two divisions:—

MAJOR-GENERAL McDOWALL'S TROOPS FROM COLOMBO.

Date.		Encamped at.	Distance.		Total.
1803.			m. f. y.		m. f. y.
Jan.	31	... Near Mootwal	4 2 0	...	—
Feb.	2	... Jaellé (crossed the Mootwal in boats)	8 6 0	...	13 1 0
,,	3	... Karrenagamme (crossed the Dandegamme in boats) ...	8 2 0	...	21 3 0
,,	4	... Halpy	10 0 0	...	31 3 0
,,	5	... Allagoolie	9 0 0	...	40 3 0
,,	6	... Katadenia	4 1 0	...	44 4 0
,,	10	... Malgamooa (crossed the Maha Oia, fordable)	7 1 0	...	51 5 0
,,	11	... Dambadenia	4 7 0	...	56 4 0
,,	16	... Kadroo Ellie	10 1 0	...	66 6 0
,,	17	... Bank of the Magroo Oia (almost dry)	6 0 0	...	72 0 0
,,	18	... Bank of the Dick Oia (then very shallow)	15 4 0	...	87 4 0
,,	19	... Gerriagamme (temple) ...	5 3 0	...	92 7 0
,,	20	... Kallagostollie	7 4 0	...	100 3 0
,,	21	... Palace of Kandy (crossed the Mahaville Ganga on rafts)	2 6 0	...	103 1 0

LIEUTENANT-COLONEL BARBUT'S FORCE FROM TRINCOMALEE.

Date.		Encamped at.	Miles.		Total.
Feb.	4	... Tambegamme (by water)	12	...	—
,,	5	... Candelay	16	...	28
,,	6	Halt.			
,,	7				
,,	8	... Jungle	8	...	36
,,	9	... Allataville	10	...	46
,,	10	... Jungle	10	...	56
,,	11	... Minary	10	...	66
,,	12	... Jungle	12	...	78

DATE.		ENCAMPED AT.	MILES.	TOTAL.
Feb.	13	Gonava	12	90
„	14	Halt.		
„	15	Neragoma	8	98
„	16	Nalande	10	108
„	17	Allola	15	123
„	18	Allatchie Graville	10	133
„	19	King's Garden	8	141
„	20	Wallapooloa	1	142

On the approach of the troops the King immediately abandoned the town, after setting fire to the palaces and temples. The force had to be constantly on the alert while quartered in Kandy. Parties of Kandians hovered continually around the outposts. They concealed themselves in the jungle, and fired upon the sentries during the night. Whenever any unfortunate stragglers fell into their hands they put them to death in a most barbarous and shocking manner. A soldier of the regiment, several Malays and Sepoys were found mangled in a most cruel way.

The Governor, on hearing from Lieutenant-Colonel Barbut that the inhabitants of the north and east of the Kandian frontiers had expressed a strong desire to place Prince Mootoo Saumy on the throne, Colonel Barbut was accordingly despatched with the Malay Regiment from Minary, whither the Prince had accompanied him, with orders to convey him to Kandy, and on the 2nd March 300 men of the Nineteenth marched from the capital to meet Colonel Barbut's detachment, and they all, including the escort, entered the town on the 4th March, Mootoo Saumy being formally crowned four days later with all the forms of Eastern ceremonial.

No sooner had this been done than articles of convention were entered into between him and the Governor, in which great concessions were made to the English, thus dividing the Kandian territory without having subdued the reigning

King! In the meantime the new monarch met with no adherents and he remained in the palace without any court but his own domestics, and supported by no other power but the British Government.

Communication between Trincomalee and Kandy was now entirely interrupted by the enemy, who had murdered small parties of coolies going up with provisions. The mail which left Colombo on the 16th March was taken near Dambadenia, when two Sepoys were killed and one dangerously wounded, from which period no correspondence from the Presidency reached Kandy till the 30th.

The policy of Pilimi Talawa, the Adikar, in opposition to Mr. North, was crafty and successful. He urged the General to make an attempt to capture the fugitive King, who was at Hangarankellie, about eighteen miles from Kandy, and requesting that two strong detachments might be sent by different routes, so as to arrive simultaneously at the Palace of Hangarankellie.

Accordingly two detachments, one of 500 men, under the command of Colonel Baillie, Ceylon Regiment, and another of 300 under Lieutenant Colonel Logan, 51st Foot, set out on the 13th March by different routes. The country they passed through was excessively strong by nature; batteries were erected on every eminence which commanded the paths through which our men had to pass, and marksmen were placed in ambush in the thickest coverts of the jungle. On the 15th the troops, having burned the palace, commenced their return march to Kandy, the King having made his escape. They met with the same opposition; the loss incurred by killed and wounded[1] was considerable, and there is little

1 Lieutenant F. Hankey, 19th Foot, Brigade-Major to the King's troops in the division of Colonel Baillie, was severely wounded.

doubt that the Adikar had planned the destruction of the troops.

After the return of this expedition Colonel Barbut went with a detachment for a short distance on the road to Trincomalee, with a view of inducing some of the headmen of the country to come in and declare themselves in favour of the Prince Mootoo Saumy. Several men were killed in the detachment, and the attempt to conciliate the natives met with no success.

On the 19th March, Captain Herbert Beaver, 19th Foot, commandant of Fort Negombo, hearing that a large body of Kandians had built a redoubt and taken post at Moohooroogampelly within the British territory, determined to march with what troops he could collect and drive them away. His party consisted of one sergeant and twelve men of the grenadiers of the 65th Foot, and a few Sepoys. On their approach the Kandians fled precipitately without even firing a shot.

As the rains had already commenced in Kandy, and were soon expected to fall in torrents from the mountains, it was determined that the greater part of the troops should return to their former stations, and that Colonel Barbut with 1,000 men should remain as a garrison in Kandy.

On the 1st April, therefore, General McDowall left Kandy, taking with him the 51st Regiment, the Ceylon Native Infantry, and part of the Bengal Artillery. At the same time the grenadier and light companies of the Nineteenth were despatched to Trincomalee.

The garrison now consisted of 300 men of the Green Howards, 700 men of the Malay Regiment, some of the Bengal and Madras Artillery, besides a considerable number of sick in hospital who could not then be safely removed.

By the end of April the climate was beginning to operate powerfully in favour of the Kandians. There were many deaths and a number of sick, and Colonel Barbut had to return to Colombo, where he shortly afterwards died.

On the 16th May General McDowall, having recovered from his indisposition, left Colombo for Kandy, where he arrived on the 25th. He remained for some days, and although he procured an interview with the King without success, he had a conference with the Adikar, who seemed to be actuated by the greatest friendship for the English. Considering tranquility for the present secured, the General returned to Colombo, taking with him the sick officers and men of the garrison, which had now begun to suffer severely from jungle fever and dysentery. Major Adam Davie, of the Ceylon Regiment, was left in command, with 200 men of the Green Howards, 500 Malays, and a few artillerymen.

The troops accomplished their respective marches to Colombo and Trincomalee, having met with no opposition. On the 15th April Captain E. Madge (19th) had marched from Kandy to Fort McDowall, a post about sixteen miles on the road to Trincomalee, with fifty-five rank and file of his regiment.

On the 13th June Major Davie received a letter from the Adikar, saying that he was in disgrace with the late King, owing to his endeavours to serve the English, and requesting him to undertake another expedition as the only means by which peace could be obtained. Major Davie was aware of the Adikar's duplicity, saw his intention of decoying the remainder of the troops, and concluded he had promised more than he was able or willing to perform. By the 20th June eleven of the Malays had deserted, and the English were dying at the rate of six men a day. Paddy was almost the only food which remained for their subsistence, and in their sickly state they were unable to perform the labour of beating it into rice.

Preparations were at the same time being made by the Kandians to attack the garrison, but Major Davie was ignorant as to whether these threatened hostilities were intended as an infraction of the treaty signed by the Adikar or whether their object was to forward its execution. Prince Mootoo Saumy, trembling at his situation, which he saw every day becoming more perilous, would gladly have renounced for ever all pretensions to the sceptre of Kandy could he have obtained a safe conduct to his old station at Jaffnee.

Before this time news arrived at Colombo concerning the state of the garrison at Kandy, and produced considerable uneasiness at headquarters. After consultation the Government determined that steps should be taken for the speedy evacuation of the Kandian country. In General Orders of the 17th June a reinforcement, consisting of sixty men of the Ceylon Native Infantry, was directed to proceed to Kandy without delay, but the want of coolies to carry their rations delayed their departure until the 20th June. In the meantime 120 men of the Nineteenth were sick in hospital at Kandy, besides smaller numbers at the different outposts, and it would have required over 1,000 coolies to carry them back to Colombo had it been possible to obtain that number then and there.

On the 23rd June the Kandians took by surprise the two strong posts of Gerria Gamma and Galla Gegedra, which were garrisoned by a few men of the Malay Regiment. By this success the communication between Colombo and Kandy was now entirely cut off. The Kandians daily increased in numbers, and poured in towards the capital from all parts of the surrounding country. As the British garrisons grew weaker, the enemy gained courage and became resolute in their determination to attack the garrison of Kandy.

Having received information to this effect, the Commandant took measures for the defence of the town, and four field pieces were placed in different positions for its protection. About 4 o'clock in the morning of the 24th June the Kandians attacked the guard posted on the hill which commanded the back of the palace where the British troops were quartered. The guard consisted of six Malays and four gun Lascars, placed in charge of one three-pounder, which the enemy took and made the guard prisoners. Soon afterwards, about 5 o'clock, a strong party of Kandians, headed by Sanguylo, their chief, attempted to force the palace at the eastern barrier, where likewise one gun was posted. They were opposed by Lieutenant Blakeney and a few men of the Nineteenth.

Sanguylo crossed the stockade, and was immediately seized by Lieutenant Blakeney; they struggled and fell together, and whilst lying on the ground Sanguylo stabbed his opponent to death with his creese. Lieutenant and Adjutant Plenderleath and a private of the regiment immediately ran two bayonets through the body of the Kandian chief. The second in command, who had followed Sanguylo in the attack, was shot outside the door of the palace. The alarm was sounded, and Captain Humphreys, of the Bengal Artillery, coming up, loaded the field piece with grape and shot down twenty-four of the enemy. Being intimidated by this loss they withdrew, and manned all the rising ground, from which they galled the garrison by the fire from their grasshopper guns. On this occasion Lieutenant Plenderleath was mortally wounded, and one man Bengal Artillery and two Malay soldiers were killed.

After this an incessant fire was kept up on both sides until 2 p.m. The officers of the garrison were exhausted by fatigue. There were only twenty Europeans fit for duty, and almost every man of the Nineteenth was sick in hospital. A torrent of Kandians pressed upon the palace. The European officers of the Malay Regiment represented to Major Davie that the place could not be much longer tenable, and entreated him to enter into a capitulation with the Kandians.

After some consultation a white flag was displayed by the garrison, and firing ceased on both sides. Many of the Kandians then approached, and Major Davie and a native Malay, Captain Noorodeen, went out to converse with them, after which they all repaired to the quarters of the Adikar. It was then stipulated that Kandy, with the stores and ammunition in it, should be immediately given up to the Kandians; that all the British soldiers should march out of Kandy with their arms on the road to Trincomalee; that Mootoo Saumy should be permitted to accompany them, and that the Adikar should take care of the sick and wounded, and supply them with provisions and medicine until such time as they could be removed to Trincomalee or Colombo.

The above articles were written, signed and exchanged between Major Davie and the Adikar, and the latter gave Major Davie a passport to enable him to proceed without molestation to Trincomalee.

Accordingly, preparations were made for the removal of the garrison, excepting such of the sick as were incapable of walking, and at 5 p.m. the troops, consisting of 14 European officers, 20 men of the Green Howards, 250 Malays, 140 gun Lascars, with Prince Mootoo Saumy and his attendants, marched out of Kandy to Watapologo, on the banks of the Mahavelle Ganga, about a mile and a half on the road to

Trincomalee. Here they were obliged to halt for the night, as the river was not fordable, and there were neither boats or rafts. It rained very hard, and the party remained on the summit of a little rising ground, exposed to the inclemency of the weather.

Next morning, the 25th June, the troops were employed in endeavouring to make rafts, but the rope could not be carried across the river, owing to the depth and rapidity of the stream. About 7 a.m. many of the armed Kandians assembled near them, and others made their appearance on the opposite side of the river. Four headmen came up to Major Davie and informed him that the King had been greatly enraged at the Adikar for allowing the garrison to leave Kandy, but if they would deliver up Mootoo Saumy they should be supplied with boats to cross the river and receive every assistance to enable them to accomplish their march to Trincomalee. Major Davie replied that he could not deviate from the articles of capitulation, which both parties were bound to observe. Two hours later another party of Kandian chiefs waited on Major Davie, spoke to him in a very mild and friendly manner, and solemnly declared that the King was desirous to see and embrace Mootoo Saumy, and that he would receive and protect him as a relation.

Major Davie now consulted his brother officers, and replied to the chiefs that he would not part with Mootoo Saumy without permission from Colombo. On this they again departed, but returned soon afterwards, declaring that if Mootoo Saumy was withheld the King would send his whole force and seize him, and prevent the British troops from crossing the river.

After another conference with his officers Major Davie told Mootoo Saumy that he had not sufficient power to keep him any longer, but that the King had promised to treat him kindly. On his arrival at Kandy he was taken in front of the King, who upbraided him for having attempted to deprive him of his crown, and gave orders that he and two of his relations should be immediately executed.

Such is Dr. Marshall's account, but Colour-Sergeant Calladine, of the 19th, who was in Colombo in 1815, says in his diary: "Shortly after the capture of Kandy a great many prisoners were brought down to Colombo, and one of the King's relatives called Mooto Sawmy, who was a very stout and fierce-looking man, with a terrible pair of whiskers and beard, in fact he looked more like a wild man than anything else."

About 4 p.m. some Kandians joined the troops, and appeared to make preparations for enabling them to cross the river. Night, however, came on before anything was completed, and they went away, promising to return with boats in the morning. The next day armed Kandians began to assemble in great numbers, but no boats appeared, nor was any assistance given in forwarding the preparations for crossing the river. Captain Humphreys had, however, succeeded in getting a warp across at 10 a.m., but a raft was not ready, and the Kandians on the opposite bank soon afterwards cut the rope.

About this time, as has been stated by Corporal Barnsley (19th), the Malays and gun Lascars began to desert in small parties to the enemy. About 100 Kandian Malays and 80 Kaffrees, followed by a great crowd of undisciplined natives, posted themselves at 11 o'clock within a hundred paces of the troops. A dissave or headman came up to Major Davie, saying that the King wished that all the garrison should return to Kandy unarmed, and that if they refused they should be surrounded and put to death.

Seeing the enormous odds against him, Major Davie thought his only safety lay in acquiescence, and after another consultation with his officers agreed to the enemy's proposals, and ordered the Malays to ground their arms.

They then all proceeded towards the town, accompanied by the Kandian Malays and a crowd of armed natives. They had advanced only half way when the Kandian force was drawn up on each side of the road, and the British troops told to move to the centre of the line. The latter were then ordered to halt, and the men of the Malay Regiment were desired to march on. A Kandian chief asked the Malays if they were willing to enter the service of the King of Kandy. Some of them answered that they were already the sworn soldiers of a great King, and that they could not serve two Governments. The chief immediately ordered those who had given this reply to be bound and handed over to the Kaffrees. He then asked the rest of the Malays whether they chose to suffer death or enter the Kandian service. They all answered that they would serve the King of Kandy, and were at once conducted towards the town.

As soon as they were out of sight the English officers were separated from the privates, and all led out two by two at a distance from one another, when the Kaffrees, by order of the chief Adikar, perpetrated one of the most barbarous massàcres which history records. The only Englishmen selected for preservation were Major Davie and Captain Rumley, of the Malay Regiment, who were carried to Kandy after the massacre was completed. During the confusion Captain Humphreys, of the Bengal Artillery, laid hold of the arm of a sub-assistant surgeon of the Malay Regiment and managed to roll down with him from where they were standing to the hollow into which the dead bodies were thrown. They contrived to conceal themselves for several days. The latter escaped to Colombo, the former died a prisoner at Kandy. Previous to the massacre of the above, all the sick in Kandy, numbering 120, most of whom belonged to the Nineteenth, were murdered in cold blood, as they lay there incapable of resistance in the hospital.

Corporal George Barnsley, already mentioned, who was left for dead in the general slaughter, found means to make his escape. In turn he was led out with his companions, knocked down by the butt end of a musket and desperately wounded by a sword across the neck. Finding himself recover, he crept into a thicket, where he lay during the day, and in the night swam across the river, arriving at Fort McDowall early on the 27th June, when he communicated the dreadful intelligence to Captain Madge (19th), who was commanding the garrison of the fort.

Captain Madge immediately determined to retreat to Trincomalee, a distance of 126 miles, before the enemy could be aware of his intentions. The garrison at Fort McDowall consisted of himself as Commandant, Captain Pearce (19th), Mr. Gillespie, Assistant-Surgeon, and Lieutenant Driburgh, of the Malay Corps, 32 men of the Nineteenth, of whom 19 were sick and unable to march, and 22 Malay soldiers.

Having resolved to abandon the sick, he spiked the guns and evacuated the fort at 10 p.m. on the 27th June. For four days he was exposed to the hostile attacks of the enemy, at the end of which period he was met by a detachment of 150 men of the Malay Regiment proceeding from Trincomalee to reinforce the garrison of Kandy. The combined force retreated to Trincomalee, which they reached on the 3rd July.

The promptitude with which the retreat was undertaken and the skill and courage with which it was effected, thereby rescuing a part of the troops from the melancholy fate of their devoted associates, reflects the highest credit on the military talents of Captain Madge. He must, however, have left the nineteen sick with great regret, being well aware that they would be forthwith put to death when they fell into the hands of the Kandians.[1]

Howell, in his life of Bombardier Alexander, states:—

"On Barnsley's near approach to Fort McDowall the sentinel was struck with terror at his emaciated figure and ghastly look; he was conducted to Captain Madge, commander of the fortress at the time, who was thunderstruck at his appearance and the melancholy tidings he bore. The first words he said were: "The troops in Kandy are all dished, your honour." Captain Madge, in astonishment, required an explanation, which was too easily given, when he immediately ordered the guns to be spiked, and arrangements made for evacuating the fort, which was done about 10 o'clock after the moon had sunk behind the hills. All the sick were left to the mercy of the enemy, who had already shown that they had none. The lamps were left burning, and the march was commenced in silence; this, however, was soon discovered, and those of the sick who were most able followed the line of march until they dropped. Poor Barnsley, after having his ghastly wound dressed by the surgeon, marched on, supporting his head by his hands, as he had done all along, and arrived, with those who were able to keep up, on the Cottiar shore where the man-of-war boats were stationed, who took them on board and brought them to Trincomalee, which they reached on the 3rd July."

---

1 Marshall's Ceylon.

The following is a list of the officers who were stationed at Kandy when the last return was received: —

BENGAL ARTILLERY.—Captain R. Humphreys.

H.M.'S 19TH FOOT.—Lieutenants William Blakeney, Martin Byne, Peter Plenderleath and Hector Maclaine, Ensign Robert Smith, Quartermaster John Brown, Assistant-Surgeon William Hope, and about 150 non-commissioned officers and men.

H.M.'S 51ST FOOT.—Lieutenant T. Ormsby, Acting Assistant-Commissary; Sergeant R. Stuart, Acting Provost-Marshal, with the rank of Ensign; and about 30 men who were sick in hospital.

H.M.'S MALAY REGIMENT.—Major Adam Davie, Captain Rumley, Lieutenant Mercer, Ensigns R. Barry, Fanthorne and Gaupil.

EAST INDIA COMPANY'S SERVICE.—Garrison-Surgeon H. Holloway.

Lieutenant Blakeney, as before mentioned, was killed in action, and Lieutenant Plenderleath died of his wounds in Kandy.

Major Davie[1], Captains Humphreys and Rumley all died as prisoners in the hands of the Kandians. By this massacre they gained possession of upwards of 1,000 stand of arms, and took nearly 400 Malays prisoners.

---

1 Major Adam Davie was a son of the late Mr. John Davie, well known in Edinburgh by the cognomen of *Sooty Davie.* Mr. Davie having become acquainted with the celebrated Dr. Hutton, they formed a copartnership, and engaged in the manufacture of *sal-ammoniac* from coal-soot, which was carried on in Edinburgh for many years with considerable success. Adam Davie obtained a lieutenancy in the 75th Regiment in 1787, and in 1793 he was promoted to a company in the same corps. Having been employed recruiting in Edinburgh, he did not join the 75th Regiment until he had obtained the rank of Captain, consequently he was not present with the corps during its distinguished services in India. He appears never to have seen any active service while he was in the 75th Regiment. In 1801 he was promoted to a majority in Major Champagne's Infantry, a regiment of Malays. He is described by those who knew him as being a well-disposed,

Corporal Barnsley, on becoming convalescent, requested a friend, Bombardier Alexander, Royal Artillery, to draw up a narrative of his experiences.

Shortly after his recovery and return to duty he was promoted to the rank of sergeant, but a few months later he was tried by court-martial for being drunk on guard and reduced to the ranks. He did duty as a private in the Nineteenth until 1805, when he was sent home with other invalids. On being transferred to a veteran corps, the last heard of him was that he was doing duty at Fort George in 1811.

There is some difficulty in accounting for the 300 men of the regiment who were left as part of the garrison of Kandy on the 1st April. The Rev. Mr. Cordiner gives the following numbers in his history of Ceylon:—

Massacred at Fort McDowall	19	158
" " Watapologo	19	
" in Hospital	120	
Escaped to Trincomalee from Fort McDowall ...		13
" " " " Kandy		1
Died from 1st April to 24th June		128
		300

The regimental records say that seven officers and 172 men were massacred at Kandy, which would make altogether

inoffensive man, without any practical experience of hostile military operations. While he was in India he was tried by a court-martial for "absconding," and acquitted. Having had a short leave from the headquarters of the 75th Regiment, then stationed on the Malabar coast, and being at Bombay, he obtained leave of absence, and permission to proceed to Europe from the Commander of the forces, without any reference having been made to the Commanding Officer of the regiment. He embarked at Bombay, but as the ship touched at one of the ports on the coast, he was taken on shore, and placed under arrest by order of the Commanding Officer of the 75th, and subsequently brought before a court-martial.—Marshall's Ceylon.

Major Davie died near Kandy in 1812, a prisoner in the hands of the King.—ED.

191 men, counting the nineteen at Fort McDowall, and this would reduce the number of deaths between 1st April and 24th June to ninety-five.

The following General Order was published at this time:

"Headquarters, Colombo,
"13th July.

"Oppressed as the Governor is by grief and indignation at the atrocious act of perfidy and cruelty lately committed at Kandy, he has sincere satisfaction in noticing the spirited and successful conduct of the garrisons of Fort McDowall and Dambadinia. The Governor requests Captain Ed. Madge, of the 19th Regiment, to accept his thanks for his gallant defence of Fort McDowall and the judicious manner in which he brought off his garrison.

"He also desires to express to Captain Blackall, of the 51st Regiment, his perfect approbation of the vigour and activity with which he conducted the relief of Dambadinia, and to Lieutenant Nixon, of the 19th Regiment, and Ensign Grant, of the Malay Corps, his high sense of the spirit with which they defended the post under circumstances of extraordinary distress."

At the same time that Kandy was taken, Dambadinia, to which the latter part of this order refers, was in a state of blockade. It was a small redoubt garrisoned by fourteen convalescents of the Nineteenth, who were on their way to Colombo from Kandy, in addition to twenty-two invalided Malays. Lieutenant George Nixon (19th) was in command of the fort, and was repeatedly summoned by the Kandians to give up the post, but declined to listen to their proposals. On the 2nd July the garrison was relieved by some troops from Colombo, under Captain Blackall, 51st Foot.

The Kandians followed up their successes by endeavouring to seduce the native subjects of the British Government from their allegiance, and to excite dissatisfaction in various parts of the maritime territory. By these means it is presumed they hoped to overcome the remnant of the troops. Towards the end of July all the Kandian frontiers were threatened by warlike assemblies, nearly all at the same time.

The enemy made their first appearance in the district of Matura, and about the 30th July twenty-five English soldiers and fifty Sepoys were sent to strengthen the principal military stations in that province.

On the 20th August, in consequence of alarming intelligence from Matura, 150 men, English and Sepoys, were sent to still further reinforce the garrison, and Captain Herbert Beaver (19th) was ordered to take command. During the week following he made nightly excursions into the country with invariable success. On the 27th August he set out to retake Tengalle, a small fort about twenty miles north of Matura, which had been evacuated. He encountered some opposition on the way, but the fort was found to be abandoned and nearly destroyed by the Kandians.

Captain Beaver was mentioned in General Orders of 2nd September in the following terms.

"The Governor has observed with peculiar satisfaction the rapid series of well judged and well executed operations by which Captain Herbert Beaver, of the 19th Regiment, has hitherto proceeded in recovering the important province of Matura from the Kandians and in bringing its deluded inhabitants to their duty. The indefatigable activity, zeal and ability which that officer has displayed since his assumption of the command in that district has fully justified the high opinion which his Excellency had formed of him from his former services, and which induced the Governor to appoint him to that arduous station in a time of such extreme difficulty and discouragement."

At this time all the other British settlements in Ceylon were either threatened or attacked by the enemy. On the morning of the 21st August they took the little fort of Hangwelle, twenty miles from Colombo. When this was discovered Lieutenant Mercer, 51st Foot, with fifty-five men, was ordered to reinforce the troops stationed at Kadavilly, and to proceed with them to attack the enemy. In the evening of the 21st he came upon an advanced party of the Kandians, and checked their progress, killing three of them. Next day he attacked and stormed a battery near Hangwelle, when the enemy lost forty as prisoners, after which he marched on and took possession of Hangwelle. The same day he was joined by Lieutenant Worsley, R.A., who brought up one coehorn and fifty-five men.

Captain Frederick Hankey[1] (19th) was now ordered to take command of all the troops at Hangwelle, and rode out by himself before dark. He lost no time in driving the Kandians out of that part of the British settlements. He proceeded against them with great speed and activity, and took possession of a strong post at Avisavillie without the loss of a man. The force was constantly sniped during a march of nearly twelve miles, and when it arrived on the banks of the river, a branch of the Kalanyganga, a slight resistance was made. The men advanced with eagerness, and the Kandians fled in great trepidation. Several of them

1 Captain 19th Foot 24th December, 1802, from 88th Foot. Major 50th Foot 29th September, 1808. Severely wounded near Kandy in January, 1803. Afterwards Sir Frederick Hankey, G.C.M.G., for many years Secretary to the Government of Malta. Died in Montague Square, London, 13th March, 1855, aged 81.

were bayoneted in the water, and they left ten dead on the opposite bank.

Captain Hankey afterwards invaded the Kandian country, where he burnt a large village. He then retired to Hangwelle, handed over the command to Lieutenant Mercer (51st), and returned to Colombo.

On the 24th August a body of the enemy, under a dessave or headman, were sent by the King to attack Putlam. Brevet-Major Evans (19th), the Commandant, made a vigorous sortie in the night with a party of Malays in the disguise of Kandian peasants, when the Kandians were surprised and retired in great disorder.

At this time many of our other posts were being attacked, and a determined effort was again made on the 6th September by the grand army of the Kandians to retake Hangwelle. They were beaten off with heavy loss by Captain Pollock (51st), who had taken command of the post. The King of Kandy was with the army, but retreated with precipitation as soon as the firing began. In his flight he was overtaken by two of his chief officials, both of whose heads, in the violence of his indignation, he ordered to be immediately struck off, and left their dead bodies unburied in a river near Royberg. Nor did his cruelty stop here: for a number of corpses that passed down the river proved that the execution of his own subjects had been considerable and indiscriminate.

At midnight on the 6th September a reinforcement, under Captain Hankey, of thirty-five British soldiers was sent to Hangwelle. These were all the men in Colombo capable of marching. Invalids and pensioners now performed garrison duty, and it was not an uncommon sight to see a sentry with only one leg on guard.

On the 9th September Captain Pollock marched forward,

accompanied by Captain Hankey, Lieutenant Mercer and Lieutenant Worsley, with a detachment of eighty rank and file (British) and seventy rank and file (Sepoys), together with two small coehorns, and some Bengal Lascars. Having driven the enemy from the strong post of Kallvagille, where they were stationed in considerable force, under the command of the Dessave of the four Corles, Captain Pollock passed Royberg without opposition and halted for the night at Aloot Ambulam, six miles beyond Hangwelle.

In the meantime Captain Buchan (Ceylon Regiment) with a detachment marched from Negombo, and was opposite Captain Pollock's party on the Kalanyganga at Minnangodda. On the 12th September both detachments marched forward by different routes, and on Captain Pollock arriving at Ruanelle the further bank was found to be lined with batteries and several pieces of cannon, from which the enemy kept up a heavy fire of round and grape shot as well as musketry. Not being sufficiently acquainted with the state of the river to attempt fording it immediately, the detachment was here obliged to halt for a few minutes. When a ford was at last discovered, Captain Hankey and Lieutenant Mercer with the advance instantly crossed, and Captain Buchan with his detachment appearing at this moment on the enemy's right flank, they fled in all directions.

A General Order, issued by the Governor on the 13th September, "requests Captain William Pollock, of H.M.'s 51st Regiment, to accept his thanks and to communicate them to Captain John Buchan, of H.M.'s Ceylon Native Infantry, Captain Frederick Hankey, of H.M.'s 19th Regiment, and all the officers who have so nobly seconded his exertions."

Captain Beaver having heard that Hambantotte was attacked, marched from Catoone on the 29th August to relieve the place. He arrived there on the 6th of September, but the blockade had been raised by a judicious sortie of the garrison, supported by the co-operation of Captain McNichol, commander of the "Snow Minerva."

The Kandians invaded amongst other places the province of Batticolao in September, and raised a general insurrection amongst the inhabitants. Two months later they were driven to the frontiers by Lieutenant Arthur Johnston (19th Foot). He seized many of the rebels, and the greater part of the inhabitants returned to their homes.

The Pandara Wannian, a chief of one of the British provinces, who had been once pardoned for rebellion, again revolted, and overran the northern district. A small party of the Ceylon Regiment, stationed at the village of Cottiar, found it necessary to retreat on the approach of the Kandians. But this important tract of country was almost immediately recovered and the enemy driven beyond the frontier by the light company of the Nineteenth, which was detached for this purpose from Trincomalee.

Another party of rebels penetrated the province of Jaffna and surrounded the small redoubt at Elephant's Pass, killing one British and two native soldiers. They remained there for a day and a night, but retired on the approach of Lieutenant John Jewell (19th) with a detachment from the garrison of Jaffnapatam, of which place he was then Commandant.

The Pandara Wannian came down in person towards Vertivoe with a great force, but retreated on the approach of Major William Vincent (19th) with a part of the Manaar independent company. By his active and judicious exertions the district of Manaar was soon entirely cleared of the enemy.

The Kandians on the 25th August attacked the Government House at Moletivoe, and the garrison, consisting of

Malays, withdrew to Jaffnapatam. The enemy were driven off by a detachment sent from Trincomalee, under Captain Edward Madge (19th), in company with another detachment, under Lieutenant Jewell, who seized a large quantity of cattle and burnt some forts on the way.

In the desultory warfare that followed and which continued during several months, there is nothing very special to refer to with the exception of the gallant conduct of a detachment, under the command of Captain Arthur Johnston, of the Ceylon Regiment. As this officer had previously served in the Nineteenth and had under his command two officers and sixty-eight men of the regiment, it may be interesting to give a short account of his expedition, of which he published a full narrative.

The forces in Ceylon having been strengthened by the arrival of the 65th Foot from England, as well as reinforcements from Madras and Bengal, it was resolved in August, 1804, to again invade the Kandian country. It was thought advisable, owing to the difficulty in procuring coolies from any one district, to divide the expedition into six different detachments, which were to meet at a given point in the vicinity of Kandy. Captain Johnston's force was composed of the following troops:—

		EUROPEANS.				NATIVES.					
CORPS.		Lieutenants.	Sergeants.	Drummers.	Privates.	Sulidar or Captains.	Jemidan or Lieutenants.	Havildas or Sergeants.	Drummers.	Privates.	Grand Total.
Royal Artillery ...		—	1	—	6	—	—	—	—	—	7
H.M.'s 19th Foot ...		2	3	1	64	—	—	—	—	—	70
,, Malay Regt.		1	—	—	—	1	1	4	—	46	53
1st Bn. Bengal Vol.		1	—	—	—	—	1	9	2	75	88
2nd ,, ,,		2	—	—	—	1	1	5	2	76	87
Pioneers and Coolies ...	550	6	4	1	70	2	3	18	4	197	305

N.B.—One one-pounder, and one $4\frac{2}{5}$ Coehorn.

His orders at first were to march on the 20th September, and to effect a junction with the other five columns on the heights of Kandy on the 28th and 29th of the same month. These orders were afterwards changed, and it was owing to his misconception of them that makes Captain Johnston's expedition so full of incident.

The detachment embarked on the 20th September at Batticoloa, and proceeded the same day up the river to Surcamony, a village on its banks, twenty-seven miles distant. On the 27th Sanbapelly was reached, after a very fatiguing march of over seventy miles, during the last sixty of which not a house or a single human being was to be seen.

At Sanbapelly the country became more open, and several villages were observed. On the 28th, as the advanced guard was descending into a deep valley, they were fired on by the enemy, and one of the latter, who was wounded, was taken prisoner. Although by this time they had marched 124 miles, this was the first native they had been able to speak with. He said he had heard nothing of the other detachments, which was rather disconcerting news to Captain Johnston, as by this time he ought to have been in touch with one of the columns. However, on the 29th they continued on their way to Kandy, and after a march of sixteen miles reached the river, which runs past the town and at this particular spot was 150 yards wide. The enemy were in large numbers on the opposite bank, but after a few discharges of round shot they retreated, and the passage was effected successfully by means of rafts.

On the 3rd October they again had to cross the stream, which is very circuitous in its course, and on the day following the enemy came to closer quarters, killing a man of the regiment and wounding some coolies. On the 5th they again attacked, killing two Sepoys and wounding several, but were repulsed with heavy loss. The troops were much exposed, both in camp and on the march. On the right ran the river, lined on the opposite bank by the enemy, and nowhere fordable. On the left was a steep mountain, which confined their march to the river bank.

After progressing flanked in this way for three miles, they came across a loop-holed house filled with Kandians. This they stormed and captured with the loss of a bombardier of the Royal Artillery.

On the 6th firing was heard in the neighbourhood, which they concluded must be some of the other columns crossing the river. This greatly raised their spirits, and as they were attempting to make a raft one of the sentries on the hill saw a boat about a mile off up the stream. Lieutenant Vincent and the men of the Nineteenth were at once ordered to secure it, but when they reached the spot the boat had gone across. Two privates, however (Simon Gleeson and Daniel Quinn), gallantly volunteered to swim and bring it over, which they did, under cover of the fire of their comrades. Lieutenant Vincent, with about twenty of his men, instantly crossed, and the Kandians fled in all directions.

The advance was at once continued, and after burning the Palace of Candalay, about five miles from Kandy, the expedition entered the town, and our troops once more took possession of the capital.

The 7th October passed without any news of the other detachments, but towards evening a Malay officer and some soldiers formerly in the English service, but forced into that of the Kandians after the surrender of Major Davie, arrived, and stated that a fortnight before a rumour had prevailed that six English divisions had entered Kandian territory, but that they had been driven back, that the Kandians were in great force in the neighbourhood, and were only waiting for the climate to enfeeble the force before attacking it.

Captain Johnston could not believe that the English columns had been driven back by the cowardly Kandians, yet their non-appearance caused him great surprise and uneasiness. His rations were greatly reduced, and much of his ammunition expended.

The situation made a great impression on the troops, both English and native. Many of the Nineteenth had served in Kandy under General McDowall, and the massacre of their comrades was still fresh in their memory. They saw displayed in savage triumph in several apartments of the palace the hats, shoes, canteens and accoutrements of their murdered comrades, most of them still marked with the names of their ill-fated owners.

Though strongly prompted by his feelings to continue following up what he deemed the object of his orders, Captain Johnston regarded the safety of his detachment as paramount to every other consideration. He therefore determined to remain no longer in the face of the enemy unsupported, and marched his force out of Kandy on the 9th of October. On the outside of the town they passed a number of skeletons hanging on trees, the remains of the massacred officers, and they shortly reached the scene of the massacre itself, and found the ground strewn with the bones of the victims. The river not being fordable, they had to encamp near this place, and commenced to make rafts, which they completed in the afternoon. In the evening a small party of the Nineteenth crossed and drove off the enemy from the opposite bank, the latter losing one of their chiefs, who was bayoneted.

The next day the troops crossed the river, but only with difficulty, and the tents had to be left behind. The route taken was the Trincomalee road. On arriving at the pass the Kandians were found to be posted on the hills that commanded it, and the force being under fire for several hours, at last gained the bottom of the pass with the loss of five men of the regiment, eight Sepoys, and thirty followers killed and wounded, amongst them being Lieutenant Berkeley Vincent (19th), who was shot in the thigh.

The march was continued, but very slowly, owing to the casualties. On the 12th, Lieutenant H. L. Smith (19th) was badly wounded in the chest, and the same afternoon two more men of the regiment were killed, besides five Sepoys.

The next day the enemy showed great determination, and being led on by Malays and gun Lascars, who the year before had deserted to them, they attacked the coolies, who, perfectly panic stricken, rushed in amongst the troops. These they threw into such confusion that two wounded men, a sergeant of the Artillery and a private of the Nineteenth, fell into their hands.

Since the departure of the expedition from Kandy on the 9th October their only food consisted of raw rice, which had become musty and mildewed. They had been engaged in one continual skirmish, exposed alternately to a scorching sun and violent rain. The latter generally set in in the afternoon and continued incessantly the whole night.

For the next few days the enemy's fire decreased, and the force was not molested. On the 16th they came up with the advanced guard in charge of the wounded, which had been lost sight of for three days, but Lieutenants Vincent and Smith, as well as two men of the regiment, were missing.

They had all either died of their wounds or fallen into the hands of the Kandians. "Thus," says Captain Johnston, "were lost to the service two excellent officers in the prime of life, who had conducted themselves throughout this arduous expedition with a degree of zeal, intrepedity and perseverance highly creditable to themselves and consolatory to their friends. I shall ever regret the loss of these meritorious young men from whose conduct I had on so many occasions derived considerable aid."

The advanced guard said they had lost their way in the woods, and were nearly starved, that the coolies had completely deserted them, and that they had no means of carrying the sick, whom they were under the necessity of abandoning, that they were without guides, and had found their way to the village, where they were met with by mere chance.

Captain Johnston, in common with the rest of his men, had marched nearly all the way barefooted. His feet had become so tender from constant wet, and so swollen that he could not put them to the ground without great pain, and in this condition, emaciated by fatigue and labouring under severe dysentery, he was carried for the last two days in his cloak fastened to a stick.

On the 19th October they arrived at Tambalagam, where they were met by some officers from Trincomalee.

No words can express Captain Johnston's surprise when he heard it was never intended he should proceed to Kandy, and that the divisions were simply meant to lay waste the country adjacent to the enemy, and his having gone to Kandy was a direct disobedience of orders.

His conduct, however, was made the subject of a court of enquiry, which entirely acquitted him of so grave a charge, owing to his orders not having been explicit enough in the first instance.

The success of so small a force in penetrating unsupported to the Kandian capital and afterwards effecting its retreat created great surprise in the island. The capital had never before been attempted by so inconsiderable a force. The troops under General McDowall in 1803 exceeded 3,000 men, and these the flower of the Ceylon garrison. Besides, since then the enemy had gained to their service 400 well disciplined Malays and Sepoys, with a number of gun Lascars, and 1,000 stand of English arms.

Major Forbes, in his work on Ceylon, says: —

"The gallantry of Captain Johnston and his party taught the Kandians a respect for British troops which they had not felt before, and afterwards reluctantly admitted, and one of the chiefs who harassed Captain Johnston's retreat assured *him* that the commander of that party must have been in alliance with supernatural powers. His personal escape while passing through such a continual ambush, and his superior judgement and energy were accountable unless this explanation was admitted."

Return of killed, wounded and missing of the detachment under command of Captain Johnston:—

	R. A.		19TH REGIMENT.					MALAY REGT.				BENGAL SEPOYS.			
DETAILS.	Sergeants.	Bombardiers.	Lieutenants.	Sergeants.	Corporals.	Drummers.	Privates.	Officers.	Sergeants.	Corporals.	Privates.	Havildars.	Naigues.	Drummers.	Privates.
Killed ...	—	1	1	—	1	—	4	—	—	—	3	1	1	1	9
Wounded	1	—	—	1	2	—	2	—	—	—	4	1	1	—	27
Missing ...	—	—	1	—	—	—	2	—	—	—	—	1	—	—	12
Total ...	1	1	2	1	3	—	8	—	—	—	7	3	2	1	48

Captain Johnston concludes his narrative in the following words:—

"To the exertions, indeed, and animating example of the officers in general, and the persevering courage of the soldiers, particularly those of the Royal Artillery and Nineteenth, may be principally attributed the safety of the detachment."

AUTHORITIES.—Cordiner's History of Ceylon; Percival's History of Ceylon; Description and Conquest of Ceylon, by Dr. Henry Marshall; A Military Expedition to Kandy, by Captain Arthur Johnston; Howell's Life of Alexander Alexander; Regimental Records; Eleven Years in Ceylon, by Major Forbes; United Service Journal, 1829; Fortescue's History of the Army; Cannon's History.

OFFICERS 19TH FOOT.

Ceylon, 1802.

RANK.	NAME.	REMARKS.
Lieut.-Col.	George Dalrymple, Col.	Appointed Lieut.-Col. from 42nd Foot 31st Dec., 1794. Served in American War with 42nd. Also in the Campaign in Germany. Died at North Berwick 19th January, 1804.
,,	Edwin Hewgill, Col.	Appointed Lieut.-Col. from Coldstream Guards. Served in Germany 1794-5. Major-Gen. 25th September, 1803. Died at Hull 25th December, 1809.
Major	Dunbar James Hunter, Lieut.-Col.	Died in Ceylon 3rd September, 1803.
,,	William Vincent ...	Lieut.-Col. 14th May, 1804. Commanded 5 companies destined for the siege of Seringapatam. Retired 25th Sept., 1806. Died at Nice 11th March, 1822.
Captain	Herbert Beaver ..	Major 3rd September, 1803. Served in Travancore Expedition. Died at Colombo 19th April, 1809.
,,	William Grant ...	Half-pay in February, 1804.
,,	Robert Honner ...	Major 18th April, 1805. Served in Kandian War 1803-4. Cashiered 13th January, 1806.
,,	Hon. Frederick Forbes	Retired 30th November, 1803. Died 2nd February, 1817.
,,	John William Evans, Bt.-Major	Served in Kandian Campaign 1803-4. Died 1st January, 1805.
,,	Alexander Blackwood Vilant	Major Royal Staff Corps 9th November, 1805.
,,	Charles Pearce ...	Served in the Kandian War 1803-4. Died 6th August, 1808.
Capt.-Lieut. and Capt.	Alexander William Lawrence	Lieut.-Col. 4th Garrison Battalion 11th June, 1812. Served in the campaign against Tippoo Sahib 1791-2. Also at siege of Seringapatam with 77th Foot. Father of Lord Lawrence of the Punjab. Died at Bristol 7th May, 1835.
Lieutenant	James McNab ...	Bt.-Lieut.-Col. 4th June, 1814. Served in Travancore War with 19th Foot. Died at Trincomalee 4th Jan., 1818.
,,	Edward Henry Madge	Served in Ceylon Campaign 1803-4. Resigned 23rd March, 1807.
,,	John Jewell	Served in Ceylon War 1803-4. Retired 28th August, 1805.
,,	Thos. Wetherall Ottley	Major 65th Foot 2nd July, 1805. Died in 1806.
,,	Alexander Dalrymple	To 46th Foot 1st December, 1804.

RANK.	NAME.	REMARKS.
Lieutenant	Arthur Johnston ...	Captain 1st Ceylon Regiment 17th August, 1804. Commanded an Expedition to Kandy 1804. Died June, 1824.
,,	James Robertson ...	Captain 1st Ceylon 7th April, 1804. Died at Matura, Ceylon, 22nd July, 1815.
,,	Robert Percival ...	Captain 18th Foot 9th July, 1803. Author of A History of Ceylon.
,,	Richard Young ...	Captain Ceylon Regiment 1st April, 1804. Died 4th December, 1858.
,,	George Nixon ...	Captain 3rd Ceylon Regiment 7th April, 1804. Retired in 1816. Served in Kandian Campaign 1803. Died May, 1839, at Chapelizod, Dublin.
,,	John Kerr, Adjt.	Died at Colombo, 18th January, 1803.
,,	Berkeley Vincent ...	Killed in action near Lake Minary, Ceylon, 14th October, 1804.
,,	James Macdonnell ...	Afterwards 85th Foot, 78th Foot and Coldstream Guards. Commanded latter at Waterloo. General 20th June, 1854. G.C.B. 1855. Served in Sicily 1805-6. Egypt 1807. Peninsula 1812-14. Died 15th May, 1857.
,,	Donald Macdonald ...	Retired as Bt.-Lieut.-Col. 25th April, 1826. Served in the 1818 Rebellion in Ceylon. Died in 1847.
,,	Hon. George Turnour	Served in the Kandian War 1803-4. Died in Ceylon 19th April, 1813.
,,	Hugh Dalrymple ...	Captain 1st Ceylon Rifles 1st November, 1802.
,,	John Campbell ...	Captain 21st Foot 3rd August, 1804.
,,	Randall Chetham ...	Half-pay 46th Foot 17th Feb., 1803.
,,	William Blakeney ...	Killed in action at Kandy 24th June, 1803.
,,	James Moore ...	Died at Colombo 11th July, 1803.
,,	Martin Harland Byne	Murdered at Kandy 26th June, 1803.
,,	Alexander Robson ...	Bt.-Major, half-pay 25th October, 1821. Served in the Travancore Campaign, 1809. Died 24th February, 1836.
Ensign	Hector McLaine ...	Murdered at Kandy 26th June, 1803.
,,	Thomas Ajax Anderson	Half-pay 60th Foot 8th April, 1819. Served in the Rebellion of 1815 in Ceylon. Died 8th January, 1824.
,,	Peter Plenderleath ...	Killed at Kandy 24th June, 1803.
,,	Robert Ball	Died in Ceylon 17th June, 1811.
,,	William Vaughan ...	Resigned 15th August, 1803.
,,	George Stewart ...	3rd Ceylon Regiment as Captain 25th October, 1809.
,,	Robert Smith ...	Murdered at Kandy 26th June, 1803.

RANK.	NAME.	REMARKS.
Ensign	Thomas Adderley Jones	Served in the Travancore Campaign, 1809. Died at Batticoloa, Ceylon, 18th April, 1818.
,,	Richard Chambers ...	Retired by sale as Captain 29th March, 1838.
Qr'master	James Le Hey ...	Died 3rd April, 1802.
Surgeon	William Andrews ...	67th Foot 15th September, 1808.
Asst.-Surg.	Thomas Duncan ...	Surgeon 1st Battalion Army Reserve 9th July, 1805. Deputy- Inspector-General 17th July, 1817. Died 3rd November, 1848.

CHAPTER VII.

Cessation of hostilities in Ceylon—The Travancore War, 1809—Notes on Dress—The Kandian War, 1815—Rebellion of 1818—Death rate in Ceylon, 1796-1820—Return to England.

BETWEEN 1805 and 1815 no active operations were undertaken against the King of Kandy, and there was a mutual suspension of hostilities. Though it was well known that the King had no intention of having any intercourse with the British, a renewal of the war was not considered prudent by the Government, with the limited troops at its disposal. Experience had shown how difficult it was to subjugate a country which nature itself had made so formidable, in addition to the unhealthiness of the climate. Frequent insurrections of the inhabitants against the cruelty of the King added weight to these considerations, and the time rapidly approached when the natives of the interior were to make common cause against their oppressor, and solicit the help of the British Government.

We have shown how small detachments of the Nineteenth successfully checked the incursions of the Kandians into British territory after the massacre, but the regiment had become very weak in numbers, and at the end of 1803 was only 500 strong, its proper establishment being double that total. The headquarters were at Trincomalee, with detachments at outlying stations.

Lieutenant-Colonel G. Dalrymple retired towards the close of the year 1803, the command falling to Lieutenant-Colonel Edwin Hewgill, who had exchanged from the Coldstream Guards with Lieutenant-Colonel Wroth Palmer Acland in May, 1800. He had served with the Guards in the campaign in Holland in 1794-95, and had been for some time private secretary to H.R.H. the Duke of York. On the 25th September, 1803, he had been promoted a Major-General in the army, but in those days this did not prevent anyone from holding the command of a regiment.

In October, 1805, the Nineteenth embarked for Colombo, and ten months later, on the 13th August, 1806, sailed for Negapatam, in Madras. After a year's sojourn in India the regiment returned to Ceylon, landing at Osnaburgh, near Trincomalee, on the 28th September, 1807. Here it remained till the end of October, when it embarked for Point de Galle, arriving there on the 4th November. Four days later the regiment marched to Colombo, there to be stationed.

On the departure from the island of the 51st Foot in January, 1807, the Nineteenth received an augmentation from it of 300 men.

On the 9th September, 1808, Lieutenant-Colonel Hewgill was appointed to the command of the York Light Infantry Volunteers, and was succeeded in the Lieutenant-Colonelcy by the Honourable Patrick Stewart, who had been promoted into the 19th from the 96th Foot in September, 1806, *vice* Lieutenant-Colonel William Vincent, who retired. Lieutenant-Colonel Stewart had obtained his first commission in the 2nd Life Guards in September, 1793, and after being appointed Adjutant in June, 1794, was promoted Captain in the 27th Light Dragoons two years later, and in 1803 got his majority into the 96th.

Some disputes of a tedious and complicated character had taken place between the British and the Rajah of Travancore, and finally an expedition against him was resolved on. The Green Howards were ordered to form part of the force, and accordingly embarked on the 31st January, 1809, in patamars, and landed at Quilon on the 7th February.

Lieutenant-Colonel Stewart, from the apparent tranquility of the country, could not be persuaded of the proximity of an enemy, so on the evening of his arrival he was permitted to advance to the front with his guns, some shrapnel and a detachment of the regiment to reconnoitre. He had not proceeded more than three miles when he was attacked and compelled to rejoin the main body with the loss of several killed and wounded.[1]

The whole force now at Quilon was under the command of Colonel Chalmers, and in addition to the Nineteenth, consisted of the 12th Foot and four battalions of native infantry.

At daybreak on the 21st February an advance was made against the enemy in two columns, one consisting of the 12th Foot and a battalion of Sepoys, under the command of Lieutenant-Colonel Picton, and the other of the Nineteenth and a battalion of Sepoys, under Lieutenant-Colonel the Hon. P. Stewart. By a combined attack they captured the enemy's stockade and extensive breastworks in gallant style, the casualties of the Green Howards being nine men killed and wounded.

After some further operations the refractory Rajah made overtures for peace, but it was not till December that the Nineteenth disembarked at Berbery, on the west coast of Ceylon, from whence it marched to Colombo.

On the departure of the regiment from Madras the following order was published by the Governor:—

"Fort St. George,
"Madras.
"15th January, 1810.

"The Right Honourable the Governor in Council considers it to be an act of public duty to express on the return of

1 Colonel Bayley's (12th Foot) Diary.

H.M.'s 19th Regiment to the establishment of Ceylon, the sentiments of approbation and respect with which he regards the services of Lieutenant-Colonel the Honourable P. Stewart and the 19th Regiment while employed under the Residency of Fort George. The 19th Regiment arrived at Travancore at a very important period of time, it eminently contributed to the suppression of the disturbance in that country, and its conduct has invariably been distinguished by a high spirit of zeal, gallantry and discipline.

"By order of the Right Honourable the Governor-General in Council.

"J. H. PEILE,
"Secretary to Governor."

On the arrival of the regiment at Colombo the Lieutenant-General commanding the troops in Ceylon ordered that as a mark of his approbation "of the conduct of this old and distinguished corps that the right wing of the 19th Regiment be struck off all duty on Monday, the 12th February, and the left the following day, and that in lieu of the extra allowance granted on former occasions, the Commandant of Colombo give directions that the regiment be on those two days entertained in tents pitched for the purpose at the public expense."

For some years past many changes had taken place in the dress of the officers and men, a few remarks on which may not be out of place here.

In 1786 the officers were ordered to wear swords instead of espontoons and black cloth gaiters, uniform with those of the men.

Three years later all officers when on duty were ordered to wear their swords slung upon their right shoulder. An oval silver breastplate with the regimental number engraved upon it occupied the centre of the shoulder belt.

The black three-cornered hat was adorned on the left side with the usual black cockade of the House of Hanover.

By an order dated 10th December, 1791, effective field officers were to wear an epaulette on each shoulder. The officers of the grenadier and light companies who already wore two epaulettes were to have the addition of a grenade or bugle horn embroidered on each.

In 1792 sergeants of infantry were instructed to make use of and to be supplied with pikes instead of halberds, which were laid aside. The sergeants of the light company, however, retained their fusils.

Towards the close of the century the coats of all ranks were fastened at the waist, and completely hid the sleeve waistcoat, which afterwards became for the rank and file an undress or fatigue coat. The illustration shows an officer in full dress about the year 1792.

The cocked hat worn by the men was discontinued in 1800, and a cylindrical shako of the "stove-pipe" pattern took its place. By General Order dated 20th October, 1806, the shako of the Waterloo period was brought in, but the officers did not give up wearing their cocked hats till late in 1811. Two years afterwards the oval breastplate was changed to one of a square pattern.

Chevrons for non-commissioned officers were adopted in 1802; sergeant-majors to have four, sergeants three, and corporals two upon the right arm.

Powder for the hair was discontinued in 1796. Wigs went quite out of fashion with the French Revolution, and frizzing, plastering and powdering the hair till it was uglier than the wig had become the mode until 1796. Stiff curls were worn on each of the temples and a long tail behind. A great authority on the dress of the soldier tells us that—

"The officers, perhaps, could afford pomatum, but the privates used the end of a tallow candle to keep this wonderful regulation head-dress in order. The army was tormented with this preposterous and most unwarlike method of dressing the hair, varying from club tails to macaroni tails and pig tails." Subsequently the tails were reduced to seven inches in length, and finally in 1808 the queue was dispensed with, and the order directing its abolition said that the hair was to be cut short in a neat and uniform manner.

In 1812 the rank of a colour-sergeant was instituted. His distinguishing badge was a regimental Colour supported by cross swords, and was worn above the chevron.

The rifle in use at this period was the old flintlock musket known as the "Brown Bess." It had a bayonet attachment, and superseded in 1700 the first flintlock musket, which was used in 1640 and the following sixty years. The great disadvantages of the flintlock were the tendency of the powder to get wet and the friability of the pyrites.

In 1807 a clergyman called Forsyth patented a priming of fulminate powder, which when struck with any metal exploded. This was a most important invention, and enabled the flintlocks in use in the army to be altered to the percussion musket introduced in 1842.

On June 25th, 1810, Lieutenant-General Sir Hew Dalrymple, from the 37th Foot, was appointed to the Colonelcy of the Nineteenth *vice* General Samuel Hulse, who was removed to the 62nd Foot.

Sir Hew had been commissioned as Ensign in the 31st Foot in 1763, Captain in the 1st Foot in 1768, and Major in 1777. He was promoted to the command of the 68th Foot in 1781, and exchanged two years later to the Grenadier Guards.

He served throughout the campaign of 1793, and was at the battle of Famars, the siege of Valenciennes and the action before Dunkirk. On being promoted Major-General in 1794, he went on the staff of the Northern District. In 1796 he was appointed Lieutenant-Governor of Guernsey, was gazetted a Lieutenant-General in 1801, and after serving on the staff in England was removed to Gibraltar.

On receiving orders to take command of the British army in Portugal in 1808, Sir Hew Dalrymple arrived there in time to become responsible for the Convention of Cintra, by which treaty the French army evacuated that country. He was promoted General in 1812, and created a baronet in 1815. He died on the 9th April, 1830, in London, aged 80 years.

Major General Tomkyns Hilgrove Turner, from the Cape Regiment, succeeded Sir Hew Dalrymple, after the latter had been a little over a year at the head of the regiment. He was appointed on the 27th April, 1812, and was an officer of some distinction. Joining the 3rd Foot Guards in 1782, he served with them in the Flanders campaign of 1793, where he was present at the battles of St. Amand and Famars and the siege and capture of Valenciennes. He was also present at the action of Lincelles and the siege of Dunkirk.

In 1794 Turner was repeatedly engaged; he accompanied the Duke of York's army to Tournay, and was with it in its memorable retreat towards Holland.

In November, 1794, he had been promoted Captain and Lieutenant-Colonel, and in 1801 as Colonel took his regiment to Egypt, where he was present at the battle of Alexandria. He received a medal from the Grand Seigneur, and returned to England in charge of the Egyptian antiquities, including the famous Rosetta stone.

Colonel Turner was promoted Major-General in 1808, Lieutenant-General in 1813, and General in 1830. He was Lieutenant-Governor of Jersey 1814 to 1816, and Governor of the Bermudas from 1825 to 1831. His decorations included the Grand Cross of the Royal Hanoverian Guelphic Order, the Order of the Crescent of Turkey, and the Order of St. Anne of Russia.

On the 24th September, 1812, there was another change in the senior ranks of the regiment, when Lieutenant-Colonel H. S. Eyre, the second in command, exchanged to the Grenadier Guards with Lieutenant-Colonel William Henry Rainsford. The latter had joined the Guards in 1793, and had been Adjutant to his battalion from 1803-1807. He had served in the campaign in Holland from September, 1794, till the evacuation of that country in 1795, and had also taken part in the Walcheren expedition of 1809. During the absence on leave of Lieutenant-Colonel Stewart this officer must have commanded the Nineteenth with success, as after an inspection in February, 1814, the Commander of the Forces says that "his thanks are due to Major McNab and the other officers, all of whom appear to have been animated with a laudable desire to support Lieutenant-Colonel Rainsford in rendering the 19th Regiment one of the most respected in His Majesty's service."

On the 26th February, 1815, the regiment embarked for Trincomalee, where it arrived during the first week in March.

The King of Kandy had for several years shown so much cruelty to his subjects and evinced such a bloodthirsty disposition, that he became absolutely unbearable to them. There was a total insecurity of life and property, and individuals were constantly being deprived of both at the caprice of the King. Things were brought to a crisis by the following circumstances:—

During the month of November, 1814, ten cloth merchants, subjects of the British Government, had been seized by the King of Kandy and, for no possible reason, were mutilated in a most shocking manner; seven of them died on the spot, but the remaining three arrived at Colombo with their arms, noses, and ears cut off, presenting a most pitiable spectacle.

The second circumstance was the arrival in British territory of one of the Kandian chiefs. This man, who was governor of a province, had been summoned to appear at Kandy; but he, expecting that the sacrifice of his life and the seizure of his property was intended, thought the best thing he could do was to seek the protection of the British.

The Governor of Ceylon, General Brownrigg, considered the time had now come for invading the kingdom of Kandy, and the troops accordingly advanced at the beginning of February, 1815, in eight divisions, which were composed as follows:—

DIVISION.	NO. OF MEN.	COMMANDED BY	TO MARCH FROM
1st or Advance	430	Major Hook[1] (2nd Ceylon Regt.)	Colombo
2nd or Reserve	780	Lieutenant-Colonel O'Connell (73rd Regt.)	Colombo
3rd or Advance	402	Major Kelly (4th Ceylon Regt.) ...	Galle and Hambantotte
4th or Reserve	642	Colonel A. M. Murray (4th Ceylon)	Galle and Hambantotte
Advance Guard	—	Major McDonald (19th Foot) ...	Galle and Hambantotte
5th or Advance	500	Major Mackay (3rd Ceylon Regt.)	Trincomalee
6th or Reserve	500	Major-General Jackson, under him Colonel Rainsford (19th Foot)...	Trincomalee
7th	205	Captain T. A. Anderson (19th Foot)	Batticolao
8th	340	Captain de Bussche, A.D.C. ...	Negombo

1 See Roll of Officers 1820.

The only divisions of this force that met with the enemy were the first, second and eighth, and the few skirmishes that took place can hardly be called fighting.

The 5th Division, which consisted of 250 men of the Nineteenth, 200 of the 3rd Ceylon Regiment, and some artillerymen, started from Trincomalee on the 1st February, and reached Callepitty on the 17th. The following table will show the different halts and distances marched by the detachment, and though they cannot be called lengthy, yet it must be remembered that Ceylon was not opened up a hundred years ago and there were practically no roads in the country. The men went through great fatigue in having to clear a way for the transport and guns, as well as having a great deal of wet weather and occasional shortness of rations.

1st February,	Kappetine	8	miles
2nd „	Tambalagam	14	„
3rd „	Halt	—	„
4th „	Kandely	26	„
5th „	Parmamoodoo ...	38	„
6th „	Halt	—	„
7th „	Caravilla	46	„
8th „	Lake Minary	60½	„
9th „	Halt	—	„
10th „	Lake Moodagallumdda	66½	„
11th „	Pecolem	74	„
12th „	Angineul Palapy ...	85	„
13th „	Wallevilla	93	„
14th „	Nalende	102	„
15th „	Nanemgamme ...	105½	„
16th „	Panama	114½	„
17th „	Callepitty	120	„

On the way to Kandely "the road was miserable beyond description," writes a doctor attached to the expedition, "for the greater part we had to wade through thick, adhesive mud, which required all the strength a person could exert to pull one foot after another; but we were fortunately shaded from the hot rays of the sun by the thickness of the brushwood. At 11 o'clock we arrived on the banks of the extensive and beautiful lake of Kandely, and as we intended to remain here for the night, we pitched our tents along the banks of the lake; the situation was certainly swampy and damp, but it was the best we could find. The rear guard did not come up for three or four hours after the division, as some of the commissariat bullocks were nearly knocked up, and with the greatest difficulty were brought in."

On the next march to Parmamoodoo the guns were only got along with the greatest difficulty, as many old trees which had fallen across the route had first to be removed. The men were now in Kandian territory for the first time. The village was quite deserted and everything movable carried off.

On arrival at Callepitty, a village about ten miles distant from Kandy, Major Mackay received a despatch from General Brownrigg mentioning an engagement having taken place between the 3rd Division and the King's adherents in which the latter were routed, and a number of the King's relatives taken prisoners, together with his wives and children and a considerable quantity of treasure. Major Mackay was ordered to halt in the meantime, and send out parties to prevent the escape of the King, who was suspected to be lurking in the neighbourhood. He was also directed to send back word to stop the march of the 6th Division, with which was the main body of the Nineteenth, under command of Lieutenant-Colonel W. H. Rainsford.

In accordance with these orders, Major Mackay sent out a party of the 3rd Ceylon to search the villages of Kinggamine and Aretane as well as the surrounding country, whilst another party of the Nineteenth, consisting of twenty-five men of the light company, under Lieutenant T. Raper, proceeded in a different direction to endeavour to intercept the flight of the hunted King.

On the 19th February the division marched $3\frac{1}{2}$ miles further on to the mouth of the Aitgalle Pass. The pioneers were set to work to clear a path through the pass, as it had been shut up by the King since 1804, and none of his subjects, on pain of death, were allowed to enter Kandy that way, but were obliged to go round by the Ballacadia Hill, very much further, and a worse road. This was done with the intention of rendering it more difficult for the English to reach Kandy, should they ever again make the attempt. "The pioneers," says the officer in medical charge, "on their return to camp from their daily labour, had more the appearance of having returned from a bloody engagement than clearing of jungle, being covered with blood from head to feet, from being attacked by the leeches. These leeches are not so large as the common ones. They are chiefly found in marshy situations in the jungle; they will immediately fasten upon any part of the body they can get hold of, and will rather suffer their heads to be pulled off than quit their hold. The only plan therefore is rather to allow them to remain till they have filled their bellies with blood and fall off, or otherwise drop some limejuice upon them, which makes them loosen their fangs immediately. The bites of these animals degenerate into very bad ulcers when not attended to."

News reached the division on the 21st February that the King of Kandy had been captured. He was immediately sent under escort to Colombo in charge of Major Hook, 2nd Ceylon Regiment, and remained there till January, 1816, when he was deported to the fort of Vellore, in Southern India.

On the 24th February Major Mackay received a report from Lieutenant Raper of his having taken seven of the King's women and some children, with a quantity of jewels of various descriptions. He came upon them in a retired house in a most wild part of the country, where they were under the care of some men who were well armed, and at first made a show of resistance, but when they saw themselves surrounded by British soldiers their courage evaporated and they quietly gave themselves up.

With the exception of the light company, all the men of the Nineteenth were now ordered to return to Trincomalee, under charge of Lieutenants Raymond and Layton, and the remainder of the force marched through the Aitgalle Pass. Lieutenant Halliday, with a small party, was left in rear to command the pass.

The division entered Kandy on the 1st of March, and the day following the troops were all drawn up in square before the Palace, where the convention was read in the audience hall before the head men of the country and the officers of the garrison. After the ceremony a *feu-de-joie* was fired on the Quapion.

The following is an extract from the General Orders published at Kandy, 20th February, 1815:—

"The Lieutenant-General sincerely congratulates the troops on the successful result of their labours. It is to the zeal and ardour displayed by every officer and soldier that under Providence is to be imputed the speedy and happy issue of the campaign terminated at the end of forty days from the time of the first division entering the Kandian territory.

"The most formidable obstacles which the nature of this extraordinary mountainous country presents have been overcome with a spirit of cheerfulness and perseverance which has never been surpassed.

"The few Malabar and other adherents of the King whenever they have made a shew of resistance have been put to flight, and the greater part are now prisoners of war. Thus the conquest of the Kingdom of Kandy has been obtained, and the British army, who have the good fortune to contribute to it, will be entitled to the thanks of their Sovereign and the British nation, for the acquisition of a possession which in richness and value of its productions, the importance of its geographical position as relating to British India, and natural beauty will vie with any of the Colonies belonging to the Crown of Great Britain."

On the 2nd November, 1815, Lieutenant-Colonel W. H. Rainsford was gazetted to the command *vice* Lieutenant-Colonel the Hon. Patrick Stewart,[1] who exchanged to the 15th Foot with Lieutenant-Colonel Alexander Milne.

During the year 343 recruits joined the regiment from England.

On the 1st February, 1816, the establishment was reduced from 1,208 to 1,008 rank and file, and two years later it was still further reduced to 800.

The regiment continued to get great praise from the General Officer Commanding whilst stationed in Ceylon, to which the following order testifies:—

"Trincomalee,

"17th October, 1817.

"The Commander of the Forces has great satisfaction in expressing his entire approbation at the appearance of the 19th Regiment in the field this morning. The regiment may

---

1 Major-General 1821. Lieut.-General 1837. Colonel 60th Rifles 1838, 44th Foot 1843. Governor of Malta 1843-1847. G.C.M.G. 1843. General 1851. Died at Eagles' Carnie, co. Haddington, 7th February, 1855.

justly be stated to be in a perfect state of discipline, and the highest praise the Lieutenant-General can bestow is due to Colonel Rainsford and the officers under his command.

"(Sd.) HENRY HARDY,

"Deputy-Q.-M.-General."

On the 1st November Lieutenant-Colonel Rainsford[1] exchanged to the 89th Foot with Lieutenant-Colonel Donald McBean,[2] and the command was now taken over by Lieutenant Colonel Alexander Milne, the senior of the rank, whose first commission had been an Ensigncy in the 15th Foot in 1798.

The Nineteenth embarked for Colombo on the 18th of February, 1818, on board the "John" and "Brilliant" transports, and landed there on the 23rd. On the day of the departure the following order was issued by Major-General Jackson, commanding at Trincomalee:—

"Major-General Jackson must here express his warm commendation of the general good conduct of the officers and men of this valuable corps, and whose merits he has had ample opportunity of appreciating for the years they have been under his command.

"In the 19th Regiment he has always observed soldierly zeal when service was in view, and a steady well conducted deportment in the discharge of garrison duties.

"He parts with them with every good and sincere wish for their welfare.

"(Sd.) H. W. LYTTLETON,

"Fort Adjutant."

Early in 1818 a new claimant to the throne of Kandy arrived in Ceylon from India, and being supported by several

1 Promoted Major-General 19th July, 1821. Died at Madras in May, 1822.

2 Died at Point de Galle, Ceylon, 5th November, 1819.

native chiefs who were hostile to the British interests, a rebellion took place.

On the 28th February the Kandians attacked Major McDonald (19th) at Paranagamme. He had not more than eighty rank and file to repulse the enemy, whose force it is conjectured amounted to not less than 6,000 or 7,000 men. The attack was renewed at intervals until the 7th March, when the contest was relinquished and the insurgents separated for the time.

The officers and men of the regiment were now employed in small columns in pacifying the country. Some desultory fighting occurred, but much greater loss of life was occasioned by the climate than by the swords of the insurgents.

Although the Kandians were not successful in attacking posts, they were with much difficulty repulsed by troops proceeding from one post to another.

On the 2nd March Major Delatre, 1st Ceylon Regiment, marched from Trincomalee for Kandy with a detachment of 100 rank and file. He was harassed on his flank by the enemy for four days until he was within a few miles of Nalende, ninety-six miles from Trincomalee. He himself was wounded, and two of his men were killed and four others wounded severely.

After remaining ten days at Nalende he proceeded to Kornegalle, carrying with him the garrison of Nalende, and had four casualties on the way.

No information having reached Kandy as to Major Delatre's movements, it was determined to open up communication with Nalende. Accordingly a detachment of 100 men, under Captain Raper (19th Regiment), marched from Aitgalle, seven miles from Kandy, on the 27th March, but so great was the opposition of the rebels that Captain Raper did not reach the bottom of the Aitgalle Pass until daylight the next morning. The distance from the pass to Nalende is twenty miles, and along the whole road he was attacked by the enemy, who killed or wounded eight rank and file and two coolies.

On reaching Nalende, which he did at 7 p.m., and finding the post evacuated, Captain Raper made the men cook and rest until midday, when he commenced his march back to Kandy. The following evening he reached the Aitgalle Pass, remained there all night, and the next morning forced his way through the enemy. He arrived at the small post of Aitgalle the succeeding morning, after an absence of fifty-six hours. Before one week had elapsed every man of Captain Raper's detachment was admitted into hospital.[1]

Consequent on the disturbed state of the country, the headquarters of the Nineteenth marched from Colombo to Kandy on the 2nd April.

From July to September the troops were chiefly employed in pursuing fugitive chiefs, together with their followers, and in capturing or destroying property belonging to the inhabitants. Two of the most important chiefs were taken on the 30th of October and beheaded, and with their death the rebellion terminated.

Great hardships and privations were undergone by the troops in this campaign, of which the Green Howards had their fair share. On account of the wooded character and almost impenetrable nature of the country, progress was very slow, and detachments were often followed by a few of the enemy who, from their knowledge of by-paths, could fire upon the party without being seen and keep up with it on the march. When a man was killed the detachment was halted

1 Marshall's Ceylon.

and the body burnt to prevent the Kandians impaling the head close to a British post, as was their custom.

By marching at night there were fewer casualties, but the fatigue of the troops was aggravated. This can be easily imagined in such a country where natural obstacles were still further increased by the ingenuity of the Kandians. In several of the stations there was great scarcity of food, and the daily ration was sometimes not more than a quarter of a pound of salt meat and about a pound of uncleaned rice. Consequently, there was much sickness amongst the soldiers, the regiment losing during the year no less than five officers and 114 men.

Dr. Davy, in his History of Ceylon, states as regards the military operations:—

"We suffered most from the harassing nature of the service; from fatigue and privation; and from the effect of these and of night marches and of an unwholesome climate producing disease. Our loss from disease alone amounted to nearly one-fifth of the whole force employed, viz., 1,000 men.

"The loss of the natives killed in the field or executed, or that died of disease and famine can hardly be calculated; it was probably ten times greater than ours, and may have amounted to perhaps 10,000."

The mean strength of the European troops in the island during 1818 was 2,863 and the mortality 678, being in the proportion of 236 per thousand, or nearly one-fourth.[1]

A few weeks prior to the close of the rebellion the Green Howards returned to Colombo and on the 25th October marched to Point de Galle.

After having served off and on for over twenty-four years

1 Marshall's Ceylon.

in Ceylon, the battalion now received orders to embark for England at an early date. During this long period the following table will give some idea of the ravages made on the strength of the regiment owing to casualties on active service and the extreme unhealthiness of the jungles of Ceylon. The return shows that the mean mortality per 1,000 in the regiment for the time specified was 76

The annual average of men invalided was ... 27

Total annual decrement per 1,000 ... 103

RETURN OF CASUALTIES, 1796-1819.

Years.	Mean Strength.	Deaths.	Invalided.	Officers. Strength.	Officers. Deaths.	Proportion of deaths per 1,000	Stations.	Remarks.
1796	1035	20	—	39	—	19	At sea, Cape of Good Hope and Colombo.	
1797	1030	46	—	42	—	44	Colombo ...	
1798	1002	26	22	42	3	25	Colombo ...	Ensign Charles Hume, Lieuts. Randolph McDonald, John Nairn, Sir James Colquhoun, Bt., Jasper Nixon, Captain Thomas Alexander Kennedy
1799	941	90	2	39	3	95	Colombo, 5 Companies in India	
1800	882	72	13	35	—	81	Trincomalee ...	
1801	854	39	—	25	1	46	Trincomalee ...	
1802	905	46	21	32	2	50	Trincomalee ...	Qr.-Mr. James Le Hey
1803	843	338	—	20	10	400	Kandy and Trincomalee.	Lieuts. William Blakeney and Peter Plenderleath killed in Kandy, Lieuts. Martin Harland Byne, Hector McLaine, Ensign Robert Smith, Qr.-Mr. John Brown and Asst. Surgn. William Hope and 172 men murdered in Kandy. Lieut.-Col. Dunbar James Hunter, Lieut. and Adjt. John Kerr
1804	642	128	12	29	3	199	Trincomalee ...	Bt.-Major John William Evans, Lieut. and Adjt. John Crooks, Lieuts. Berkeley Vincent, Henry Littleton Smith and Charles Douglas
1805	528	44	27	27	3	83	Trincomalee ...	
1806	451	12	12	26	—	26	Colombo and India	
1807	793	20	7	24	—	25	India and Colombo	
1808	828	17	9	26	1	20	Colombo ...	Captain Charles Pearce
1809	796	45	42	31	2	56	India	Major Herbert Beaver, Lieut. Thomas James Rodney
1810	776	39	29	36	2	50	Colombo ...	Lieut. and Adjt. Richard Phepoe Nixon, Lieut. Arthur Saunders
1811	729	17	15	35	1	23	Colombo ...	Captain Robert Ball
1812	785	14	27	37	—	17	Colombo ...	
1813	938	20	11	41	1	21	Colombo ...	Captain Richard Parsons
1814	910	67	45	41	2	73	Trincomalee ...	Qr.-Mr. Thomas Blake, Lieut. John Bowyer Edenson

Years.	Mean Strength.	Deaths.	Invalided.	Officers. Strength.	Officers. Deaths.	Proportion of deaths per 1,000	Stations.	Remarks.
1815	1142	114	63	41	5	99	Trincomalee ...	Captain Hugo Weymss, Captain James Peter Fitzgerald, Lieut. Robert Gardiner Lieut. Francis Goodall and Ensign Adam Callander drowned in the wreck of the 'Arniston'[1]
1816	1019	38	60	42	2	37	Trincomalee ...	Captain Harry Beaver, Ensign William Thornton
1817	930	91	57	34	2	97	Trincomalee ...	Captain Robert Brown Duke, Lieut. George L. Halliday
1818	748	114	26	28	5	152	Kandian Provinces	Bt.-Lieut.-Col. James McNab, Captains James Aldersey Jones and James Edwin McGlashan, K.H., Lieut. Fortescue Hatherley and Asst.-Surg. Richard Hooper
1819	598	41	57	22	2	58	Kandian Provinces and Galle	Lieut.-Col. Donald McBean, Captain John Gore-Langton, Lieut. Donald Campbell and Paymaster Robert Nicholls
Total	20,105	1,498	...	...	50	...		
Average	873	63	...	...	...	76		

✠ ✠ ✠

" And though no stone may tell
Their name, their work, their glory,
They rest in hearts that lov'd them well ;
They grace their country's story."

Before leaving Ceylon 220 men were drafted into other regiments serving in the island, out of which 170 went to the 73rd and the remainder to the 45th and 83rd Regiments.

On January 7th the grenadiers and four companies under command of Captain Dobbin embarked for England at Point de Galle on board the "Maister," but did not sail till the next day.

---

[1] The "*Arniston*" was wrecked off the Cape of Good Hope on the 30th May, 1815. Only six men who happened to be in a boat when she struck, were saved. She went down immediately.

NOTE.—This return has been compiled from the regimental records. The names of the officers have been added by me, but do not quite agree with the totals.—M.L.F.

"It was a happy thing," says Colour-Sergeant Calladine,[1] in his diary, "that we remained only till the next day, as we had such a number of black women coming alongside who were left behind, some with three or four children, and although they were only blacks, still I conceive they felt as keen a sorrow as if they had been white. I suppose the 19th Regiment left more children than any regiment leaving the island, as it was between twenty-four and twenty-five years there. Some of them were grown up and girls were married, while boys who had been brought up at the Government School at Colombo were filling respectable places as clerks or otherwise had entered the army. We had a number sent to the school when the regiment came away."[2]

On January 22nd the headquarters and remaining five companies, under the command of Brevet-Colonel Donald McDonald, embarked for England in the "Dick" transport.

ADIEU TO CEYLON.

Written on board ship by Captain Thomas Ajax Anderson, 19th Foot.

Farewell ye mountains of Ceylon !
Fast fall the twilight shadows down,
And as the vessel cuts the brine
Each moment hides some charm of thine ;
Soon from my raptur'd sight will glide
Thy cocoa forests waving wide,
No more when morning clears the skies
Their dew-bright foliage charms my eyes,
Thy cloud-crown'd hills no more shall raise

---

1 Col.-Sergt. George Calladine wrote an account of his life in the 19th Foot from 1812-1837, which has been published in "The Green Howards' Gazette." A distinguished officer well versed in military lore, described it as the best-written experience of a non-commissioned officer he had ever read.—M.L.F.

2 At about this period there were 14 sergeants and 102 men married to or living with native women, with permission to live out of barracks.
Inspection Returns War Office.

Their tow'ring summits to my gaze,
Where mighty nature reigns alone,
Majestic on her ancient throne!
Farewell ye rocks, and glens and streams
Long will ye haunt my waking dreams,
And deeply must thy mountain charms,
Delight the breast whom nature warms.
Regret still mingles with the adieu
That now, fair isle! I bid to you.

AUTHORITIES.—Regimental Records; Col.-Sergt. G. Calladine's Diary; Cannon's History; Journal of the March of the 5th Division to Kandy in 1815; Description and Conquest of Ceylon by Dr. Henry Marshall; Letters on Ceylon by Captain L. de Bussche; Narrative of Events in the Island of Ceylon, by Edward Jeffery; Fortescue's History of the Army; An Account of the Interior of Ceylon, by Dr. John Davy; Eleven Years in Ceylon, by Major Forbes.

OFFICERS—1820.

RANK AND NAME.	RANK IN THE REGIMENT.	REMARKS.
Col. Sir Tompkins Hilgrove Turner, K.C., l.g.	27 Apr., 1811	See text.
Lieut.-Col. Alexander Milne	2 Nov., 1815	Died at Georgetown, Demerara, 5 Nov., 1827.
Major Donald McDonald, l.c.	11 Jun., 1812	Half-pay 25 April, 1826. Died at Preston, Brighton, in 1847.
,, Lionel Smith Hook, l.c.	5 Jan., 1818	To 16th Foot 8 March, 1821. Died at Benares when Lieut.-Col. 16th Foot 7 Feb., 1834.
Captain Alexander Robson,[1] m.	31 Dec., 1804	Half-pay 25 Oct., 1821. Died 24 Feb., 1836.
,, Edmund Lockyer, m.	29 Aug., 1805	Major 57th Foot 21 Oct., 1824. Died at Sydney, N.S.W., 10 June, 1860.
,, Henry Hardy, l.c.	30 Jan., 1809	Died at Trinidad when in command 16 April, 1835.
,, Thomas Dobbin[1] ...	10 Apr., 1813	Lieut.-Col. half-pay unattached 10 June, 1826. Died at Albany Street, Regent's Park, 13 Nov., 1862.
,, William Nihill ...	23 Dec., 1813	Retired 8 Feb., 1821.
,, William Cox[1] ...	21 Sep., 1815	To 54th Foot 21 Nov., 1822. Died in 1859.
,, Philip C. Lanphier..	19 Apr., 1818	Half-pay 25th Foot 6 Oct., 1825. Died at Kinsale, 1860.
,, Ronald McDonald[1]	4 Feb., 1819	Retired by sale 24 Aug., 1821.
,, Montgomery Cairnes	8 Apr., 1819	To 81st Foot 17 Jan., 1821. Died in Dublin, 1877.
,, Hubert Farquharson	31 Dec., 1819	Major unattached 8 Apr., 1826.
Lieut. Victor Raymond[1] ...	16 Nov., 1809	Half-pay 40th Foot 15 Sep., 1825. Died in 1859.
,, Lionel H. Hughes ...	16 Jan., 1810	Died as Major 22 Jan., 1837.
,, Crosbie Christian ...	6 Feb., 1811	To 34th Foot 15 June, 1820.
,, Henry O'Shea[1] ...	7 Feb., 1811	To 87th Foot 26 Dec., 1822.
,, Robert Cormack ...	29 Aug., 1811	Half-pay 29 June, 1820. Died in 1822.
,, James Martin Sullivan	17 Mar., 1814	Half-pay 17 March, 1814. Died 26 Nov., 1856.
,, Charles Stewart ...	22 Apr., 1814	To 4th Foot 24 Jan., 1821.
,, Charles McDonald ...	25 May, 1815	Retired half-pay 86th Foot 8 April, 1826.
,, Charles Forbes ...	3 Jun., 1815	Captain half-pay 24 Oct., 1821. Died at Aberdeen 7 May, 1843.
,, Thomas T. Hamilton	19 Oct., 1815	See 1840 Roll.
,, Thomas Taloe ...	30 Nov., 1815	Died at Gravesend 17 June, 1820.

1 Served in the Travancore Campaign of 1809.

RANK AND NAME.	RANK IN THE REGIMENT.	REMARKS.
Lieut. David Burns ...	25 Jan., 1816	See 1840 Roll.
,, Henry Hawker, Adjt.	22 Feb., 1816	Retired on half-pay 10 Jan., 1834.
Ensign Stephen G. Bagshawe	25 Jan., 1816	To 89th Foot 25 Jan., 1821.
,, John Stirling ...	6 Jun., 1816	See 1840 Roll.
,, Francis Tydd ...	7 Sep., 1816	Died at Trinidad 11 Feb., 1835.
,, Edward Chas. Soden	19 Dec., 1816	To 2nd West India Regiment 27 June, 1824.
,, Alexander Scott ...	2 May, 1818	See 1840 Roll.
,, Thomas Bayley ...	3 May, 1818	To 20th Foot 14 June, 1821.
,, Isaac Bryne Robinson	17 Dec., 1818	Half-pay 8 Feb., 1834. Died in Jersey 30 May, 1881.
,, Edward C. Mathias	22 July, 1819	To 44th Foot 20 Nov., 1822.
Paymaster Robert Nicholls	14 Aug., 1803	Died at Point de Galle 28 Aug., 1819.
Surgeon William J. B. Parker	2 Aug., 1815	Retired half-pay 26 Oct., 1826. Died at St. Helier's, Jersey, 10 Oct., 1845.
Asst.-Surg. Wm. McDonnell, M.D.	20 Sep., 1810	Died at the Hotwells, Bristol, 8 Sep., 1822.

Facings green, lace gold. Agent, Mr. Fitter, Silver Street, Golden Square.

CHAPTER VIII.

Stationed at Chatham and Weedon—Coronation of King George IV. Service in Ireland—Mullingar, Dublin, Athlone and Boyle—Still hunting—Castlebar and Limerick—Service abroad—Demerara, 1826—Notes on Dress—St. Vincent—Tobago—Trinidad and Barbados—Ordered home—Ireland 1836-1842—Roll of Officers, 1840—Malta and Ionian Islands, 1843-1845—West Indies and Canada to 1851—England—The Duke of Wellington's Funeral—Ordered to the East.

THE Green Howards disembarked at Gravesend and marched to Chatham on the 19th May, 1820. After a few weeks' stay there the route was received for Winchester, where they arrived on the 17th June. Here they found the depôt which had left Jersey for the purpose of joining the regiment on its return from Ceylon. It had first of all gone to Hull, expecting the battalion to disembark there, and had been nearly wrecked on the passage by running on a sandbank. It then received orders to return and disembark at Southampton, and had arrived at Dover at the time of Queen Caroline's return, going ashore there as a guard of honour to Her Majesty.

The depôt was now consolidated with the regiment, and as it was the stronger of the two there were many more young faces in the ranks than old ones. The duty at Winchester was very easy, which made it a good place for a young regiment, and there was nothing to do but drill.[1]

In July, when the Assizes came on, the Nineteenth marched to Romsey, eleven miles from Winchester, and remained there in billets for ten days.

In September orders came for a return to Chatham, and

1 Col.-Sergt. Calladine's Diary.

leaving Winchester on the 25th, the regiment marched into the town on the 4th October.

A short time after its arrival a sham fight took place before the Duke of Clarence, who had come down from London, accompanied by Sir H. Torrens, to inspect the garrison, The Veteran Battalion and the recruits of the East India Company occupied St. Mary's Battery, whilst the Marines and the Nineteenth formed the storming party. The latter led the attack, with the Marines close on their heels, and were soon ascending the scaling ladders in front of the works. After the supposed enemy had been driven from their posts at the point of the bayonet, the men were formed up inside the batteries and were inspected by the Duke of Clarence. Finally, the whole force took off their caps and gave three cheers, the massed bands at the same time playing "God Save the King."

The garrison and other duties at Chatham were very severe, and some of the guards were a considerable distance from barracks. At this period it was the receiving depôt for invalids from all regiments on foreign stations, who on their arrival were quartered at Fort Pitt until they proceeded to Chelsea to pass the Invaliding Board there. It was also the receiving depôt of the East India Company's recruits, who were generally pretty numerous. Regiments, too, were either embarking for or disembarking from foreign service, and when the route came for the Nineteenth to march to Weedon in relief of the 86th Foot there was much satisfaction.

Passing through London, Barnet and Dunstable, the right wing arrived at its destination on the 26th February, 1821, followed by the left wing on the 12th March. A detachment of three companies, under Brevet-Lieutenant-Colonel McDonald, was quartered at Northampton.

In August the establishment was reduced from—

	F.O.	Capts.	Subs.	Staff.	Sergts.	Crpls.	Drs.	Ptes.
	4	10	20	5	35	30	22	620
to	4	8	16	5	29	24	12	552

"The 19th July being appointed for the Coronation of King George IV., the regiment marched to Daventry, four miles distant, and from a hill there in view of the town fired a *feu-de-joie*, after which line was formed in the principal street of the town, when there was repeated cheering. In the afternoon all the officers and men sat down to different tables in the barrack square, which were supplied with all the good things obtainable.

"The non-commissioned officers and their wives sat at separate tables, the band and drums at another, and the whole of the artillery and their wives and families at another. After dinner was over Colonel Milne came to the iron railing that surrounded the elevated space where the officers dined and proposed the first toast, which was 'The King,' all the men immediately rising and giving three cheers. After this he gave a number of toasts and sentiments which were received with loud acclamations. During the intervals there were toasts drunk and songs sung at the different companies' tables, and all was the height of hilarity during the remainder of the day."[1]

Having received orders for Ireland, the Nineteenth set out from Weedon for Liverpool on the 25th October, and arrived there on the 5th November. They embarked in a small packet on the 7th, and landed at Pigeon House Fort, Dublin, the next day. After a short stay at the Royal Barracks the regiment marched out of the city on its way to Mullingar on the 10th November, where it arrived three days

1 Col.-Sergt. Calladine's Diary.

later. Detachments were found at Longford, Tullamore, Phillipstown, and Maryboro'.

This part of the country was greatly disaffected at the time, and murders, robberies and incendiarism were of frequent occurrence. The men had the ordinary jail duties to perform, in addition to sending out parties in search of arms, attending executions, escorting prisoners, etc.

In March there was an execution of three men at Johnstown, distant some eleven miles from Mullingar, which the headquarter companies were ordered to attend as well as the two at Longford. A squadron of cavalry from Newbridge was also brought over for the occasion, in addition to all the county police, as there was some expectation of an attempt by the inhabitants to rescue the prisoners. It was thought advisable to march this force to the spot where the execution was to take place, near where the crime had been committed and also in the neighbourhood of the residence of the men's relations.

The soldiers moved from Mullingar very early in the morning, escorting the prisoners in a carriage and conveying the gallows on a car. The number of followers increased every mile till the arrival of the party near Castletown Devlin, when the crowd was so great that there would probably have been a rescue had not the strength of the military and police prevented the possibility of any success. "Montford, the principal character of the prisoners, had a great number of friends in this part of the country, and it was certainly an awful sight to see three young men cut off in the prime of life, for the same crime that perhaps one-half of the surrounding multitudes had been guilty of. While the gallows was erecting and the finisher of the law was performing his part of the execution, we were formed in a circle round the elevated piece of ground where the gallows stood, facing outwards. Every hill and elevated object for a great distance away was covered with men, women and children, and when the car was drawn from under them there was a tremendous scream set up which made the country resound for miles around."

"After the bodies had been suspended the usual time they were taken down, and we returned to Mullingar, where we arrived very late in the evening without any disturbance taking place, being well tired with our day's work."[1]

On the 15th July, 1822, the various detachments and the headquarters marched to Dublin, where they arrived on the 18th, and took over quarters from the 71st Regiment at the Royal Barracks.

On the 10th March, 1823, the regiment left Dublin for Athlone, in the centre of Ireland, arriving there on the 13th, after passing through Maynooth, Kilcock, Kennegad, Tirril's Pass, Kilbeggan and Moate. The only detachment here furnished was a small one at Moydrum Castle, on Lord Castlemaine's property. In addition to the Nineteenth there was a detachment of dragoons and artillery in the garrison, which was commanded by Major-General Sir John Elley, K.C.B.

This officer was "very curious in his ways," according to Colour-Sergeant Calladine, "never letting it be known beforehand, even when we had the half-yearly inspection. He sent the order to the Deputy-Adjutant-General's office overnight with directions for it not to be opened till nine o'clock the next morning, when at the same time it was for our inspection at eleven the same day. Several times he called us out suddenly, which was the means of keeping us always ready for anything.

1 Col.-Sergt. Calladine's Diary.

He remained at a pleasant residence about a mile and a half from the town. On one occasion we had a five days' exploit on the banks of the Shannon, about three miles above Athlone, where Lord Castlemaine and family were on the river in his yacht, and other gentlemen of the neighbourhood in their pleasure boats. The pleasure was theirs, but the fatigue was ours."

In May the route was received for Boyle, and the regiment commenced its march there on the 15th, arriving two days later. A number of detachments were found in the counties of Roscommon, Leitrim, Sligo and Mayo. Headquarters and two companies were at Boyle and two companies at Sligo, under Major Bloomfield,[1] one at Carrick, Ballyhadareen, Strokestown and Roscommon, with small detachments of a subaltern and twenty men at Drumnsa, Tobercurry, Castlereagh, Lanesborough, Mohill, Ballinamore and Grange.

In addition to their ordinary duties the men were frequently engaged in still hunting, very fatiguing and harassing work, and eight to ten miles had often to be covered before the search commenced. The parties selected for these expeditions generally left barracks in the middle of the night, accompanied by the revenue officers, so as to surprise the farm where the still was at work. But if any warning was had of their danger, the people would carry off the still and sink it in a lake or boghole. The soldiers had simply to follow the revenue officer and his men and prevent them from being assaulted. Extra pay was given to the officers and men so employed.[2]

1 William Bloomfield, promoted Major 19th from the 16th Foot 8th March, 1821. Retired 22nd July, 1824. Died at Boulogne, February, 1828.

2 The remuneration that a soldier gets is 1s. 8d. per 24 hours, but if the party only goes half a mile from barracks they are entitled to the same allowance. The corporal is allowed 2s. and the sergeants 2s. 6d., the officer half a guinea, and whenever any seizure is made we are entitled to an extra allowance for every full seizure of stillhead and worm, and in proportion for part seizures.—Col.-Sergt. G. Calladine's Diary.

On the 24th May, 1824, the headquarters and out-stations left Boyle for Castlebar, where they all arrived by the end of the month. The headquarters of all the companies were stationed in Castlebar, but small detachments of an officer, or sergeant, and from fifteen to twenty men were sent to Ballycastle, Newport, Eskey Bridge, Killala, Clare, Crossmolina, Westport, Ballina and Foxford.

In the middle of August the route was received for Limerick, the regiment leaving Castlebar on the 16th and arriving there on the 21st. For a few months the companies were all together, but in December a number of detachments were sent into the country, two companies going to Rathkeale and Ballingarry, under command of Major Lenn,[1] which constituted second headquarters.

The South and West of Ireland was in a very disturbed state owing to the outrages of the Whiteboys, an illegal agrarian association whose object was "to do justice to the poor by restoring the ancient commons and redressing other grievances" (Lecky). The members of the association assembled at night with white frocks over their other clothes, threw down fences, levelled enclosures, and destroyed the property of harsh landlords or their agents, the Protestant clergy, the tithe collectors, and any others who had made themselves obnoxious to the association.

One night towards the end of February, the detachment at Rathkeale was ordered out by the magistrate, as some of these ruffians had broken into a farmer's house named Byrne, had dragged him away from his family, and having beaten him, left him dying in a ditch, where he was found by some

---

1 Exchanged to the 57th Foot with Major Edmund Lockyer 2nd May, 1822. Died at Cork 24th October, 1825.

men of Captain Rose's[1] company. Whilst the attack was being made on Byrne a house on the common was set on fire so as to attract the attention of the police and soldiers in that direction.

In March, 1825, an augmentation of two companies took place, raising the establishment to 704 privates, with a corresponding increase in non-commissioned officers.

On the 10th June the detachments were relieved by the 61st Regiment, and all the companies were together once more in Limerick, where steady drill was the order of the day. Here they were inspected by Lord Combermere, commanding the forces in Ireland, shortly after their arrival, who expressed himself well pleased with the discipline and state of the regiment.

On the 2nd and 3rd of August the Green Howards marched in three divisions for Buttevant, previous to proceeding to Cork. They arrived at Buttevant on the 5th, sending detachments to the Magazine at Cork and Spike Island to relieve detachments of the 79th, ordered on foreign service. Leaving Buttevant on the 24th of August, and passing through Mallow the regiment marched into Cork the next day.

On the 1st May, 1826, the battalion was inspected by Major-General Sir George Bingham, who reported on it as follows:—

"This is an exceeding good regiment, and fit for any service it may be ordered on. The Commanding Officer, Lieutenant-Colonel Milne, is a most zealous, active officer,

---

1 Hugh Henry Rose, Ensign 19th Foot 8th June, 1820. Lieutenant 24th October 1821. Captain 22nd July, 1824. Major, unattached, 30th December, 1826. Afterwards Lord Strathnairn, G.C.B., G.C.S.I. Commander-in-Chief India 1860-1865.

and it is impossible to speak of him in too high terms of praise. He is much beloved by officers and men, and by his firmness has prevented his regiment being so encumbered by women and children as some other regiments now are, and whenever the 19th is called on to embark it will exhibit few of those scenes of misery these occasions so frequently afford.

"If I was to particularize any part of the system of this regiment beyond the rest, it would be its steadiness under arms, the alacrity and correctness of movement in the field. It is impossible for a regiment to move better than the 19th, either alone or with other troops, and both officers and men are very well instructed in the movements of Light Infantry."[1]

In the autumn of 1826 the regiment was divided into six service and four depôt companies, and the former embarked at the Cove of Cork in three divisions between the 11th and 18th of October for Demerara, under command respectively of Lieutenant-Colonel Milne, Major Donald Campbell,[2] and Captain Lionel Hughes, where they all arrived by the 11th December. Before they left Cork, Major-General Sir George Bingham issued a very flattering farewell order, in which he mentioned "the high sense he entertained of the behaviour of the regiment since it had been under his command. The steady conduct of the men in quarters had kept pace with the high discipline, brilliant appearance and correct movement in the field, and their alertness on guard had been noticed by field officers of other regiments."

The marching-in inspection of the battalion took place on the 19th December, and called forth the following

1 Inspection Returns War Office.

2 Promoted Major 19th Foot from the 17th Foot 29th August, 1826. Retired 16th June, 1837. Died 28th November, 1865, and buried at Inverary.

complimentary order from Major-General Sir Benjamin D'Urban, commanding the troops in the colony:—

"District Office, Demerara,

"19th December, 1826.

"The Major-General Commanding the District has been very highly gratified at his inspection of the 19th Regiment. The soldier-like appearance of the men, the correctness and precision of the field exercise and movements, and the arrangement and cleanliness of the barracks, all deserve his unqualified approbation, and he desires to assure Colonel Milne that he will not fail to report to the Commander-in-Chief the high order in which the regiment has been found, notwithstanding all the disadvantages of a long and recent voyage.

"(Sd.) F. SHEA,

"Major of Brigade."

On the 5th of November, 1827, Lieutenant-Colonel Alexander Milne died of fever, to the great grief of the regiment. He was much lamented and highly respected by all ranks. His last dying wish was that the Colours should be wrapt round his body and buried with him, and this request was complied with. "No man," said a contemporary writer, in giving an account of his funeral, "had ever been buried in the colony so universally and sincerely regretted."

Major Donald Campbell, pending the appointment of a Lieutenant-Colonel, now assumed temporary command.

On the 18th January, 1828, the headquarters and one division embarked for Barbados, followed on the 21st by the second, under Captain L. H. Hughes, where they both arrived a week later.

On the 17th February Lieutenant-Colonel Henry Hardy was appointed to the command of the regiment. He had served with it as a Captain in Ceylon, where he gained two brevets for service on the staff. He was now promoted from Major in the 9th Foot.

The new Commanding Officer had taken part in two campaigns in India, with the 12th Foot in 1804 and with the Nineteenth in 1809. He was also present in Ceylon during the operations in 1815 and 1818, where, in addition to his brevets, he had been three times mentioned in General Orders.

In December of this year the officers' shakos were altered in height to six inches, and the gold lace taken off as well as the Hanoverian cockade. The only ornament now on the head-dress was a universal gilt shako plate, star-shaped, with crown over, some five and a half inches high. All officers wore white feathers twelve inches in height, except those of the light company, who wore green ones.

By a warrant issued in February, 1829, the double-breasted coatee was authorized, which, with little alteration, remained the dress of the officers until the Crimean War. Two epaulettes were worn by all officers.

More changes came in 1830. The gorget was abolished. The shako feathers were reduced to eight inches; musicians were dressed in white, instead of the hitherto fancy costumes generally worn. Officers of the regular army wore gold lace only. Fusils replaced the pikes, which had so long been carried by the sergeants.

NOTE.—In 1829 men enlisted for life; this continued until 1847, when enlistment for ten years was introduced. In 1825 the highest flogging sentence on record in the British Army was passed on a private soldier. He was awarded one thousand nine hundred lashes by a court-martial, and received one thousand two hundred. In 1830 a regimental court-martial could award three hundred lashes, and in 1831 a district court-martial five hundred lashes. In 1838 no corporal punishment was to exceed two hundred lashes. The following were the approximate rates of pay issued from 1800 to 1867:—Sergeant-Major, 1s. 6¾d.; Quartermaster-Sergeant, 1s. 6¾d.; Sergeant, 1s. 0¾d.; Corporal, 8¾d.; Drummer, 7¾d.; Private, 6d.—after deducting 6d. a day for rations.—"The Army Book for the British Empire."

On the 15th January, 1830, the left wing, consisting of three companies, under the command of Major Timothy Raper, embarked for the island of St. Vincent, followed the next day by the remainder of the regiment, under Lieutenant-Colonel Hardy, destined for the island of Grenada, where they disembarked respectively on the 18th and 22nd January. On the 24th April, 1831, the headquarters removed to St. Vincent.

Before Christmas the left wing, consisting of 6 officers, 12 sergeants, 3 drummers and 233 rank and file, sailed for Tobago, where they arrived early in January, 1832. There they disembarked two companies the next day in relief of the Royal Scots, and the remaining company (the light) proceeded to Trinidad. The headquarters and right wing had arrived at Trinidad on the 10th, after sending a detachment, consisting of one officer and sixty-three rank and file, to Naparina.

On the departure of the headquarters from St. Vincent the following resolution of the House of Assembly in that island was passed and published in the "St. Vincent Gazette":—

"At a meeting of the House of Assembly on Tuesday, the 6th March, Mr. Munro moved and it was resolved—'That the thanks of this House be given to Lieutenant-Colonel Hardy, the officers, non-commissioned officers and soldiers of the 19th Regiment for their meritorious, peaceful and excellent conduct during their residence in this colony. That the inhabitants of Kingstown will ever remember with the deepest gratitude the prompt and ample protection afforded to themselves and their property by the detachment of that gallant regiment sent by Colonel Hardy, when the town was almost devastated and all the shipping wrecked by the hurricane on the 11th August last. Nor will the poor forget to offer their prayers and thanks for the prosperity and happiness of their benefactors who generously contributed to their relief when bereft of their homes and means of subsistence by the furious elements.'"

The Nineteenth continued to get high praise from all the General Officers under whom it served in the West Indies. Writing of his inspection in September, Sir Lewis Grant, K.C.H., mentions "his unqualified approbation of the excellent appearance of the corps, and the style in which the firing and field manœuvres were performed; the respectable and soldier-like appearance of the men at all times," etc., etc.

Again, in July, 1833, Major-General Middlemore "congratulates Colonel Hardy on the truly excellent appearance of the regiment, and also upon the unanimity and cordial co-operation which he meets with generally, in enabling him to preserve that discipline and regularity which is so apparent, and which the Major-General has no doubt will never cease to prevail in this admirable corps."

The regiment suffered a great loss by the death of Lieutenant-Colonel Hardy, which took place at Trinidad on the 16th April, 1835.

The Governor of the colony on hearing of it at once wrote to Lieutenant-Colonel Doherty, commanding the 1st West India Regiment, as follows:—

"Government House,

"16th April, 1835.

"Sir,—In consequence of the death of Lieut.-Colonel Hardy, of the 19th Regiment of the Line, the command of His Majesty's troops in this Colony has devolved on you.

"The first duty in that capacity I am anxious you should discharge is to express in words the estimation in which as

Commander-in-Chief I held that excellent gentleman. His Majesty has lost in him one of his most loyal subjects, one of his most valuable military officers.

"The officers of the 19th Regiment have lost their friend, their adviser, their hospitable, cheerful companion, whose courteous manners and moral example secured the well-being, and much contributed to establish the character of that corps, for all that is correct and gentlemanlike.

"The non-commissioned officers and privates of the 19th Regiment have lost in the lamented death of Lieut.-Colonel Hardy a humane protector and charitable reliever of their wants and difficulties and a generous contributor to and superintendent of the education of their children.

"Society at large has been deprived of a truly honest and honourable member, and I have to deplore the loss of a sincere friend.

"I well know your liberal mind, and am confident you will feel pleasure in giving publicity to these sentiments in which you fully participate.

"I have, etc.,

"(Sd.) G. F. HILL."

The Author of "Truths from the West Indies" refers to Colonel Hardy as "One of the best men that ever breathed—one of the most loyal subjects King ever possessed—one of the most chivalrous spirits that ever adorned the ranks even of the British Army."

On December 18th, 1835, Lieutenant-Colonel Timothy Raper arrived and took over command. He had been appointed Ensign in the Nineteenth in 1810, and had served in the campaigns of 1815 and 1818 in Ceylon with the light company of the regiment. He had command of a flying column in 1818, made up from the light companies of various corps, and was several times noticed in General Orders for his services.

On the 29th March, 1836, the headquarters embarked at Trinidad, touched at Tobago, took on board the two companies stationed there, and arrived at Barbados on the 31st March.

The left wing, consisting of three companies, under the command of Captain Potts, embarked at Trinidad on the 9th April, and arrived at Barbados three days later.

The regiment only remained a few months at Barbados, orders being received for it to return to England. Prior to its embarkation Major-General Sir Lionel Smith, K.C.B., commanding the troops, issued a most flattering order, stating in no measured terms his approbation "of the sober, orderly and soldier-like conduct of the regiment during its service of nearly ten years in the West Indies."

On the 26th July the Green Howards, under the command of Lieutenant-Colonel Raper, embarked at Barbados for Cork, where they arrived on the 27th August. The regiment then marched to Buttevant, where it was quartered.

The marching-in inspection by Lieutenant-General Sir Edward Blakeney, K.C.B., Commanding the Forces in Ireland, took place in October, when he was pleased to express himself in the most flattering terms of the discipline, steadiness under arms, and soldier-like appearance of the men.

Orders were here received for the handsome regimental lace worn by the rank and file to be discontinued, but the drummers retained it till 1866, when the universal crown pattern was introduced.

Changes of station were now of frequent occurrence. On the 21st March, 1837, the headquarters and a wing proceeded to Cork, and remained there till the end of May. They then moved to Templemore, where they were joined by the remainder of the regiment.

After being a little over a year at Templemore the Green Howards marched to Kilkenny in July, 1838.

On Lieutenant-Colonel Raper[1] retiring by sale of his commission on the 28th December, the command was given to Major Thomas Talbot Hamilton, who had been transferred from the 57th Foot in 1815.

The regiment moved to Dublin in April, 1839, and whilst there the establishment was augmented to 39 officers, 47 sergeants, 40 corporals, 14 drummers and 700 privates.

Orders were received on the 28th December to embark on board the packets "Ballinasloe" and "Express" for Bristol. This was only a temporary move, as the regiment was back again in Dublin in March, 1840.

---

1 Died at his residence, Hoe Court, Herefordshire, on the 22nd July, 1862.

OFFICERS—1840.

RANK AND NAMES.	DATE OF COMMISSION.	REMARKS.
Col. Sir Tompkins Hilgrove-Turner, G.C.H., K.C.	27 Apr., 1811	See text.
Lieut.-Col. Thos. T. Hamilton	28 Dec., 1838	Retired 30 Aug., 1842.
Major Charles Craufurd Hay	16 June, 1837	Retired as Lieut.-Col. 14 Apr., 1854. Afterwards Lieut.-Gen. Died at Freshwater, Isle of Wight, 27 Sept., 1873.
,, Studholme J. Hodgson	28 Dec., 1838	Lieut.-Col. unattached 8 Aug., 1845. Afterwards General and Colonel 54th and 4th Foot. Died at Torquay 31 August, 1890.
Captain John Stirling ...	13 May, 1826	Died at Malta 25 Jan., 1841.
,, Charles H. Potts, m.	5 Feb., 1829	Retired as Bt.-Colonel 28 Nov., 1854.
,, David Burns ...	20 July, 1830	Retired as Bt.-Major 12 Oct., 1849. Died at Kingstown, co. Dublin, in 1830.
,, Thomas Beckham...	3 Dec., 1830	Served in the Peninsula with 43rd Foot. Retired as Major unattached 9 Nov., 1846.
,, Francis Price ...	7 Feb., 1834	Retired 11 Sep., 1840. Died at Charlton Kings, Gloucestershire 14 Sep., 1863.
,, John Semple ...	5 Sep., 1834	Retired 7 Aug, 1846, as Major. Died at Moorside House, Neston, 21 April, 1891.
,, Alexander Scott ...	23 Jan., 1837	Retired 20 May, 1842.
,, Robert Lovelace ...	16 Jun., 1837	Retired 4 Sep., 1840.
,, Thomas Hilton ...	30 Mar., 1838	Retired 4 April, 1845.
,, John Duke Simpson	28 Dec., 1838	Retired 6 August, 1847.
Lieut. William Bernard ...	25 Jan., 1828	Retired on half-pay unattached 14 Feb., 1845.
,, Frederick Deacon ...	8 Feb., 1824	Captain unattached 7 March, 1845. Died at Lordswood, Southampton, 10 Aug., 1864.
,, Mark A. H. Tuite ...	21 Feb., 1834	Succeeded to his father's bar'etcy in 1841. Retired 14 Aug., 1848. Died at Kilruane, Nenagh, 10 March, 1898.
,, Richard A. M. Franklin	17 Apr., 1835	Half-pay 55th Foot 25 Oct., 1844.
,, Anthony Walshe ...	23 Jan., 1837	Paymaster 2nd Foot 10 Sep., 1841.
,, Francis Seymour ...	16 June, 1837	To Scots Fusilier Guards 21 Jan., 1842. Gen. 1 Oct., 1877, and Col. Devonshire Regt. Created a baronet 28 Oct., 1869. Died 10 July, 1890.

Rank and Names.	Date of Commission.	Remarks.
Lieut. George R. Langley	4 Aug., 1837	Captain unattached 20 Aug., 1844. Served in the Peninsula with the waggon train.
,, George A. Stanley ...	26 Dec., 1837	Captain unattached 10 Sep., 1841.
,, James T. Bowdoin ...	30 Mar., 1838	Captain 4th Dragoon Guards 16 Feb., 1844.
,, James Cochrane ...	28 Dec., 1838	Captain unattached 5 Oct., 1849.
,, Henry Calley, Adjt.	2 Feb., 1839	Retired as Major 15 Feb, 1853. Died 3 May, 1881.
,, Geo. Hannover Tuite	7 Jun., 1839	Brother of Lieut. M. Tuite. Died at Calais 1 Sep., 1853.
,, William Dillon ...	30 Aug., 1839	Retired 9 June, 1843. Died at Monkstown, co. Dublin, 18 July, 1843.
Ensign Edward J. Ellerman	11 Feb., 1837	Captain 98th Foot 6 June, 1851.
,, Henry Butler Stoney	5 May, 1837	Captain unattached 28 May, 1852.
,, Robert Sanders ...	26 May, 1837	See Crimean Roll.
,, Frederick A. Jeffreys	16 Jun., 1837	See Crimean Roll.
,, James Ker	30 Mar., 1838	See Crimean Roll.
,, Robert J. S. Mansergh	28 Dec., 1838	Retired 6 Aug., 1847. Died 31 May, 1871.
,, John Phillips ...	2 Feb., 1839	Retired half-pay 57th Foot 21 March, 1851.
,, Hugh J. M. Campbell	31 May, 1839	Retired 1851. Died in London 16 Feb., 1852.
Paymaster Thos. R. Travers	23 Nov., 1838	Died at Montreal 23 Aug., 1849.
Quarterm'r. Richard Barrett	30 Oct., 1835	Ensign and Adjt. 13 June, 1845. Captain unattached 2 Oct., 1855.
Surgeon James Young, M.D.	18 Oct., 1839	To 13th Light Dragoons 30 May, 1843. Died in 1866.
Asst.-Surgn. T. Williams, M.D.	28 Sep., 1826	Surgeon 57th Foot 26 Jan., 1841.

Facings Green.

Agents, Messrs. Cox & Co.

In the summer the Nineteenth was divided into six service and four depôt companies, and on the 20th June the former marched to Cork, where they embarked on board H.M.'s ships "Vanguard" and "Rodney," for conveyance to Malta, arriving there after a voyage of three weeks.

The next change of station was the Ionian Islands, the regiment proceeding there in the troopship "Resistance" in January, 1843, under the command of Lieutenant-Colonel

Charles Craufurd Hay. A few days after its arrival at Corfu the headquarters and three companies embarked in the "Boyne" transport for Argostoli, Cephalonia.

Lieutenant-Colonel C. C. Hay had been appointed to the command on the 30th August, 1842, on the retirement of Lieutenant-Colonel Hamilton. He had been commissioned as Ensign to the Nineteenth on the 24th June, 1824, and was a great musketry enthusiast.

On the 7th May General Sir Tompkins Hilgrove Turner, G.C.H., died at his residence in Jersey. He had been Colonel of the regiment since 1811, and he was now succeeded by General Sir Warren Marmaduke Peacocke, K.C.H., K.C., a Peninsular veteran.

This distinguished soldier was the eldest son of Marmaduke Peacocke, of London, and was born in 1766. He was appointed an Ensign in the 88th Foot in 1780, and in 1793 exchanged into the 2nd Foot Guards. He was aide-de-camp to Sir George Nugent in Ireland in 1796, and during his period of service there he took part in the operations against the rebels at Antrim and Ballynahinch.

He went with his regiment to Egypt in 1801, and was present at every affair in which the Guards were engaged. In October, 1805, Major Peacocke, as he now was, embarked with the expedition to Hanover, and in July, 1807, took part in the attack on Copenhagen. Having been promoted Brevet-Colonel he went out to Portugal in 1808, and in the following year was appointed by Lord Wellington to command a brigade. He was present at the passage of the Douro, and later on was in command of the garrison at Lisbon, a position of great labour and responsibility, as the town was the base of operations.

THE DINNER ROUND. LIGHT COMPANY. 19TH FOOT.

From the Painting by C. B. Newhouse.

In 1811 Peacocke was gazetted Major-General, was promoted Lieutenant-General in 1821, and General in 1838. His services in Egypt were rewarded with the Order of the Crescent, and he received permission to accept the Insignia of a Knight Commander of the Tower and Sword of Portugal. He was knighted by the Prince Regent, and decorated as a Knight Commander of the Hanoverian and Guelphic Order.

After nearly three years' service in the Ionian Islands the Nineteenth embarked on board the "Java" freightship for the West Indies once more, under the command of Lieutenant-Colonel Hay, and sailed on the 5th December, 1845.

On the 20th January, 1846, the regiment disembarked at Barbados, and during the year had detachments quartered at Demerara and Trinidad.

The establishment was augmented in March to 950 privates, making a total of all ranks of 1,118.

In February, 1847, the headquarters and two companies moved to St. Vincent, with detachments at Grenada and Barbados.

Previous to the departure of the Nineteenth from St. Vincent in April, 1848, the following resolution was passed unanimously by the House of Assembly:—

"That His Honour the Speaker be instructed to communicate to Colonel Hay, the officers, non-commissioned officers and privates of the Nineteenth Regiment under his command (about to quit our shores for Canada), the high sense this House entertains of the quiet and orderly conduct of the regiment during the time they have been stationed here, and that the thanks of this House and the public at large are justly due to Colonel Hay for his great kindness in permitting the excellent band of his regiment to perform in public so frequently for the charitable purpose of supporting the Colonial Hospital.

"(Sd.) JOHN DRAPE,
"Speaker."

The headquarters of the regiment left St. Vincent on the 1st April for Barbados, picking up the Grenada detachment on their way, and, transhipping there, sailed with all the service companies for Canada, under command of Major Robert Sanders. Lieutenant-Colonel Hay remained in the West Indies as Acting-Lieutenant-Governor of St. Lucia, and did not rejoin until six months later. After again transhipping at Quebec the regiment arrived at Montreal on the 20th May, 1848.

In accordance with a warrant dated 30th June, the officers' blue frock coat was ordered to be discontinued and a plain scarlet shell jacket with pointed cuffs of green cloth, gold-twisted shoulder cords, and small regimental buttons was substituted. A great coat of grey cloth took the place of a cloak, which had been worn for many years.

About this time, too, the old cross belts, worn by the men for eighty years, gave place to a shoulder belt to carry the pouch, the bayonet being suspended by a frog from the waist belt fastened by a brass union locket.

Early in 1849 the establishment was reduced to 730 privates, a total of 874 of all ranks.

In a fire which took place early in January, when the prison of St. Helen's was destroyed, the bravery of Private Samuel Baldock (19th Foot) was brought to the notice of the garrison in General Orders. The Commanding Officer was personally assured by Captain Knight, the Governor of the prison, that had it not been for the assistance rendered by Private Baldock, many, if not all, of the men confined in the cells in the top storey of the prison, in number about forty-eight, must have been burnt to death.

In the middle of July the regiment encamped on the island of St. Helen's, owing to the prevalence of cholera at Montreal, and remained there for four months.

General Sir Warren Peacocke, K.C.H., K.C., died in London on the 31st August, 1849, when he was succeeded in the Colonelcy by Major-General Charles Turner.

This officer was appointed a Cornet in the 1st Dragoon Guards in 1795, and purchased a troop in the 29th Light Dragoons three years later. In March, 1799, he accompanied his regiment to Portugal, and in December, 1800, embarked from Lisbon to join Sir Ralph Abercromby's expedition to Egypt, where he landed as Brigade-Major of the cavalry. He was present in the action of the 13th March, 1801, but on the 18th was captured in a skirmish near Lake Mareotis, and remained a prisoner in Alexandria until the fall of that town in September.

In August, 1803, Captain Turner rejoined his regiment, and shortly afterwards was appointed aide-de-camp to Lieutenant-General Floyd, second in command in Ireland. After purchasing his majority he exchanged to the 13th Foot, and embarked with it from Barbados in the expedition against Martinique, ending in the capture of the island.

After serving for many years on the Staff, chiefly in Ireland, he was promoted Major-General in 1841, and Lieutenant-General ten years later.

For his services in Egypt he received the gold medal, and a silver medal and clasp for Martinique.

By the 6th May, 1850, the headquarters and five companies of the Nineteenth had arrived at Quebec, when one company was sent to Three Rivers and small detachments to Grosse Island and Sorel.

On the 28th June, 1851, the six service companies sailed for England, and on arrival at Plymouth on July 25th took over quarters at Devonport, sending out detachments to Pendennis Castle, Dartmoor, St. Nicholas and Market Heights.

The following year, on the 8th May, the headquarters and four companies moved by rail to Exeter, from whence they marched to Winchester, being met there by the two outlying as well as the four depôt companies.

Field-Marshal the Duke of Wellington had died on the 14th September, but his funeral did not take place till the 18th November. The Nineteenth was one of the regiments ordered to attend, and it left Winchester by rail two days beforehand. On arrival in London the men were billeted in Boro' High Street, and on the day of the funeral the regiment marched to St. Paul's, where it was formed up four deep on the south side, being finally posted with the right on Dakin's tea shop and the left towards Temple Bar. The funeral car stuck on the newly-laid gravel on Ludgate Hill, and some of the men were ordered to sling their arms and help it up the rise.

"I was there," wrote the late Mr. S. M. Milne, "as a boy from school. My father and I had good seats in Ludgate Hill opposite to where the Nineteenth were stationed. The day before the authorities had placed about three inches of yellow gravel up Ludgate Hill, and possibly elsewhere also. Well, the heavy rain in the night had turned the gravel into yellow slush, and I can well understand how the heavy funeral car would require assistance at the steep part of the hill below where I was seated.

"I did not know at the time that the Nineteenth Foot was all about there, but I well remember seeing groups of officers with green facings (no doubt of the Nineteenth) moving about, looking well in their handsome full-dress uniforms, but plastered up to the knees with the before-mentioned yellow slush."

Quartermaster Hendry attended the parade with the regiment. He was 6 feet 6 inches in height and 25 stone in weight, and was a source of much interest to the spectators. He had served with the 79th Highlanders at Quatre Bras and Waterloo, and finally died a Knight of Windsor.

The next day the Nineteenth returned to Winchester, and on the 5th January, 1853, moved by rail to Portsmouth.

NOTE.—The accompanying illustration is from an original sketch by the late Mr. R. Ebsworth, who also did one of the 19th marching out of the Tower of London to embark for the Crimea.

The following remarks were written by him on the back of the funeral sketch:—

19th Foot, Winchester, 1852. As just home from Canada (many of the men and officers and part of musicians on leave) were hurried up to London, billeted in Boro' High Street to assist in keeping the ground: Duke of Wellington's funeral: Two parades in heavy marching order took place in grounds of Guy's Hospital, musicians furnished with bayonets (as drawn). Duty calls were allowed but no band permitted. Drummers had their instruments covered with crêpe: Colours ditto as per regiments in the procession.

These I had the opportunity of seeing when 19th Foot moved Winchester to Tower, February, 1854 (also I saw them when 19th attended Chobham Camp, 1853).

Two Lieut.-Cols., Hay and Unett, commanded in turn. Col. Hay, first Commandant School of Musketry.

Lace two bar. Drummers' wings white lace with red lines.

I have ventured to put more in of the 19th group really as seen. It was that dull November day, a very crowded parade of some 600 men in yard of Guy's Hospital, Boro' High Street. A convenient post offered for the stacking of the Colours, bare of honours. A similar post used to be, and perhaps still is in the colour yard St. James' Palace, so also was one in Dublin Castle for daily guard.

Before leaving the Tower, 1854, there was some mention that new Colours were to be presented, but route arriving for foreign service prevented that being done. After all they were in fair condition.

After repeated delays, left London for the seat of war one cold spring morning, along the cheerful district of Wapping, 1,000 strong. Ensign Massy (afterwards Colonel 5th Lancers) carried one of the Colours.[1]—R.E., 1890.

[1] The artist is wrong here as Ensign Massy did not join the 19th till 1855.—M.L.F.

In the middle of May four companies were sent to Weymouth in relief of the 88th Foot. At the same time the headquarters and six companies proceeded to Gosport to relieve the 65th and 88th Foot at Haslar and Fort Monckton.

A further change of station came in July, when the regiment assembled for a few weeks at Chobham Camp, where it was brigaded with the 79th Highlanders and the 97th Foot, under command of Colonel Lockyer, K.H. After this training the headquarters and ten companies marched to Woking on the 19th August, where they entrained for Deal and then marched to Walmer Barracks.

On the death of Lieutenant-General Charles Turner, which took place at Sutton Lodge, Chiswick, on the 12th June, 1854, Major-General William Rowan was selected to fill the vacancy.

This highly distinguished officer was the son of Robert Rowan, Esq., of Mullans, co. Antrim, and was born in the Isle of Man on the 18th June, 1789. He received a commission in the 52nd Light Infantry in 1803, and served with it in Sicily, 1806-7, and in Sweden in 1808. On obtaining his company he went out to the Peninsula in October, 1808, but left in 1809 to take part in the abortive Walcheren expedition. Returning to the Peninsula in 1811, Captain Rowan was present at the action of Sabugal, and from January, 1813, up to the end of the war served in the Peninsula and in France, taking part in the battles of Vittoria, the Pyrenees, Nive, Nivelle, Orthez and Toulouse, besides many minor affairs.

At Waterloo he was with the 52nd in the famous charge against the Imperial Guard, and was promoted Brevet-Colonel for his services.

FIELD MARSHAL SIR WILLIAM ROWAN, G. C. B.

FROM A DRAWING AT MOUNT DAVY'S, CO. ANTRIM.

He commanded the forces in Canada from 1849 to 1855, when the Nineteenth was one of the regiments quartered there. After going through the various grades of promotion he was finally appointed a Field-Marshal in 1877.

Sir William Rowan was created a G.C.B. in 1856, and was in possession of the war medal with six clasps for the Peninsula as well as the medal for Waterloo.

The band of the Nineteenth at this period, 1844-1854, had a great reputation, and was one of the few line regiments that possessed stringed instruments. This was due principally to the exertions of the bandmaster, Mr. James Smyth, so well known in the Royal Artillery band from 1854 to 1881.

The son of a guardsman, he was born in London in 1818, and while a child his father was promoted into the Nineteenth, the boy later on being taken into the regimental band. His abilities were so marked that he was appointed sub-conductor, and on the retirement of Mr. Brown in the early forties, Corporal Smyth was selected to succeed him.

On the outbreak of the Crimean War the Officer Commanding, not wishing to part with the bandmaster (who was not allowed to accompany the regiment) offered him the appointment of quartermaster just then vacant, but Lord Hardinge, the Commander-in-Chief, would not sanction such an unprecedented promotion as it was at the time.[1]

On the 4th February, 1854, the Nineteenth entrained for London, and on arrival took up quarters in the Tower.

Whilst it was stationed in London the negotiations with the Emperor of Russia were brought to an abrupt conclusion. Owing to his unprovoked aggression against Turkey, and his

1 The British Musician, 1900.

NOTE.—When the regiment was at Walmer the Countess of Mount Edgecumbe presented the band with a silver medal and two kettle drums as a recognition of its merits. The drums are still used by the band.—M.L.F.

rejection of the terms offered him by the principal European Powers, Her Majesty Queen Victoria was compelled to declare war.

On the 24th March the band and drums, together with two companies of the regiment, marched to the Royal Exchange, from the steps of which the Royal Proclamation was read by the Herald. At its conclusion the band played the National Anthem and the men presented arms.

The Nineteenth was one of the regiments warned for service with the Eastern Expedition, and we shall follow its fortunes in the next chapter.

CHAPTER IX.

The Green Howards embark for the East—Arrive at Scutari and proceed to Bulgaria—Land in the Crimea—Battle of the Alma.

THE headquarters and six companies of the Nineteenth, under command of Bt.-Lieutenant-Colonel Unett, embarked on the 17th April, 1854, at Blackwall, in the s.s. "Victoria," and sailed a week later for the East. Other small detachments of the regiment, under Captain Chippindall, Lieutenants G. Lidwill and Godfrey Massy, had embarked earlier in the month. In the meantime the depôt companies had separated and marched to Walmer, under command of Captain O. A. Grimston.

The total strength going out was 3 field officers, 8 captains, 16 subalterns, 6 staff officers, 47 sergeants, 40 corporals, 15 drummers, and 810 privates.

Early in May the headquarters arrived at Malta, and leaving next day, the "Victoria" anchored at Constantinople on the 10th of the month.

The regiment disembarked at Scutari and marched into camp there, being told off to the Light Division, under command of Lieutenant-General Sir George Brown, K.C.B., and posted to the 2nd Brigade, under Brigadier-General G. Buller, the 77th and 88th being the other two regiments composing it.

The Nineteenth was encamped with its right resting close to the wall of the Turkish cemetery, with the other two regiments on its left. The division lay between the cemetery and Kadikoi, a village on the Sea of Marmora, about a mile distant.

"The position of our camp in the eye of an artist could hardly be surpassed. It lay between the great Turkish cemetery of Scutari on the right and Kadikoi, a village on the Sea of Marmora, on the left, say a mile in longitude.

The 19th, as senior regiment, was on the right, the 88th, as junior, on our flank, and the 77th, as second senior, on the extreme left. As we stood we had, when in line, the great barracks of Scutari in our immediate front, an enormous building, said to be able to house 10,000 men. It was not a very picturesque pile, built no doubt with an eye more to economy than taste. Glaringly white, it was grand in dimensions. On the extreme right was the Champ de Mars of the Turks, with ancestral yews and cypress plantations sad and sombre to the view, studded underneath with monuments of wonderful and careful artistic manipulation: all had head-stones erect, beautifully and elegantly carved, painted and inscribed to the memory of the departed beneath deposited. The gilding and painting on these slabs was to us a source of interest and admiration.

"The tall, conical and sepulcral-coloured groves gave to the scene a never-to-be-forgotten sentiment as to the greatness, and now the humilitude of the original constructors of these curious depositories of the dead. To the left of our picture is seen the Seraglio and Seraglio Point, on the European side of the Bosphorus. More distant again the great dome of the Mosque of St. Sophia, while further in the distance many other structures, such as the Mosque of Sulieman the magnificent, with minarets and pillars *ad infinitum*.

"More to our left lay the tranquil Sea of Marmora, reaching as far as the eye could view towards the Dardenelles. While looking in this direction we see in the extreme distance Olympus covered with snow, and at its foot the wooded country round Broussa, and the Marmora, studded with some half-dozen volcanic-looking islands, some conical, others serrated in appearance, all more or less inhabited."[1]

1 Extract from a letter of Major Lidwill.

The 1st Brigade consisted of the 7th, 23rd and 33rd Regiments and the 2nd Battalion Rifle Brigade, under the command of Brigadier-General R. Airey.[1]

Whilst in camp at Scutari Lieutenant-Colonel Robert Sanders, who had been travelling abroad, joined the regiment and took command. He had been promoted on the 14th April *vice* Lieutenant-Colonel Hay,[2] on the appointment of the latter as Commandant of the newly formed School of Musketry at Hythe, and had received his first commission in the Nineteenth as Ensign on 26th May, 1837.

---

1 Succeeded shortly afterwards by Major-General W. Codrington.

2 Afterwards Lieut.-General C. C. Hay, Commanding Troops Cape Colony, 1868-1873. Died at Freshwater, Isle of Wight, 27th September, 1873.

Colonel Hay took with him to Hythe Sergeant John M'Kay, 19th Foot, to assist him in the work of organization, and when the School of Musketry was finally established in 1854, Sergt. M'Kay was promoted Ensign and Quartermaster of the school. He was born in Beggars' Bush Barracks, Dublin, where his father was then serving as a private in the Green Howards. On the death of the latter in the West Indies in 1832, the officers of the 19th took care of the boy until at the age of ten he was duly admitted to the Duke of York's School in 1833. When fourteen years of age he enlisted in the 19th as a drummer in 1837, and in due course was promoted to the rank of Sergeant.

He was gazetted to a Lieutenancy in the 41st Foot on the 9th March, 1855, but never joined, for within a month he was appointed a Lieutenant-Instructor, and Acting Adjutant at Hythe. His success in this appointment was so marked that on the 1st April, 1856, he was made Deputy-Assistant-Adjutant-General of the School of Musketry, an office which he held with the greatest credit for eleven years, having in the meantime been promoted to the rank of Captain on the 6th May, 1859, and Major 1st July, 1867.

On 1st August, 1867, he was gazetted to the 1st Battalion 12th Foot, and did duty as Major of that battalion until 1st May, 1871, when he was selected to command the 2nd Battalion of the regiment, then in India, with the rank of Lieut.-Colonel. It was an almost unprecedented distinction for a man to rise from the rank of sergeant to the command of a regiment in sixteen years without the expenditure of a penny. In those days first appointments to commissions and subsequent promotions in the cavalry and infantry were almost invariably obtained by purchase.

On the 1st May, 1876, M'Kay was promoted to the rank of Colonel in the Army, and on the 20th March, 1878, was selected by the Commander-in-Chief for a Distinguished Service Reward. On the 10th April, 1878, after commanding his regiment for seven years, he was appointed to the command of the 32nd Brigade Depôt at Bury St. Edmunds, known now as the 12th Regimental District. He held this post till he attained his sixtieth year, in 1882. In the following April he was promoted to the rank of Major-General, and retired from the service, carrying with him the well-deserved respect of the whole British Army. He died in October, 1887.—From "The Sons of the Brave," 1897.

The regiment remained at Scutari till the 29th of May, when it embarked on the s.s. "Medway" for Varna, and disembarked the day following. There it remained for a week, when the division marched to Alleydyn, ten miles distant, where it encamped. This was a very trying march for the men; the road was a bad one, and the trees and bushes along the route kept away a free circulation of the air. There were many cases of heat apoplexy.

Alleydyn was beautifully situated for a camping ground. A lake ran westward from Varna for twenty-seven miles, and on the northern wooded slopes were certain spacious clearings, where were scattered the tents of the regiment where opportunity existed.

Constant drills took place every morning, even though the heat was increasing daily, and the commanders of the Light Division gained unenviable notoriety for their harassing drills all the time it was quartered in Bulgaria.[1]

There was little sickness amongst the men till the cholera appeared. It broke out in the Nineteenth on the 16th June, and soon spread with marvellous rapidity.

Alleydyn not proving healthy as a camp, the division marched to Devna, over eight miles distant, on the 30th June. Devna lay open and the camp faced south, whilst close by ran a river of the purest water rising from natural springs, yet notwithstanding this, it was the unhealthiest camp of all.

1 The infantry drill, "The Red Book," as we know it, was first published in 1859. Previous to this regiments had their own drill written in manuscript, and copies could be got on payment from the orderly-room. The 19th had a very good light infantry drill up to this time written by Lieut.-Col. Timothy Raper, who had been in command of the Light Company of the regiment. The last time it was used was in Alleydyn, in Bulgaria, in 1854. The 19th then covered the front in splendid style. General Buller, commanding the Brigade, said it was not the drill of any regiment or army in Europe. "No matter," said Sir George Brown, commanding the Light Division, "what drill it is, its d——d well done," and complimented Lieut.-Col. Sanders on the smartness of the battalion.—Extract from "The Green Howards' Gazette."

On the 6th July the Light Division was inspected by Omar Pasha, commanding the Turkish army. He was accompanied by Lord Raglan and Staff, the Duke of Cambridge, Admirals Dundas and Lyons, and other military and naval officers.

Up to the 19th July the health of the men had been very good on the whole, but as the heat increased so did the cholera, and the Nineteenth, in common with other regiments, lost many men, no less than twenty dying one day in the division. It attacked the French with the greatest severity, and they sank under it at the rate of sixty to eighty per diem.

To still further avoid the cholera the regiment moved on the 24th July to Monastir, in the hope of finding a more salubrious site for their camp a few miles further west. Monastir and the country around were superb as to scenery. In the distance lay the majestic Balkan range, well wooded as far as the walnut and forest trees could get soil to mount its slopes. The camp of the regiment was in the midst of a cluster of these trees, looking north. There was no drill here, as owing to the cholera the brigade was broken up, and the 77th and 88th were sent away a mile or two in different directions.

The men's time was employed in throwing up entrenchments and being trained in the use of gabions and fascines. They also had constant practice with the Minie rifle, which was issued whilst in this camp.

On the 26th July a draft arrived, under Lieutenant A. Goren, consisting of one assistant-surgeon, fourteen corporals and ninety-eight privates.

About this time Lord Raglan gave the men permission to grow moustaches, but Sir George Brown, a soldier of a very old and rigid school, objected to the Light Division doing so. Sir George was a smart looking, well set up man of sixty-six years of age, and had served in the Peninsula forty years earlier. He was a stickler for pipe-clay, leather stocks and a clean shaven chin, but when the real work of war commenced a difference was soon found between theory and practice. Razors got rusty and blunt, and were pitched away. Instead of pipe-clay, mud was the prevailing ingredient used upon the belts when the Nineteenth settled down before Sebastopol.

On August 27th the regiment, with the rest of the brigade, marched from Monastir and encamped at Yursakova, ten miles eastwards, the same day. Karagoli was reached on the 28th and Varna on the 30th. The Crimea having been chosen as the scene of operations against Russia, the regiment embarked there on the "Courier," and on the 3rd September assembled with the rest of the fleet in Baltchik Bay.

On the 14th the Nineteenth disembarked early in the morning at Old Fort, Crimea, about five or six miles north of the Bulganak river, the first company to land being No. 6, under Lieutenant Lidwill, which had been transhipped to the "Pride of the Ocean" on the 1st of the month, the "Courier" being overcrowded.

The strength on disembarking was 3 field officers, 6 captains, 12 subalterns, 5 staff, 49 sergeants, 43 corporals, 15 drummers, and 763 privates.

The whole of the Light Division landed in about an hour. Forming up they marched in a southerly direction for about four miles, where they bivouacked on a plateau. Cossacks were to be seen at a great distance inland, mostly posted on tumuli, and an unwary convoy of corn crossing the front was captured by the 23rd Fusiliers.

The first night in the Crimea was a most miserable one, rain fell in torrents, and no tents were available, for they were not landed till the next day, and then they had to be reshipped, as there was no transport for them. The three following days were full of labour and activity applied to the disembarkation of the vast array of guns, ammunition and stores requisite for the army.

On the morning of the 19th September the allied armies commenced their march southwards towards Sebastopol, twenty-seven miles distant. The Turks on the right were close to the sea, then the French, 28,000 strong, with the English, 27,000 in all, on the left. The cavalry were on the extreme left. The English advance was in double column from the centre of divisions, the Nineteenth and 23rd marching together.

The whole of this moving mass covered several square miles, and carried with it the hopes of three nations; nor was this the limit of the picture: a splendid fleet steamed and sailed southward as the army marched, each, fleet and army watching and admiring the other.

The officers and men landed in full dress, the former carrying sword and revolver, with greatcoats rolled horse-shoe fashion over the shoulder. The men had a blanket, a greatcoat, in which were folded a few necessaries tied up with the knapsack straps, and three days' rations, together with sixty rounds of ammunition. The men's knapsacks containing their boots and shell jackets, were left on board ship, as were also the band instruments, so that there was no music on the march.

The distance to the Bulganak was only ten miles, but, as before stated, the health of the men had not been good, and their long confinement on board ship, contrary to expectation, had not improved it. The day was very hot, consequently their thirst was very great, and many fell out on the march, some in the throes of cholera, which had not yet disappeared. It was therefore a great relief to all when they reached the banks of the Bulganak, a small, slow-running stream, where they bivouacked for the night.

The following description by Major Lidwill of the advance to the Alma and the part played by the Nineteenth in the battle will give a good idea of what took place on this eventful day.

THE BATTLE OF THE ALMA.

"Though till now ungraced in story scant although thy waters be,
Alma, roll those waters proudly, proudly roll them to the sea."

The dawn of the 20th of September was remarkable for the first dense fog that the regiment experienced in the Crimea since landing, and it was long after daylight before one could leave the piled arms of the regiment and be sure of retracing his steps back again.

The hum of voices could be heard and the neighing of horses far and near, but as bugle sounds were forbidden, commands were passed by word of mouth, and comparatively silent was our great army. Not so the French, for their drums and trumpets were heard continually.

When at last a light sea breeze swept away the wall of cloud, which, rolling inward, as it were, drew up the curtain on a scene which it was the lot of many to behold and long remember.

Facing the regiment in a northern direction was the camp of the Russian army, its right wing resting on an elevated round-headed hill, called by them the *Courgan*. A similar mound would be called by the French a *mamelon*, in Ireland we would designate it a *drum*, and it being high, would probably be called *dromard*, or the high round ridge.

This hill had on it an earthwork, and the clay thrown up could be seen, though it was so elevated that such additional strengthening could add little to its importance.

Just under it was another elevated plateau between it and the river, on which was placed a much more formidable and conspicuous work, mounting fourteen guns.

The latter was the grand object of attack of the Light Division, and was the point where the blood of our army was freely shed. Beneath this high plateau ran the Alma, with its villages, orchards, vineyards, enclosures and walls. Several tall Lombardy poplars showed the course of the stream, and added variety to an otherwise uninteresting bit of landscape.

The sea lay to the right; blue and clear was the horizon studded with white sails, and the dingy smoke of moving steamers alone tainted the fresh-looking streaks of the ocean. There, and at no great distance, but obscured from vision by the nature of the ground, lay the allied fleets, the base of supplies.

It was a long time before any order to fall in was given, owing, it is said, to a want of fixed plan being agreed to by the allied generals, but eventually the companies were formed up and the brigade takes post in line of quarter columns, with the 2nd Division, under Sir De Lacy Evans, at deploying distance on the right.

The Light and 2nd Divisions form the first or fighting line, together with the 1st and 3rd Divisions, the 1st being composed of Guards and Highlanders. The 4th Division in reserve. The French are in dark masses on the right of the British with a numerous artillery.

A curious stillness and silence pervaded the army. No bugles were allowed to sound, and unless the neighing of a horse or the rocking of the wheels of a gun on the soft soil, no sound escaped our line.

At last a crowd of officers is seen having a council of war, and an order soon comes to form line of battle.

In the case of the Light Division it was deployed right and left from the front company of the Nineteenth, which deployed to its left. Then came the 88th and 77th, with the Rifle Brigade covering the front with skirmishers and supports. The 1st Brigade was deployed to our right.

The army was now drawn up in battle formation, with a front of over two miles, the French on our right probably occupying similar space, with 7,000 Turks again on their right close to the sea, and the order is given—"The division will advance in double column from its centre."

The divisions of each battalion being fifty-four files in front or 108 men in the leading grand division, one can imagine what an imposing and magnificent a machine of war the Light Division was, with its old twenty-one years' service men, seasoned and trained in the Mediterranean islands, the West Indies and the snows of Canada, for all these regiments had done their tour of service in those latitudes, and had only within a few years returned to England, and for their smartness and efficiency were selected by Sir George Brown to form his legion, ready to emulate the deeds of the old Peninsular Light Brigade, in which he had been trained.

The ground traversed was of a deep clay soil covered with a coarse dried-up vegetation with an odour like thyme. Dark clouds of dust rose at times from the tramping of the troops. Partridges and hares were disturbed by the onward flow of the army, and many were killed, being apparently dazed and not knowing whither they went. A few halts were made from time to time until the men came within easy eye-range of all in front of them.

From the Post House on the Bulganak in rear, down to the bridge over the Alma in front, ran a fine macadamized road crossing the stream and winding ziz-zag up the grassy slopes of the downs before us, out of sight to Sebastopol.

This was called the Waronzoff Road, and it divided the enemy's army into two wings. Looking to the left we see the Russians, their right posted in masses of grey columns, on a broad dome-shaped hill, flat on the top and converted into a temporary battery. This earthwork had a glacis of some 400 yards running down to the river, and extending round its flanks, with twelve to fourteen heavy guns mounted in it. Immediately in its rear was the Courgan already alluded to, the highest point of ground held by the Russian artillery. About and in rear of these works were masses of troops. Mounted men were leaving and arriving continually.

Carrying the view round towards the right, the top of the ridge was held by columns with guns quite visible, until the eye rests on a building like a lighthouse. This was a telegraph station communicating with Eupatoria from Sebastopol. Just here the position appeared to be very strongly held. Towards the sea further to the left the troops had been withdrawn from the fire of the ships. Close before us lay the tortuous Alma, its banks dotted here and there with farmhouses, and tall and graceful poplars lined its course.

After a long halt and a consultation amongst the Generals the division moves on, and soon after a dense column of smoke is seen arising from the buildings all along the front, which have evidently been set on fire to impede our onward movement.

The march continues, a few cannon shots are fired at us, but fall short, and the order comes—"The division will form line on the two leading double companies." The Nineteenth and 23rd move into line by echelon, the others only alter their line of march in column outwards to their positions. This deployment was effected by the different regiments with beautiful precision. It was while leading my company into its place that I saw the effect of the first deadly aimed cannon ball.

No. 1 was halted and dressed, when a shot passed through it, apparently killing two men front and rear rank, and striking Lieutenant Wardlaw in the leg. He instantly fell mortally wounded.

About this time the Russian skirmishers open a deadly fire, little puffs of smoke coming from the vineyards in front. We had not been long in line before the order to take ground to the left came, the reason being that the correct distance between the 2nd and Light Divisions was very badly maintained, and the deployed lines were overlapping each other.

The smoke from the burning village of Burliuk gave us some protection from the view of the enemy, and while marching in open column the cannon balls shot through the intervals of the moving companies with wonderful harmlessness.

The musketry fire on us now ceased, and the Rifle Brigade rush into the vineyards. At this time an order came for the regiment to form square on the left centre company, and mounted officers were seen galloping to the 77th and 88th, apparently with the same orders.

Away at a considerable distance were a body of Hussars on dark-coloured horses, and the Horse Artillery of our Light Division passed my left flank at a gallop towards this

THE BATTLE OF THE ALMA.

cavalry. With the great imposing earthwork obliquely on our right front, it would have been madness to have formed square in the face of such a powerful battery, and would have meant an awful sacrifice of life for no purpose whatever.

The Nineteenth therefore continued its flank movement with the rest of the division in its rear, and shortly afterwards wheeled into line, and so hot was the fire that the men had to lie down close to the ground.

In the meantime the guns had gone ahead, and were getting ready to fire on the Russian cavalry, but were anticipated. Like magic the whole force moved rapidly across our view, and out of sight round the trees and enclosures in our present front.

Had these horsemen done something, or even threatened our flank during the march, much anxiety would have ensued. As it was, General Buller, with the 77th and 88th, remained long halted, and did not partake in the subsequent advance of General Codrington's brigade. His reason was that the left of the army being threatened, he thought it only right to keep two regiments in hand until he should be supported.

The Nineteenth thus found itself associated with the 1st Brigade for the remainder of the day.

The advance was continued up to the end of the vineyard plantations, where a certain amount of cover is obtained. Where my right was posted there was a vista through the apple and pear trees, and I had a clear view of the great red clay work on the mound on the opposite side of the Alma. The position looked very uncomfortably strong; all the masses of columns on its flanks appeared a terrible barrier to force.

I had been here some ten or fifteen minutes when Sir George Brown rode up to me, alone, on his white horse. Until now I had not seen him, as he had been with Lord Raglan. He wore his big boomerang cocked hat, but the feathers of staff officers were taken off by order, and he looked more like a naval officer than a military one.

His object was to know when I last saw General Buller. I said, "Not since the line was formed; I got no orders from him, and am sure he is a long way off on the left." He was very displeased. I made him move out of where he was, as he was seen, and brought shots on himself. A little on my right a house quite sheltered him.

From the view I had I could see the French on my right climbing the cliffs and scaling their way through the broken defiles up to the white telegraph building. All was life. The sun shone brightly, and in every direction on the hill tops and on the battery near in front were the gunners busy at their deadly occupation. The noise of the artillery was grand, continuous and stupefying, and a new element of disturbance —shells—were pelted at us, so we kept under cover of the big earth bank. The dogs of the village ran about scared at this novel change in their affairs, and pigeons darted and dived on the wing through the blue smoke in evident terror of their lives.

After a long delay and being in a state of nervous inactivity, we at length get an order. It was given me by Captain Pearson, A.D.C., to cross the vineyard and river "at a rush" on a given signal. Just before this I distinctly saw Major Norcott attacking the right flank of the battery. He was with the supporting companies of the Rifle Brigade, and they were covered by skirmishers, who were much hidden by their own smoke. It looked a daring, dashing bit of business to see a mere handful of men going slap at a multitude. The supports reinforced the firing line, and galled the columns and gunners almost with impunity.

At last the signal is given, and over the bank we climb, and now are up to our waists in the vines. They are planted in holes some four or five feet broad and three feet deep. The sides act instead of poles, that are not to be had in this land of steppes. Falling and tumbling about as they advance, the men soon reached the river bank, where they were treated to a hot fire of shot and shell, but plunging into the river, which at this point was three or four feet deep, they soon crossed to the south side and got good cover on the opposite bank. The light company and No. 7 had shallow water where they crossed, with no high banks, comparatively speaking, to break their formation, and consequently were better dressed than the others when they advanced up the heights. Neither General Buller nor any of his staff were with the regiment, which instinctively put itself under the command of General Codrington, and took his orders.

The Colonel had to leave the regiment and go to the bridge-ford of the Alma, a very considerable distance, before he could rejoin, and the regiment had to advance without him.

Onward we go, and though we are not side by side with our old friends the 88th, who under Wallace did such wonders in Spain, where fighting and the use of the blue cold steel was carried to perfection, we with light hearts advance to measure our strength with men who even rolled back the tide of war on the great Napoleon.

The shot strike us, many fall, and we begin a general file firing against the earthwork. Bayonets are fixed, and the advance continued, the rear rank carrying their arms at the port or slope. The ringing of the steel bayonets when struck by the grape is novel and curious. On we tramp up the slope, firing all the time, until we are almost clear of the left front of the earthwork, which has greatly slackened its fire. We were now about one hundred yards from the nearest battalion of the enemy, and I could see the heads and glittering bayonets of any number of men in column. Their strange drab coats and ugly heavy-looking helmets appeared anything but sightly, though perhaps martial. Between these columns and the earthwork were skirmishers kneeling, and one very big man used to come out of the battery, kneel and fire as I thought at myself, so I shouted for Sergeant Owen, who was a splendid shot, and I pointed him out what I wanted. Owen straightaway fired, and this troublesome fellow fell forward. The effect of our fire was very noticeable now. I could see the men in the great column wasting away, whilst we were having little loss. The strength of the enemy seemed enormous, and having advanced closer, looked like the dense crowd as you see it at Epsom on Derby day. The assembled mass before us now was, however, of very different material. Just at this moment, as we were closing on the enemy, the advance came somehow to a halt, and I momentarily dropped to the rear to find out the reason. I saw Colonel Sanders on his horse in the centre of the regiment waving his sword and encouraging the line to advance. He was hit soon afterwards in the leg, and I saw him being carried away.[1] Ensign Stockwell, carrying one of the Colours, was also about this time shot dead. The ball passed through his forehead, making a very small aperture.

1 Lieut.-Colonel Sanders' horse was killed at the same time. He was invalided home and did not return to the Crimea. Sir George Brown, writing to him some time after, said: "Nothing could have been more spirited than your conduct, and that of your gallant corps, at the Alma, and no one had a better opportunity of observing it than I had. Many a time since I wished to see you again at the head of the 19th."

The fire on both sides was now pretty hot, and no further advance was made, as the Russian columns had come to the "port," and were making a slow, steady advance down the hill. We had been stepping back some ten or fifteen paces, firing as we did so, till a sergeant, calling me by name, said: "Sir, the right of the line has halted," so I shouted, "Halt, halt," and we were all steady again, and kept up a warm fusilade on the enemy.

The Grenadier Guards, who were in support of this part of the line, had now come up, and they passed through the Nineteenth, and the Russians seeing these reinforcements advancing in perfect formation, turned and fled, throwing away their arms and all impedimenta in their hurry to get away. Their knapsacks were found to contain a good supply of clean and neatly folded linen, which our men were not slow to annex. In this manner a lot of drums fell into our hands belonging to the Vladmir and Minsk Battalions, five of which are now in possession of the Nineteenth.

Hardly had the Grenadiers passed through my front when the word ran—"Quarter distance column on No 1." No. 1 was retired a little, and in a few minutes the column was formed and told off. Our men being smart, didn't take long to get into their places. Colonel Unett now came and took command. The 1st Brigade was kept in column, and marched up the slope, whilst we were presently joined by the 77th and 88th, under General Buller.

The battte was over and won, and a wild shout and hurrah rang over the field that must have astonished the beaten army. The 7th, 23rd, 33rd and Nineteenth did not look half what they had been in the morning. My company, fifty-four, in front, was now only thirty-six, and the others were about the same.

The division was formed in line of contiguous quarter distance columns on the heights of the Alma, facing south, as soon as the battle ceased. When I arrived on the first steppe I could see the enemy retiring in a cloud of dust in the distance. Our light cavalry was at hand, but made no attempt to accelerate their rout or throw them into confusion. Had they done so all their guns would have fallen into our hands at the river beyond the plains called the Katcha, some three or four miles off where there was only one bridge.

The Nineteenth passed the river opposite the great battery, as we may call it. How the 7th and 33rd got over I can't tell. The Nineteenth went across the vineyard at a most rapid rate, and there being no room for the 23rd, it moved along our rear, and eventually in advancing up the glacis tacked on to our left, where I had been all the time expecting to see the 88th.

It has been said that the enemy's skirmishers lay on the south bank of the Alma and fired down on us. Not a word of truth in it. They skedaddled, and some were shot escaping up the hill. Above where the Nineteenth forded the river it was easily crossed. It was a rapid, stony-bottomed trout stream, and was no obstacle to either the 88th or 77th. The Rifles and Russian cavalry crossed it easily, and so too must have the Guards and Highlanders. How our Nineteenth mounted officers got over I don't know. I never at the time thought of asking them, but the Adjutant of the Grenadiers told me he had to go to the ford at Burliuk and gallop up to meet his battalion on the other side. I presume Colonel Sanders did so too. The words of command were passed at the river bank by captain to captain in the regiment, and I did not notice Colonel Sanders until just as he was struck down. Cardew, our Adjutant, was early hit on the line advancing. They said he was dead. The 95th Regiment of the 2nd Division took part in the attack with us, but such is the extraordinary confusion of battle that I was not aware of it till next day. They suffered a similar loss to ourselves.

The chief part of the fighting took place in front and on the left of the great battery, for here next day lay the dead of both sides in numbers. The 23rd, owing to their close formation, came off worst. Eight officers fell, and are buried opposite the salient. They had a very warm time of it. They were on the left rear as I stood when the Grenadiers relieved us.

They had not received assistance as we had, but that assistance had a moral effect, for Bell, one of their captains, although he had fired off all the rounds in his revolver in self-defence, jumped over the parapet and ran at the drivers of a gun trying to escape, and presented his unloaded

---

NOTE.—Lieut.-Colonel Unett had some difficulty in crossing the river with his mare, as will be seen from the following extract from a letter he wrote immediately after the battle.

"We were ordered to advance, and did so to the best of our ability in line, across stone walls and a vineyard. Here the plunging fire from grape, round shot, shells, minie muskets, etc., was awful, and also across the stream through which we made our way with the greatest difficulty; more particularly us mounted officers. My mare got into a hole in the water, and was all under for some time except her head and neck. I dismounted and got her to the side, and attempted the high bank, which was nearly perpendicular and very slippery. I struggled to get up, sticking my fingers into the grass, and she by desperate struggles, came up by my side as I had hold of the bridle. Shot falling all this time very thick. I mounted and pushed on, and we got into something like a line under the crest of the hill."

Ensign and Adjutant A. M. Cardew, who was badly wounded both in the leg and face, wrote home a few days after the battle:—

"We had a splendid fight on Wednesday, the 20th inst. We found the enemy entrenched strongly on the summit sides of a very steep hill at the foot of which flowed the River Alma. This river winds so much that we had to cross it twice in many parts of the line. My horse was struck by a spent ball almost the first thing, but I managed to keep him up until we got to the river a second time, when he was struck in the hock and fell, giving me a glorious sousing. I pulled him up on the opposite bank, and left him tied to a tree, and racing on on foot, we then (that is the 7th, 23rd, 33rd, and ourselves) came under a most awful fire of grape, canister and minie balls, which cut us up tremendously."

revolver at them. They jumped off on the other side of their horses, and left two guns a prize to the 23rd.

The proper right flank of the Russian battery had no fighting at all. The 88th and 77th, had they been brought into action, would have had their work here.

I walked to the telegraph station where the French fight took place. It was a trumpery affair. There were not fifty Frenchmen killed, and perhaps fewer Russians. When I returned to the regiment the rear-guard had come in, and grog was drawn for 119 or 120, studiously concealing the fact that out of this number a very considerable proportion were not to the fore to claim rations. Their share was devoted to what was called "Drinking the dead men's health."

To still further explain the part played by the Light Division in the battle, we cannot do better than quote from Kinglake:—

"At length the whole Light Division, together with the additional regiment which had strayed into its company, was upon the Russian side of the river. But as yet the troops only stood upon the narrow strip of dry ground at the water's edge, and such of them as were in the centre or towards the right were penned back by the rocky bank which rose steep and high above their heads. The soldiery were a crowd, a crowd shaped and twisted by the winding of the river's bank, yet with some remains of military coherence; for although the enclosures and the fording of the river could not but destroy all formation, the men of every company had kept together as well as they were able.

"Though forming part of the next brigade, the 19th Regiment was suffered ere long to associate itself with

Codrington's advance, so with this and the other stray regiment, which clung to it, Codrington's brigade was swollen to a force of five battalions. These five battalions were extended in a broken chain at the foot of the bank on the Russian side of the river, under the close fire of the skirmishers who crowned the top.

"Codrington was at this time between the 33rd Regiment and the 23rd Fusiliers. He strove to do something towards restoring the formation of the troops, but the crowd jammed together, twisted into a fantastic shape by the bend of the river's bank, could hardly even try to perform an evolution requiring free space and time. His immediate power over the disordered masses around him was confined within the range of his voice, but lifting himself in his stirrups he spoke to the men in his clear ringing voice, and ordered them (all who could hear him) 'To fix bayonets, get up the bank, and advance to the attack.'

"The Russian column descending from the eastern flank of the work marched against that part of our line which was formed by the 19th Regiment and some of the left companies of the 23rd. In their English way, half sportive, half surly, our young soldiers seemed to measure their task; and many of them still holding betwixt their teeth the cluster of grapes which they had gathered in the vineyard below, began shooting easy shots into the big solid mass of infantry which was solemnly marching against them. Soon the column was ordered or suffered to yield. Then the 19th and left companies of the 23rd began as they advanced to head towards their right, and became a part of the force which was storming the Great Redoubt.

"Then the five regiments, all of whom had become broken up and more or less mixed, drove the enemy up the slope, and notwithstanding the heavy fire of twelve guns of heavy calibre, captured the redoubt, the enemy retreating in confusion before them.

"But now an unfortunate incident occurred. Just as the men were reforming and preparing to continue their advance, the 'cease fire' was sounded, and the enemy taking advantage of the hesitation which ensued, advanced some heavy columns and drove the Light Division from the redoubt with heavy loss.

"The first division now came up, and advancing through the Light, retook the redoubt at the bayonet's point, the Light Division forming a second line.

"The key of the position being thus again forced, the Russians again retreated, and the day was won."

The losses sustained by the regiment in the battle of the Alma were:—

Killed: One subaltern (Ensign Stockwell), 1 drummer and 36 privates.

Wounded: Two field officers (Lieutenant-Colonel Sanders, severely, Major H. E. McGee, slightly), two subalterns (Lieutenant L. D. Currie, severely, and R. Wardlaw, mortally), 1 staff (Ensign and Adjutant A. M. Cardew, severely), 6 sergeants, 13 corporals, 2 drummers, and 174 privates, many of whom afterwards died of their wounds.

The allied forces bivouacked on the ground they had won, and the Nineteenth were employed for the remainder of the day in collecting the dead and wounded.

COLONEL ROBERT SANDERS, C. B.

FROM A PAINTING AT CHARLEVILLE PARK, CO. CORK.

CHAPTER X.

The flank march to Sebastopol—The Siege—Battle of Inkerman—The great Storm—Privates Evans and Lyons—Assault on the Redan, 18th June, 1855—Final assault, 8th September—Honours and Rewards—Memorials—Roll of Officers.

THE Light Division with the rest of the army moved forward on the 23rd September. On all sides was evidence of the hasty flight of the Russians, as arms, accoutrements and cooking utensils lay strewn about in every direction. During the absence of Lieutenant-Colonel Sanders the Green Howards were under the temporary command of Brevet-Lieutenant-Colonel Thomas Unett. The delay in not immediately advancing after the Alma was due to indecision and want of cohesion between the French and English commanders, and judging from the state of panic that the Russians were in, some vigorously concerted action might have brought about the fall of Sebastopol.

Having crossed the heights of the Alma, a march of six miles brought the Allies into the smiling valley of the Katcha, where the Light Division bivouacked for the night near a small village which had been forsaken by its inhabitants, but only for a time, as by degrees they came back on discovering that no harm was being done to their homes and property.

The next day the march was resumed, and there from the hills, half-way between the Katcha and the Belbec, the men looked down on the town of Sebastopol, which they were not to enter for so many weary months.

Early on the 25th the flank march round the east side of Sebastopol was begun. The orders were to march south—south-east, and to rendezvous at McKenzie's farm. This was a very tiring day for the regiment, and in common with others they advanced but slowly. Cholera still clung to it, and Ensign Phipps, who had carried the Queen's Colour at the Alma, was attacked and died. The route took them through a forest, which was so dense that they could preserve no possible line of formation, and at times were compelled to go in Indian file.

Every officer and soldier had his own story to tell of the progress of his regiment. Now diverging too far to the west and coming into dangerous proximity to Sebastopol, now losing the track altogether through a maze of trees and undergrowth, and when the army emerged on the other side, regiments, and even brigades, were hopelessly mixed. During this march the Nineteenth came suddenly on the rear-guard of some Russian troops with their baggage, who were making for Simferopol. A few shots were fired, and the whole convoy was captured. The soldiers ran to the waggons, and some time was taken up in arranging the captured stores and catching the horses. A large quantity of gunpowder was seized and blown up, the effect of which must have been seen and heard in Sebastopol, in fact all over the inhabited Crimea as well.

The march was continued, and it was a great relief to all when they came to the banks of the Tchernaya, where the bivouac was selected. Here a picquet of the Nineteenth and another of the 88th were thrown out across the river to the south. Next morning saw the army again in motion, and soon the pretty village of Kadikoi was reached, the people of which said that Balaclava was undefended, but Lord Raglan not trusting their statements, sent the Light Division and some artillery to occupy the heights, shortly after which the town surrendered.

Captain Lidwill with his company was placed in charge of Balaclava till the 29th September, when the division marched away, and the sickly men of each regiment were made up into a battalion and left there as a garrison.

The Light Division formed the line of investment on the south side of the fortress, where it bivouacked on the left of the position taken up by the Allies, and marched on the 30th to the ground it occupied during the siege of Sebastopol.

For the first fortnight the men were employed in revetting and entrenching Gordon's 21-gun Battery, and the right Lancaster Battery was also constructed by Buller's Brigade.

On the 17th October the first bombardment of the town took place. Two hours before daybreak the left wing of the Nineteenth paraded with three days' rations, and under command of Major McGee, marched off to the Woronzoff ravine to form a guard to the recently constructed earthworks of Gordon's Battery. The embrasures were all masked with bags of grass and weeds to prevent the enemy from seeing that everything was ready for action.

On a given signal this battery opened fire, then Chapman's on the left, and further on again the French fire was heard. Gordon's Battery in its advanced position received several shots from the Flag Staff Battery, but no harm was done. Fragments of shell, sky high, came down with little or no result. The sides of the quarry were struck with splinters and the grass set on fire, but beyond this there was no great damage.

Not very long afterwards an artillery tumbril was seen coming with powder from the camp: the driver, instead of keeping out of the line of fire, neglected to do so, and a shell suddenly struck the tumbril, blowing up man, horse, ammunition and everything to atoms. Fortunately not an officer or man of the Nineteenth who were close by were injured, and their salvation seems to have been owing in some degree to their proximity, as the force of the discharge was directed chiefly upwards. On the parapet of the Barrack Battery the Russians were standing, shouting and cheering, but on fire being brought to bear on them they disappeared. The tables were soon turned, for the magazine of this very battery exploded not long afterwards with a rumbling like an earthquake. The men fixed bayonets, but no order came to storm as was expected. Had this been done the Redan might have fallen, and then the Malakoff would have been taken in rear. But it was not attempted.

On the morning of the 18th the left wing was relieved by the right, under Lieutenant-Colonel Unett. The damage done to the Redan had been repaired in the night, and the firing was continued with unabated energy.

From this time up to the 5th November the artillery duel between the Allies and their enemy never ceased by day.

Shortly after the battle of Balaclava, which was fought on the 25th October, the Russians concentrated an overwhelming force of 50,000 men to attack the scanty British force on the plateau of Inkerman.

On the 5th November the Nineteenth were detailed for trench guard, and fell in a few hours before daybreak. Colonel Unett was field officer of the picquets, and marched away two companies of each regiment.

As day dawned a smart rifle fire was heard from the direction of Inkerman, and the 77th and 88th, together with No. 6 company, which had come off trench duty, at once stood to arms. They were afterwards joined by Captain Ker's company and later on by the details in camp, who were formed into a third company, under Captain Bright. These three companies, which had moved on to the Victoria Ridge, were held in readiness to reinforce on the right bank of the middle ravine, and though they were continually under fire did not actually take part in the action which raged all day.

A heavy rain had continued to fall during the whole of the previous night, and towards morning a dense fog settled in the valley of Inkerman and completely enveloped the height, rendering it difficult to see objects distinctly for more than a few yards.

Favoured by the mist and taking advantage of the darkness, at early dawn the Russians surprised our picquets, and were fast drawing near to the British lines ere the troops received warning to collect and dispute the enemy's advance. Column after column, closely pressing on each other, ascended the plateau, and protected by a tremendous artillery fire, assailed the position at all points. Everywhere our troops found themselves outnumbered by the enemy's forces, and it required efforts and gallantry of no ordinary kind to maintain their ground. Such a struggle has rarely, if ever, been witnessed in modern warfare. For nearly seven hours 8,000 English, subsequently joined by 6,000 French, not only sustained a hand to hand fight against upwards of 45,000 men, supported by a most powerful artillery, and succeeded in holding their own, but they forced two or three times their number to retreat beyond the Tchernaya with considerable loss.

The casualties in the regiment were Captain Ker and one private killed, Sergeant-Major Madden and two privates wounded, all of whom afterwards died of their wounds.

From the 5th November the Nineteenth were on trench guard, advanced trench guard, and the Malakoff picquets up to Christmas. Three companies were on picquet opposite the Malakoff, one from each regiment of the brigade, and were relieved every morning. There were frequent sorties. Early in 1855 a French corps took over the Malakoff picquets, and the Light Division had then nothing but guards and working parties to furnish.

After the battle of Inkerman the regiment became so weak, not only numerically but physically, that there was considerable difficulty in finding men in sufficient numbers to do duty in the trenches.

There was no firewood, and the men had to search for the roots of bushes to make some sort of fire with which to cook their salt pork or roast their green coffee beans, for which they used pieces of tin taken from the lining of their ammunition pouches.

The weather at this time became very unsettled, rain fell and sudden storms arose. On the 14th November one of the most terrible hurricanes ever experienced swept over the Chersonese. About noon the storm was at its height, and sleet and snow fell blindingly. Roofs were stripped off the houses, all the tents in the camp were blown down and torn, and in the open no one could keep their feet.

The men-of-war in the harbour of Balaclava rode out the storm, but nearly all the other vessels grounded near the Belbec and went to pieces. Most of these had as cargo ammunition, warm clothing for the troops, and other necessaries for the winter.

Sir Evelyn Wood, in his reminiscences, says:—

"It is impossible to describe the scene of misery, but some idea of it may be realized if my readers will imagine they are on the bleakest of the Surrey hills, 800 feet above the sea, without even a tree for shelter, and exposed to the wildest storm of wind, rain and sleet they have ever experienced. There were indeed two or three hovels near our camp, but they sheltered only a few, and were crowded with wounded soldiers, and in many cases these dwellings lost their roofs. Horses broke loose from their picket ropes, and, wild with terror, careered over the upland. Waggons were upturned, and to some it seemed that the end of all things had come."

On 21st November a draft arrived from England of 1 sergeant, 3 corporals and 98 privates, and during January the battalion was still further increased by 5 officers (Captains O. A. Grimston and H. F. Massy, Lieutenants Kirke, Evans and Warburton), 1 corporal and 65 privates.

The nights in November and December were very long. Those who were not on trench duty or on guard remained in their tents; there was no kind of light provided, and the men improvised lamps for themselves, which were made out of their blacking boxes with the fat of the salt pork as oil and a piece of shirt as the wick.

They took a turn of twelve hours in the trenches at a spell, and suffered considerably from the effects of cold, wet and exposure. The prevalent diseases were fever and dysentery, and owing to the scarcity of vegetables, scurvy too broke out amongst the troops. In the Light Division, on which a large share of the work fell, there were 350 men on the sick list. By the end of November their clothes were threadbare, and not fit to resist rain or cold.

During November the 97th Regiment arrived, and was attached to the division, as was also the 90th, which landed in the first week of December.

From the 1st to the 12th January the weather was exceptionally severe, and the working parties were engaged almost exclusively in clearing out the snow and mud from the batteries and trenches. The frost was so hard at times that no impression could be made in the ground except with difficulty, and at others pelting rain inundated the trenches, making the earth heavy, so that it took double the work to clear it out. But though it was the hardest worked, the Light Division was reported to be the healthiest in the army.

In February Lieutenant-Colonel Unett proceeded to England to join the depôt, and the command of the regiment was assumed by Major George Rooke.

The 9th April saw the second bombardment of Sebastopol, which was continued until the close of the 18th, and though the Allies worked their guns all this time both by day and night with destructive effect, the enemy never failed before morning to repair their broken defences and restore their artillery power.

On the 13th April part of the regiment was on duty in the trenches, and the conduct of three men, viz., Privates James Calligan, Thomas Keating and Samuel Evans, was so marked for intrepedity and gallant courage, that Major-General Jones, commanding the Royal Engineers, made a special report of it.

This was the subject of a very flattering divisional order relative to the regiment, part of which is here quoted:—

"These incidents have appeared to the Lieutenant-General so creditable to the character of the regiment and to the individuals concerned, that he deems it but justice to them and their gallant corps not only to notice their exemplary conduct in divisional orders, but purposes also to recommend to the Field-Marshal Commanding the Forces that they may be granted a pecuniary gratuity as an encouragement to others."

These men volunteered to repair an embrasure under a very heavy fire, and afterwards received a gratuity of £5 each, Evans being subsequently granted the Victoria Cross.[1]

Not alone were there occasional sorties and fighting, but during the siege the men in the trenches were constantly under the fire of the batteries, and were killed and wounded at all hours.

At daybreak on the 20th April four men of the Nineteenth and four sappers volunteered to go out and destroy an advanced rifle pit, which they did successfully. The night before this the Russian rifle pits were attacked and carried by the 77th, who, however, suffered heavy loss, and Lieutenant-Colonel Egerton, who led the attack, was killed.

On the 5th June a draft, under Lieutenant R. Molesworth, consisting of 1 sergeant, 1 corporal and 49 privates, joined from Malta.

Next day the third bombardment commenced, and on the 7th the Mamelon was captured by the French and the Quarries by the British. The Green Howards had a working party of fifty men employed on this occasion, and Lieutenant E. W. Evans was severely wounded.

On the night of the 10th June the conduct of a working party of the regiment on duty in the trenches again called forth the approbation of the officer under whom they served, and the bravery of one man in particular, Private John Lyons, was so conspicuous that Lord Raglan was pleased to express himself in high terms of his gallantry as well as that of his regiment. He awarded a gratuity of £5 to this soldier for throwing a live shell out of the trenches, and thereby

1 Pte. Evans died on the 4th Oct., 1901, and in his will left his Victoria Cross and other medals and decorations to the Officers' Mess 1st Battalion.

most probably saving many of his comrades' lives. This latter must have been a greater satisfaction to him than either the gratuity or his subsequent promotion to corporal. He was afterwards awarded the Victoria Cross and Legion of Honour.[1]

On the 18th June (Waterloo Day) the first assault of the Great Redan took place, the Light Division leading, the French at the same time assaulting the Malakoff Tower.

As soon as the troops had shown themselves beyond the trenches they were assailed by a most murderous fire of grape and musketry. After endeavouring to advance without success they had to retire, but not before almost all the senior officers had been killed or wounded.

The British casualties were 100 officers and 1,444 other ranks, but the Nineteenth, under command of Major Rooke, formed part of the reserve and suffered no loss.

The French took the Mamelon and two redoubts on the right. They were driven out of the former, but ultimately retook it and held it. They advanced up to the Malakoff, but could not enter.

On the 28th June, ten days after this lamentable failure, Lord Raglan died. He was very popular with the army, and his death caused much despondency amongst the different regiments. He was succeeded in the command by Lieutenant-General James Simpson.

On the 12th July a draft, under Lieutenant W. G. D. Massy, consisting of 1 sergeant, 2 corporals and 28 privates, joined from Malta, followed nearly a month later by another of 2 subalterns (Lieutenants H. Browne and G. Nutting), 1 sergeant, 3 corporals and 57 privates from England and Malta.

1 Both these decorations are now in the Officers' Mess 1st Battalion.

Bt.-Lieutenant-Colonel T. Unett rejoined from England on the 28th July, and took over command from Major Rooke.

After the failure of the first assault on the Redan on the 18th June, hopes ran high in the British lines that the second attempt, which was fixed for the 8th September, would prove successful. General Simpson had arranged with the French that they, with 30,000 men, should attack the Malakoff, and when they got possession they were to hoist the tricolour on the fort, which was to be the signal for their allies to storm the Great Redan. The British attack was under the direction of General Codrington, and his division (the Light) and the 2nd, under General Windham, were chosen for this honour. The covering party, ladder party, working party, and the main body numbered 1,700; the supports were 1,500 strong, and the remainder of the two divisions, numbering 3,000 men, were in reserve in the third parallel. Lieutenant-Colonel Thomas Unett, C.B. (19th Regiment), in the absence of the brigadier, was commanding the leading troops of the Light Division, and the question arose between him and Colonel Windham as to which party should show the way. It was decided by tossing with a shilling, and Colonel Unett won. The 2nd Brigade Light Division was detailed for the storming, the Nineteenth Regiment, 420 strong, under command of Major Robert Warden, forming part of the supports.

The French were successful after a hard fight, and the tricolour was seen fluttering on the Malakoff. This was at about 12 noon, and a few minutes after that hour the Light Division stormers rushed out from the fifth parallel into the open and came immediately under a heavy fire from the guns of the Barrack Battery. Colonel Unett fell badly wounded before he had reached the abattis, and the men, rushing rapidly over the 200 yards of fire-swept space, suffered heavily from the round, grape and case shot, as well as the musketry directed on them from every available point. Those who got across first, passed with ease over the battered rampart, and entered the work, but the majority clung to the edge of the ditch, and began firing alongside the covering troops. The Nineteenth, with the rest of the supports, now pressed forward on the heels of the stormers, and under the same withering fire reached the ditch at the projecting angle of the Redan, which at this point was about fifteen feet deep. The Green Howards, led by their officers, leaped in, and, scrambling up the other side, some of them entered the work with little opposition. The few Russians who were in front retired to the rear of the work, and our men took cover behind the traverses and parados which the enemy had erected to protect them from the heavy fire of our batteries. Unfortunately the men seemed to think that as long as they held the ground they had captured all would be well, but this method of action proved really the death knell to success. The more they crowded in the more was the disorder, and the more they suffered from the enemy's fire. The reserves had come up, but in moving through the trenches towards the point of attack, were obstructed and discouraged by meeting the numbers of wounded men and their bearers, who of necessity returned by the same narrow route. Colonel Windham, who had brought up the 2nd Division, saw that reinforcements were an absolute necessity, owing to the tremendous losses, and sent Ensign Young (19th Regt.) back to the nearest trench to order more men up, but Young was badly wounded on the way, and, not returning as soon as was expected, Colonel Windham left the Redan, and passed over the same ground, in order to get more support.

Every moment our men were diminishing in numbers, and the Russians reinforced by parties driven from the Malakoff, were now in such strength, that unless some wonderful chance attended the British efforts, the capture of the Redan was well-nigh hopeless.

The enemy now had collected several thousand men in rear of the breastwork, and seeing our men scattered and confused behind the traverse, crossed the breastwork and charged the broken groups with the bayonet. The combat that ensued was short, desperate and bloody, and the struggling band was forced back by the Russians, who moved on like a heavy wedge, driving our men pell mell into the ditch, which was already choked with the dead and dying of the Light and 2nd Divisions. Into this mass of mixed soldiery the enemy poured a merciless fire, and even rolled stones on to them, and a scene of chaos presented itself seldom seen in modern warfare. Most of their officers having now been either killed or wounded, and the men themselves falling fast, the sorry remnant of the two divisions was at last forced to retire, having participated in one of the most unfortunate attacks in which British soldiers had ever been launched.[1]

In the Light Division there were 73 officers and 904 men killed and wounded. Of this the share of the Nineteenth

1 Bt.-Lieut.-Colonel Rooke, writing home on this failure, said:—"The failure of cur attack on the Redan must be attributed mainly to the old thing, "mismanagement." The distance from the 5th parallel to the salient angle is 284 yards. Over this distance the men had to advance under a crushing fire of canister, grape and musketry from heavy flanking batteries, which would have thrown the best soldiers of our best days into confusion, and, of course, caused great disorder among the attacking parties of young recruits, for the majority of men in our regiments out here are but little better. Our regiment, I am told, advanced in perfect order, but on arriving at the ditch got mixed up with and broken by the disorderly mass of the attacking columns. Young, Massy, and Goren are still in a precarious state, all the other wounded officers are doing very well."

was 192, or 45 per cent. of its strength at the commencement of the action. At the Alma the division had 1,001 killed and wounded. Sir W. Russell, in his History of the War, says:—

"From the above statement it will be seen that this gallant body (the Light Division), which behaved so well at the Alma, and maintained its reputation at Inkerman, suffered as severely as it did in gaining the former great victory, and an examination of the return will, I fear, show that the winter, the trenches, and a careless recruiting have done their work, and that the officers furnished a noble example of devotion and gallantry."

In the Green Howards, out of 18 officers and 420 men engaged, there was 1 officer, 3 sergeants and 25 rank and file killed, and 9 officers, 9 sergeants, 1 drummer, 16 corporals and 124 privates wounded.[1] Many of the latter afterwards died of their wounds. Missing, 4 privates, who were made prisoners, but were afterwards exchanged.

OFFICER KILLED.

Lieutenant Peter Godfrey.

OFFICERS WOUNDED.

Lieut.-Colonel Thomas Unett, C.B.[2] (severely, died subsequently).
Captain and Brevet-Major Robert Warden (slightly).
Captain E. Chippindall (slightly).

---

1 For names see casualty list in the Appendix.

2 Colonel Thomas Unett was appointed a Cornet in the 2nd Dragoons in 1820, and transferred to the 7th Dragoon Guards the year following. In 1830 he joined the 67th Foot, and exchanged to the 19th in 1845 as Captain and Bt.-Major. At the Redan he was wounded in the hip and was again hit as he was being carried to the rear. He died on the 15th September, to the great grief of his brother officers, and particularly of the privates of the regiment, whose comfort and welfare he had invariably studied, and to whom he was a father and a friend.

Memorials to his memory are in the Crimea and in St. Philip's Churchyard, Birmingham.—M.L.F.

LIEUT. COLONEL THOMAS UNETT, C. B.

Lieutenant W. G. Dunham Massy[1] (dangerously).
,, Richard Molesworth (severely).
,, Edward Bayley (slightly).
Ensign Robert Martin (severely).
,, Walter Young (severely).

The capture of the Malakoff by the French settled the fate of the war. Sebastopol was no longer tenable, further resistance was in vain, and the Russians silently and skilfully evacuated the town without the knowledge of the Allies. On the 9th of September they retreated across the harbour by a bridge of boats to the forts on the northern side, after blowing up those on the southern side and sinking their fleet. The docks were soon afterwards destroyed by the Allies, and the great Russian fortress reduced to absolute ruin.

On the 9th September the following order was issued:—

"Head Quarters, Sebastopol,
"9th September, 1855.

"The Commander of the Forces congratulates the army on the result of the attack of yesterday. The Commander of the Forces returns his thanks to the General officers and officers and men of the Second and Light Divisions who advanced and attacked with such gallantry the works of the Redan. He regrets from the formidable nature of the flanking defences that their devotion did not meet with that immediate success which it so well deserved. He condoles

1 The late Lieutenant-General W. G. Dunham Massy, C.B., Colonel 5th Lancers. Lieutenant Massy had his thigh broken and was left out all night. In the absence of his cousin he led the Grenadier Company, and was one of the first men of the regiment to get into the Redan. When the British retreated he was left where he fell, and when the works of the Redan were blown up in the night by the retreating Russians, he had his right leg fearfully crushed by a falling stone. He was found in the morning by some Highlanders and brought to his regiment almost dead from loss of blood. His conduct on this occasion won for him the name of "Redan Massy," and when he arrived in Ireland the students of Dublin University presented him with a Sword of Honour.

and deeply sympathises with the many brave officers, non-commissioned officers and men who are now suffering from the wounds they have received in course of their noble exertions of yesterday. He deeply deplores the death of so many gallant officers and men who have fallen in the final struggle of this long and memorable siege. Their loss will be severely felt, and their names long remembered in this army and by the British nation.

"By order,

"(Sd.) H. W. BARNARD,

"C. of S."

The following are the reasons generally given for the failure on the Redan. The number of troops which actually stormed the works was ridiculously inadequate. The point of attack was not sufficiently extended, and the converging thereon produced the greatest degree of confusion. The supposed mining of the Redan has also been very strongly put forward as the true reason.

On the anniversary of the Alma the regiment was inspected by Lieutenant-General Sir William Codrington, now commanding the division, and was presented with medals for the Alma, Inkerman and Sebastopol by Lord William Paulet.

After the siege the men kept very healthy, and drills, field days and inspections were common. In addition, they were very busy making roads in the Crimea. They began to settle down comfortably. Lieutenant-Colonel Unett had seen that excellent kitchens had been erected for them, and had even invented a plan for making bricks, which were manufactured out of the earth in the neighbourhood.

A terrible accident occurred on the 13th November, when the French ammunition siege train blew up, hurling shells, shot and ball for nearly a mile in every direction, and even blowing down some of the huts. There were no less than seventy-nine casualties in the Light Division, the Nineteenth having nine of these. The regiment happened to be out road making at the time or there would have been many more.

On the 19th December the regiment was inspected by Colonel D. Lysons, now commanding the 2nd Brigade, on which occasion he was pleased to express himself in very satisfactory terms on the discipline and interior economy of the corps.

New dress regulations were issued this year, when the coatee was abolished and the coat tails of the army disappeared. In its place a double-breasted tunic was introduced. The shako of black beaver was made smaller at the top than before and much lighter, with a gilt star of eight points in front, surmounted by a crown, and the regimental number in bright gold on a black centre. The sword was carried suspended to a white enamelled waist belt, fastened by a gilt union locket with number and name of the regiment. A double-breasted blue frock coat for officers was adopted for undress. Buttons hitherto of pewter for non-commissioned officers and privates, were now changed to brass and made convex.

On the 28th February, 1856, news arrived that an armistice had been arranged with the enemy until the 31st March. At the end of the month this was prolonged till further orders. On the 2nd April notification was published of peace having been signed at Paris on the 30th March.

A grand review of the British troops took place on the 17th April, when about 30,000 men were under arms. The most distinguished of the Russian, French and Sardinian generals honoured the parade with their presence.

On the 24th May the whole of the English army paraded in review order on the plain of Balaclava to witness the presentation of the French war medal for "Valeur et Discipline." Nine non-commissioned officers and men, whose names appear later, received this decoration for their distinguished service during the campaign. Shortly after this, on the 11th and 15th June, the regiment, in two detachments, embarked at Balaclava and Kazatch on board the steam transport "Imperatrice" and H.M.S. "Furious," and landed at Portsmouth on the 28th June and 24th July. On arrival they proceeded by rail to Farnborough, and thence by march to South Camp, Aldershot.

AUTHORIZATION OF CRIMEAN BATTLE HONOURS.

"London Gazette," 16th October, 1855.

"The Queen has been graciously pleased to command that in commemoration of the gallant conduct of the troops concerned, the words 'Alma,' 'Balaclava,' and 'Inkerman' be borne on the regimental Colours of the regiments specified in the accompanying list.

"Also that the several corps composing Her Majesty's army in the Crimea on 5th September, 1855, shall bear the inscription 'Sevastopol' on the regimental Standard or Colours as a memorial of the arduous and successful operations which have led to the reduction of that fortress."

19th Regiment.

"ALMA," "INKERMAN," "SEVASTOPOL."

HONOURS AND REWARDS.

THE VICTORIA CROSS.

Private Samuel Evans, 13th April, 1855.
,, John Lyons, 10th June, 1854.

COMPANION OF THE BATH.

Lieutenant-Colonel J. L. R. Rooke.
,, R. Sanders.
,, T. Unett.

KNIGHTS OF THE LEGION OF HONOUR.

Lieutenant-Colonel J. L. R. Rooke.
Brevet-Lieutenant-Colonel R. Warden.
Major R. O. Bright.
Brevet-Major E. Chippindall.
Lieutenant-Colonel H. E. McGee.
Captain G. W. H. Massy.
Lieutenant W. G. D. Massy.
No. 1651 Pte. John Lyons.

IMPERIAL ORDER OF THE MEDJIDIE, 5TH CLASS.

Brevet-Lieutenant-Colonel R. O. Bright.
Brevet-Major E. Chippindall.
Captain R. D. Barrett.
,, George Clay.
,, H. T. Uniacke.
Lieutenant Ames Goren.
Captain E. R. Ward Bayley.
Brevet-Lieutenant-Colonel R. Warden.

THE SARDINIAN WAR MEDAL.

Captain H. T. Uniacke.
,, G. A. Warburton.
Brevet-Lieutenant Colonel R. Warden.
Captain E. R. Ward Bayley.
Lieutenant and Adjutant T. Thompson.
2242 Sergt. John Sherlock.
1106 Pte. John Halloran.

THE FRENCH WAR MEDAL.

Col.-Sergt. W. J. Britts.
Sergt. William Murphy.
,, Thomas Murphy.
,, William Smith.
Sergt. Henry Strick.
,, George Robins.
Pte. H. B. Higgins.
,, James Duffy.
Pte. Samuel Evans.

MEDAL FOR DISTINGUISHED CONDUCT IN THE FIELD.

	Sergt.-Major P. Campion	Annuity of	£20
	,, J. Magner	,,	,,
1381	Col.-Sergt. D. Bell	Gratuity of	£15
1635	Corpl. J. Austin	,,	10
1643	,, H. Mitchell	,,	,,
2494	,, H. McClister ...	,,	,,
1401	,, T. Donaghoe	,,	,,
790	Pte. F. Collins	,,	£5
1963	,, P. Brown	,,	,,
869	,, J. Watson	,,	,,
912	,, M. Reneham	,,	,,
2703	,, W. Siddle	,,	,,

1007	Pte. J. Glennon	... ...	Gratuity of £5
1073	,, J. Willcox	... ...	,, ,,
1669	,, T. Mullery	... ...	,, ,,
1079	,, P. Smiley	... ...	,, ,,
1031	,, N. Barr	... ...	,, ,,

SPECIAL PROMOTION.

Sergt. Fredk. Arthur, promoted Ensign 19th Foot, 15 Nov., 1854.
,, A. W. Kirby, promoted Ensign, 5th Nov. 1854.
Col.-Sergt. Wm. Rawding, promoted Quartermaster, March, 1856.

RETURN OF CASUALTIES.

Return of casualties from its arrival in the East to the 30th April, 1856:—

	Officers.	Sergeants.	Drummers.	Rank and File.	Total.
Killed in action	1	3	1	69	74
Wounded severely	13	11	2	250	276
,, slightly	7	14	1	224	239
Died of wounds	3	3	1	61	68
,, disease	1	8	—	308	317
Invalided, not included in above	—	6	—	57	63
Deserted	—	—	—	4	4

CRIMEAN MEMORIALS.

Of all the burial grounds in the Crimea that which is considered the most picturesque, if this term can be used towards a graveyard, is that of the 2nd Brigade, Light Division (19th, 77th, 88th, 90th and 97th Regiments), situated on Cathcart's Hill, close to the Woronzoff Road. A certain foreign character was imparted to it by being laid out in walks and alleys, somewhat after the manner of "Père la Chaise." The stone is that of the country, which is generally soft in quality, although durable and remarkable for its dazzling whiteness. Most of that made use of was procured from the quarry in front of the 3rd Division camp, the wall behind the Redan, the Inkerman quarry, or the docks after they were blown up by the Allies, from whence was also procured some granite.

The following is the inscription on the 19th memorial:—

(*Left.*)

Sacred to the Memory

OF THE NON-COMMISSIONED OFFICERS AND PRIVATES BELONGING TO THE 19TH REGT. WHO LOST THEIR LIVES IN THE SERVICE OF THEIR COUNTRY DURING THE CRIMEAN CAMPAIGN.

THIS STONE IS ERECTED BY THE OFFICERS OF THE REGIMENT IN ADMIRATION OF THEIR FORTITUDE AND BRAVERY.

(*Right.*)

19TH REGIMENT

LIEUT. WARDLAW, KILLED IN ACTION, 20TH SEPR., 1854.

ENSN. STOCKWELL, KILLED IN ACTION, 20TH SEPR. 1854.

ENS. PHIPPS, DIED OF CHOLERA, 25TH SEPR., 1854.

CAPN. KER, KILLED IN ACTION, 5TH NOVR., 1854.

CAPTAIN P. GODFREY, KILLED AT THE FINAL ASSAULT OF SEBASTOPOL.

LIEUT. COLL. T. UNETT, C.B., FELL 8TH SEPR., 1855.

There is also a separate monument to Lieutenant-Colonel Thomas Unett, C.B., inscribed as follows:—

IN MEMORY OF
LIEUT. COL. UNETT, C.B.
HE FELL WHILE LEADING
THE 19TH REGIMENT
TO THE ASSAULT,
8TH SEPTR., 1855.

Some regiments, the Nineteenth, the 18th Royal Irish and the 90th Light Infantry, erected their own monuments. In other instances these, together with all public ones, were constructed by men of the Royal Engineers' Camp of the Light Division. The centre obelisk was put up in memory of the 2nd Brigade Light Division. It faces the road looking towards Sebastopol:—

TO THE
OFFICERS,
NON-COMMISSIONED OFFICERS
AND MEN
OF THE
SECOND BRIGADE
LIGHT DIVISION,
WHO FELL
1854—1855.

R. BARRET,
90TH L. I.

(Right.)

Similar inscription to above in Russian.

(Rear.)

REGIMENTS OF THE BRIGADE.
XIX.
LXXVII.
LXXXVIII.
XC.
XCVII.

(Left.)

Similar to the above in Russian.

There is also a memorial brass in York Minster erected by the officers of the regiment as a tribute of respect and esteem for their departed comrades.

ROLL OF OFFICERS OF THE 19TH REGIMENT

Who were present in the Crimea, giving details of their services, the rewards and promotions received, and the date of landing.

P—signifies present in the engagement. K—killed, or died of wounds. W—wounded.

	Rank and Names.	Date of Landing.	Present at Alma.	Present at Inkerman.	Present at Sebastopol.	Present at Redan, 8th Sept.	Decorations, Promotions, etc., as Reward.	Remarks.
1	Captain Richard Doyle Barrett ...	14 September, 1854	P	P	W	—	Medal and 3 clasps. 5th Class of Medjidie. Turkish medal.	Slightly wounded 21 Oct., 1854. Invalided 8th Feb., 1855. Did not return to the Crimea. To 100th Foot 29 Sept., 1880. Died at Blackheath 21 April, 1905.
2	Asst.-Surgeon Samuel Joseph Bayfield...	1 May, 1855 ...	—	—	P	P	Medal and clasp. Turkish medal	With Regiment to 23 July, 1856.
3	Captain Edward Robert Ward Bayley...	14 September, 1854	P	—	W	W	Medal and 2 clasps. Sardinian and Turkish medals. 5th Class of Medjidie.	Slightly wounded in the trenches 10 June, 1855. On leave 8 Dec., 1855 to 4 March, 1856. Now a Military Knight of Windsor.
4	Major Robert Onesiphorus Bright ...	14 September, 1854	P	P	P	P	Medal and 3 clasps. Legion of Honour. 5th Class of Medjidie. Turkish medal. C.B. Mentioned in despatches. Bt.-Lieut.-Colonelcy.	See text.
5	Lieutenant Henry James Browne ...	2 August, 1855	—	—	P	P	Medal and clasp. Turkish medal	Retired 25 August, 1861. Died in 1877.
6	Lieut. and Adjnt. Ambrose M. Cardew	14 September, 1854	W	—	—	—	Medal and clasp. Turkish medal	To hospital at Scutari 28 October, 1854. Returned to Crimea 9 March, 1856. Half-pay 23 October, 1857. Died at Falmouth 4 March, 1895.
7	Captain Edward Chippindall	14 September, 1854	P	P	P	W	Medal and 3 clasps. Legion of Honour. 5th Class of Medjidie. Turkish medal. Mentioned in despatches. Brevet of Major.	See text.
8	Captain George Clay	14 September, 1854	P	P	P	—	Medal and 3 clasps. 5th Class of Medjidie. Turkish medal.	Sick leave 8 Feb., 1855 to 17 Nov., 1855. Major, half-pay. 1 May, 1866. Died in London 30 June, 1878.
9	Paymaster Andrew Clendining	14 September, 1854	P	—	P	—	Medal and 2 clasps. Turkish medal ...	On leave medical certificate 8 March, 1855. Did not return to the Crimea.
10	Captain Leonard Douglas H. Currie ...	14 September, 1854	W	—	—	—	Medal and clasp. Turkish medal	Invalided on account of wounds. Did not return to the Crimea. See 2nd Battalion Roll, 1859.
11	Lieutenant James Robert Dalton ...	9 March, 1856	—	—	—	—	... ...	Joined after the siege. Died at Lymington 4 February, 1866.
12	Lieutenant Francis Davis	2 September, 1855	—	—	P	P	Medal and clasp. Turkish medal	Died at Parsonstown 20 October, 1866.

13	Lieutenant Edward William Evans ...	January, 1855	—	—	W	P	Medal and clasp. Turkish medal ...	... Severely wounded in the attack on the Quarries 7 June, 1855. See Hazara Roll.
14	Lieutenant George Forbes	1 August, 1855 ...	—	—	P	P	Medal and clasp. Turkish medal ...	... 78th Foot 15 Jan., 1861. Died 29 Dec., 1885.
15	Lieutenant Peter Godfrey	14 September, 1854	P	P	P	K	Medal and 3 clasps. Turkish medal	... Mortally wounded 8 September, 1855. Died 13 September, 1855.
16	Lieutenant Ames Goren	18 October, 1854	—	P	W	W	Medal and 2 clasps. 5th Class of Medjidie. Turkish medal.	Slightly wounded in the trenches 28 July, 1855. Retired 22 January, 1869. Died 30 November, 1910, at Brighton.
17	Lieutenant Edward St. John Griffiths...	3 October, 1855	—	—	—	—	... ...	... Joined after the siege. See Hazara Roll.
18	Asst.-Surgeon William Richard Grylls...	14 September, 1854	P	P	P	—	Medal and 3 clasps. Turkish medal	... To Inniskilling Dragoons May, 1855. Resigned 1856.
19	Captain Oswald James A. Grimston ...	5 January, 1855	—	—	P	—	Medal and clasp. Turkish medal ...	... Left Crimea 21 July, 1855. Did not return. Retired 26 October, 1855.
20	Ensign Edward Hales	28 December, 1855	—	—	—	—	... ...	... Joined after the siege. Died at sea 3 November, 1858.
20A	Ensign George Douglas Harris ...	31 March, 1856 ...	—	—	—	—	... ...	... Joined after the siege. Retired 20 Sept., 1864. Died at Ryde 14 June, 1878.
21	Asst.-Surgeon Exham Long Hiffernan...	14 September, 1854	P	P	P	—	Medal and 3 clasps. Turkish medal	... At siege up to 3 July, 1855. Died at Netley 24 April, 1881.
22	Captain George Bingham Jennings ...	7 October, 1854	—	—	P	—	Medal and clasp. Turkish medal ...	... Left Crimea on sick leave 29 Dec., 1854. Rejoined 17 Oct., 1855. Died at Fort William, Calcutta, 6 March, 1870.
23	Captain James Ker	14 September, 1854	P	K	—	—	... ...	... Mortally wounded at Inkerman and died 7 November, 1854.
24	Lieut. Edward Nassau M. Kindersley...	6 November, 1855	—	—	—	—	... ...	... Joined after the siege. See 2nd Battalion Roll, 1859.
25	Lieutenant John Henry Kirke	28 December, 1854	—	—	P	—	Medal and clasp. Turkish medal ...	... Invalided 26th July, 1885. See 2nd Battalion Roll, 1859.
26	Ensign James Knox	28 December, 1855	—	—	—	—	... ...	... Joined after the siege. See 2nd Battalion Roll, 1859.
27	Lieutenant Edward Levett	14 September, 1854	P	P	P	—	Medal and 3 clasps. Turkish medal	... To 10th Light Dragoons 7 September, 1855. Died 28 December, 1899.
28	Captain George Lidwill	14 September, 1854	P	P	P	—	Medal and 3 clasps. Turkish medal	... Invalided 30 May, 1855. Did not return to the Crimea. Retired 24 July, 1857. Died at Dromard, Tipperary, 14 October, 1908.
29	Surgeon Thomas Longmore	14 September, 1854	P	P	P	P	Medal and 3 clasps. Legion of Honour. Turkish medal.	Afterwards Sir T. Longmore, K.C.B. Died at Swanage 12 September, 1895.
30	Lieutenant Robert Connolly Martin ...	31 August, 1855 ...	—	—	P	W	Medal and clasp. Turkish medal ...	... To half-pay 22 April, 1862. Died at Cloone Grange, Mohill, co. Leitrim, 13 Dec., 1903.
31	Captain Hugh Francis Massy	22 January, 1855	—	—	P	—	Medal and clasp. Turkish medal ...	... Invalided 13 August, 1855. Did not return to Crimea. Retired on half-pay 2 June, 1865. Died at Bray, co. Wicklow, 3 February, 1900.

	Rank and Names.	Date of Landing.	Present at Alma.	Inkerman.	Sebastopol.	Redan, 8th Sept.	Decorations, Promotions, etc., as Reward.	Remarks.
32	Captain Godfrey William Hugh Massy	14 September, 1854	P	P	P	—	Medal and 3 clasps. Legion of Honour. Turkish medal.	Brother of above. Invalided 22 June, 1855. Major unattached 20 June, 1856. Died in London 4 June, 1862.
33	Lieutenant William Godfrey D. Massy	12 July, 1855 ...	—	—	P	W	Medal and clasp. Legion of Honour. Turkish medal. Mentioned in despatches.	Invalided on account of wounds 22 Feb., 1856. Did not return to the Crimea. Captain military train 1 February, 1857. Afterwards Lieut.-General and Colonel 5th Lancers 4 October, 1896. Died at Grantstown, Tipperary, 20 Sept., 1906.
34	Major Henry Edward McGee	14 September, 1854	W	P	P	—	Medal and 3 clasps. Legion of Honour. Turkish medal. Bt. of Lieut.-Colonel.	To half-pay 31 August, 1855. Died at St. Peter's Port, Guernsey, 28 April, 1866.
35	Lieutenant Henry Mitford	14 September, 1854	P	P	P	—	Medal and 3 clasps. Turkish medal ...	Invalided 28 Dec., 1854. Did not return to the Crimea. Transferred as Captain to 98th Foot 20 Sept., 1856. Died at Godesberg-on-the-Rhine 17 Feb., 1910.
36	Ensign William Henry Moffatt	8 September, 1855	—	—	P	—	Medal and clasp. Turkish medal	See Hazara Roll.
37	Lieutenant Reginald Molesworth ...	5 June, 1855 ...	—	—	P	W	Medal and clasp. Turkish medal	To 1st Dragoons 16 July, 1861. Died a Military Knight of Windsor at Windsor Castle 2 April, 1900.
38	Ensign Alexander Brooke Morgan ...	28 December, 1855	—	—	—	—	... ...	Joined after the siege. See Hazara Roll.
39	Captain Richard Fielding Morrison ...	14 September, 1854	P	—	P	—	Medal and 2 clasps. Turkish medal ...	Invalided to Scutari 28 October, 1854. To 51st Foot 17 August, 1855. Died at Larkfield, Ballybrack, co. Dublin, 19 July, 1902.
40	Lieutenant George Stewart Nutting ...	10 August, 1855 ...	—	—	P	P	Medal and clasp. Turkish medal	Invalided 23 November, 1855. Retired 18 December, 1856.
41	Quartermaster Thomas Palmer ...	14 September, 1854	P	P	P	P	Medal and 3 clasps. Turkish medal ...	Died at Nowshera 29 June, 1867.
42	Ensign Digby Cotes Pedder	7 September, 1855	—	—	P	P	Medal and clasp. Turkish medal	Retired 5 March, 1857.
43	Ensign Wm. Fredk. Hothersall Phipps	14 September, 1854	P	—	—	—	... ...	Carried the Queen's Colour at the Alma. Died of cholera 25 September, 1854.
44	Quartermaster William Rawding ...	14 September, 1854	W	—	P	—	Medal and 2 clasps. Turkish medal ...	Colour-Sergeant at the Alma. Promoted Quartermaster 7 March, 1856. Died at Barrackpore 5 June, 1858.
45	Ensign Walter Lacy Rogers	28 December, 1855	—	—	—	—	... ...	Joined after the siege. Resigned 12 Sept., 1856. Died 18 April, 1885.
46	Major John Lewis Richard Rooke ...	14 September, 1854	P	P	P	—	Medal and 3 clasps. Turkish medal. Knight of the Legion of Honour. C.B.	Died at Fort William, Calcutta, when commanding 19th Foot 27 November, 1857.

47	Lieutenant-Colonel Robert Sanders ...	14 September, 1854	W	—	—	—	Mentioned in despatches. Medal and clasp. Turkish medal. C.B.	Invalided for wounds October, 1854. Did not return to the Crimea. Retired in 1856. Died 1 November, 1860.
48	Captain Lawrence Shadwell	14 September, 1854	P	P	P	P	Medal and 4 clasps. Legion of Honour. Sardinian and Turkish medal. 5th Class of Medjidie. Bts. of Major and Lieut.-Colonel.	A.D.C. to Sir Colin Campbell. Subsequently Assistant-Quartermaster-General. Present at Balaclava. Lieut.-Colonel unattached 31 August, 1855. Afterwards a Lieut.-General. Died at Reading 16 Aug., 1887.
49	Captain Wm. Lempriere Fredk. Sheaffe	27 October, 1855	—	—	—	—	... ...	Joined after the siege from 51st Foot. Retired 6 July, 1858.
50	Ensign Geo. Dickson Thomas Stockwell	14 September, 1854	K	—	—	—	... ...	Carried the Regimental Colour at the Alma.
51	Lieut. and Adjnt. Thomas Thompson...	14 September, 1854	P	P	P	—	Medal and 3 clasps. Sardinian and Turkish medals.	Promoted from Quartermaster-Sergeant 11 August, 1854. On leave from 14 Feb., 1855, to 9 March, 1856. Died at Cape Town 18 March, 1860.
52	Ensign Henry Thompson	13 March, 1856 ...	—	—	—	—	... ...	Joined after the siege. To Indian Army 28 March, 1858.
53	Bt.-Lieutenant-Colonel Thomas Unett...	14 September, 1854	P	P	P	K	Twice mentioned in despatches. Medal and 3 clasps. C.B. 12 December, 1854. Turkish medal.	Mortally wounded 8 September, 1855, and died 15 September.
54	Ensign Alexander Fraser Unett ...		—	—	P	—	Medal and clasp. Turkish medal	Son of above. Invalided 28 October, 1854, and did not return to the Crimea. Died March, 1884.
55	Captain Henry Turner Uniacke... ...	14 September, 1854	P	—	P	P	Medal and 2 clasps. Sardinian and Turkish medals. 5th Class of the Medjidie.	See 2nd Battalion Roll, 1859.
56	Captain Geo. Alexander Warburton ...	22 January, 1855	—	—	P	P	Medal and clasp. Sardinian and Turkish medals.	To Depôt 29 March, 1856. Retired 23 December, 1858.
57	Major Robert Warden	14 September, 1854	P	P	P	P	Mentioned in despatches. Medal and 3 clasps. Legion of Honour. 5th Class of the Medjidie. Sardinian and Turkish medals. Bts. of Major and Lieut-Col.	On leave 23 November, 1855 to 29 February, 1856. See 2nd Battalion 1859 Roll.
58	Lieutenant Ramsay Wardlaw	14 September, 1854	K	—	—	—	... ...	Died after amputation of the leg on board the s.s. "Andes" 27 September, 1854.
59	Asst.-Surgeon William Marshall Webb	14 September, 1854	P	P	P	—	Medal and 3 clasps. Turkish medal ...	To Staff 15 November, 1859. Afterwards Deputy-Surgeon-General. Died 18 March, 1899.
60	Captain Walter Montiford Westropp ...	25 March, 1856 ...	—	—	—	—	... ...	Joined after the siege. Retired 7 August, 1857. Died at Marston, Oxford, 17 April, 1856.
61	Lieutenant Walter William Young ...	7 September, 1855	—	—	P	W	Medal and clasp. Turkish medal	Invalided from wounds 12 October, 1855. Presented with a sword of honour by the Corporation of Belfast, and a silver salver by the County of Monaghan. Died at Auckland, New Zealand, 21 August, 1865.

CHAPTER XI.

Royal Review at Aldershot—The 19th sails for India—The Mutiny—The Cholera—The Green Howards' Hunt—The Hazara Campaign and Black Mountain Expedition—Return to England—Roll of Officers present with the Hazara Field Force.

ON the 8th July, 1856, headquarters and three companies were reviewed by Queen Victoria at Aldershot, in addition to three battalions of Guards, 7th Royal Fusiliers, 1st Royals, 1st Battalion Rifle Brigade, 79th Highlanders (Crimean troops), and other regiments. Her Majesty addressed a few representatives of each of the battalions which had been in the Crimea, expressing her admiration for their gallantry and fortitude during the campaign in the following words:—

"Officers, non-commissioned officers and soldiers: I wish personally to convey through you to the regiments assembled here this day my hearty welcome on your return to England in health and full efficiency. Say to them that I have watched anxiously over the difficulties and hardships which they have so nobly borne; that I have mourned with deep sorrow for the brave men who have fallen for their country; and that I have felt proud of that valour which, with their gallant Allies, they have displayed on every field. I thank God that your dangers are over, whilst the glory of your deeds remain. But I know that should your services be again required, you will be animated with the same devotion which in the Crimea has rendered you invincible."

In August the depôt companies, consisting of 408 of all ranks, joined the battalion from Walmer, and were amalgamated. In October a further change in the establishment was made, and the battalion was divided into eight service and four depôt companies, the former consisting of 800 men and the latter of 200, both with a proper proportion of officers and non-commissioned officers.

Lieutenant-Colonel John Lewis Richard Rooke, C.B., was gazetted to the command on the 10th November, 1856, *vice* Lieutenant-Colonel R. Sanders, C.B., who retired on half-pay owing to ill-health. Lieutenant-Colonel Rooke had been acting in command since the death of Lieutenant-Colonel Unett at the Redan, as Lieutenant-Colonel Sanders was absent, not having recovered from his wound at the Alma. Joining the Nineteenth in 1840 as an Ensign, he had spent all his service in it, and had received the thanks of the General Officer commanding the troops in Tobago for his conduct in 1847 during a hurricane that took place there. We have already seen with what distinction he served in the Crimean War.

Early in February, 1857, the depôt companies, under command of Captain L. D. H. Currie, moved from Aldershot to Parkhurst, Isle of Wight, to form part of the depôt battalion at that station.

Whilst quartered at Aldershot the Nineteenth received frequent commendation from Major-General the Hon. A. A. Spencer, C.B., commanding the brigade, who referred to its soldier-like appearance on parade, the steadiness of the men under arms, and the general good order of the regiment.

On the 16th June the battalion moved to Portsmouth by rail, and took over quarters in the Clarence Barracks, sending out one company on detachment to Tipnor magazine.

Consequent on the outbreak of the mutiny amongst the native troops in India, the Nineteenth was warned for service in that country, the establishment at the same time being raised to 1,140 non-commissioned officers and men, with the addition of a second lieutenant-colonel and third assistant-surgeon.

Between the 22nd and 29th of July the regiment embarked for India in three detachments, consisting of 45 officers and 1,007 men, under command of Lieutenant-Colonel John Rooke, C.B.

After transhipping at Point de Galle, the last detachment reached Calcutta on the 19th December, the others having arrived a few days earlier.

To the great disappointment of the officers and men their services were not required up country, for during the few months they had been at sea affairs in India had begun to wear a very different aspect, and the mutineers had been checked at all points. For the time being, therefore, the regiment took up its quarters at Fort William, sending a detachment of three officers and 120 men to Alipore.

On the 27th November Lieutenant-Colonel Rooke was attacked by cholera, and succumbed after a few hours' illness, to the general regret of the regiment. Major J. B. Jennings, writing home the sad news to Lieutenant-Colonel Rooke's old friend and brother officer, Captain John Margitson, said:

"I need not tell you how sincerely he is regretted by the regiment. Besides, he was so good an officer and there was so much confidence in him that all were delighted at the prospect of going on service with him. There were few more promising officers in the Service, and his untimely end is not only a misfortune to the regiment, but a loss to the army in general."

The command now devolved on Lieutenant-Colonel George Valentine Mundy, C.B. This officer had been appointed second Lieutenant-Colonel in July from the 33rd Foot. Originally commissioned to the Coldstream Guards in 1835, he had been transferred to the 33rd in 1841. With the Guards he had served in the Canadian rebellion of 1838-9 and with the 33rd all through the Crimean War, where he was twice slightly and once severely wounded.

On the 5th February, 1858, the Green Howards marched to Barrackpore for duty, sending out one company to Dum Dum and leaving two behind at Fort William. In Barrackpore there were 6,000 Sepoys confined as prisoners, and the regiment had to find an outlying picquet over their camp and watch them closely. At Fort William the King of Oudh and his Prime Minister were under guard, with sentries placed over their quarters.

During the months of April and May there was a good deal of sickness amongst the men owing to the intense heat, and in June there were fifteen deaths.

In July eight officers and 230 men marched to Dacca, and were joined a few days later by fifty-four more.

Late in October the remainder of the regiment moved to Dinapore *viâ* Raneegunge, where they assembled on the 24th November.

During this year the grenadier and light companies were abolished in the army, but it was not till a few years later that they finally disappeared. In future the companies were distinguished by a letter instead of a number.

On the 22nd January, 1859, a detachment of five officers and 188 men, under command of Captain H. F. Massy, marched from Dinapore to join Brigadier-General Douglas' column in pursuit of rebels in the Monghur district, rejoining headquarters a month later.

Another detachment, under command of Lieutenant-Colonel R. O. Bright, consisting of seven officers and 311 non-commissioned officers and men, marched from Dinapore to join the Field Force, under Colonel Kelly, C.B., 34th Foot, on the Nepaul frontiers. They were followed at the end of March by the rest of the regiment, under Lieutenant-Colonel Mundy, with the exception of the three officers and 130 men left at Dinapore. On this expedition the Ganges was crossed below Patna, and in pursuit of the Sepoys the men had often to wade rivers breast high, holding their arms and accoutrements high above their heads. The only safe way to cross was for them to hold each other's hands, otherwise there was great risk of being swept away. Though the rebels were never actually met with, as they were always on the run, still the force had plenty of hard marching, and underwent all the privations incidental to active service. As one of the men aptly put it: "If they were not up to their knees in blood, they were up to their hips in water day and night."

These detachments, together with headquarters, re-assembled at Dinapore on the 18th May, and during the summer sent small parties to Muzufferpore and Bankipore.

The latter having rejoined, the regiment marched from Dinapore to Benares in February, 1860, where it arrived on the 12th March.

Lieutenant-General Sir William Rowan was transferred to his old regiment, the 52nd Light Infantry, on the 10th March, 1861, when Major-General Sir Abraham Josias Cloeté, K.C.B., K.H., was appointed Colonel in his place.

He was the second son of the Hon. Peter Lawrence Cloeté, of Capetown, and was born in 1794. In 1803 he was sent to Holland to be educated, and in 1809 to England, when the Duke of Cumberland gave him a commission in the 15th Hussars. He became Aide-de-Camp to Lord Charles Somerset, who commanded at the Cape, and was placed in

GENERAL SIR ABRAHAM JOSIAS CLOETÉ, G. C. B.

FROM A PAINTING BY GLAZEBROOK.

charge of the expedition which took possession of the island of Tristan d' Acunha.

In 1817 and 1819 Cloeté commanded a squadron of the 15th in the Mahratta War, and was employed on the staff for no less than forty-six years, serving in every possible capacity, from Aide-de-Camp to Quartermaster-General and Chief of the Staff.

In 1820 he carried out an extensive measure of emigration along the frontier of the colony, and settled a serious question of Kaffir policy with the envoy of the Zulu chief, Chaka. In 1842 he commanded the successful expedition sent to Natal to relieve a detachment besieged by insurgent Boers. Colonel Cloeté's success in this expedition, which paved the way for the final settlement of Natal, led him to be recommended for the C.B., which was subsequently given him by the Duke of Wellington in person.

During the Kaffir War of 1851-53 he served as Chief of the Staff to the force operating in the Basuto country, and commanded a division at the battle of Bereia (medal and K.C.B.).

He was in command of the Windward and Leeward Islands from 1855 to 1861, and died in London on the 26th October, 1886, aged 92 years.

Whilst the Nineteenth was at Benares a company was stationed at Raj Ghat, which was relieved every month.

On the 11th December, 1861, the headquarters and seven companies, the remaining three being still detached at Dacca, marched to Allahabad, where they arrived in eight days' time, and encamped to await the incoming of the 92nd Highlanders. Proceeding by rail to Shikirbad, the march was continued

NOTE.—With a view to facilitate the provision of cloth by contract, the number of shades of green were diminished, and from 13th March, 1861, the facings of the regiment were ordered to be "grass green."

thence to Mean Meer *viâ* Agra, Allyghur, Delhi, Umballa, Jullundur and Umritsar, which was reached on the 25th February, 1862.

Early in April the three companies which had been left at Dacca, and more recently at Senchal, joined headquarters, under command of Major G. B. Jennings, having been absent for three years and nine months.

From February till August a company was on detachment at Lahore Citadel.

Towards the end of August cholera broke out in the regiment, in consequence of which three companies, under command of Major Chippindall, moved into camp at Shadra, three companies, under Captain Moffatt, to Chubeel, and the remainder of the battalion, consisting of headquarters and four companies, under the temporary command of Major Jennings, to Neaig Beg. Up to the 18th September, when the regiment assembled again at Mean Meer, the admissions into hospital from cholera had been 131 men, 5 women and 8 children, the deaths being 64 men, 2 women and 5 children.

On the 19th March, 1863, the headquarters and four companies marched from Mean Meer to Phillour, two companies being left to garrison the fort, the remainder proceeding to Kussowlie, where they arrived on the 9th April. They were stationed there till the 4th October, when they left for Ferozepore, in relief of the 7th Royal Fusiliers.

NOTE.—In the Church at Mean Meer there is a memorial slab to those of the Nineteenth who died there during the cholera epidemic, 1862-3, with the following inscription:—

SACRED
TO THE MEMORY OF
One Colour-Sergeant, one Drummer, and 85 rank and
file, five women and eleven children of the 1st Batt.
19th Regiment, who died at this station in the years
1862—1863.
This tablet is erected as a mark of respect by their comrades
in the regiment.

On the death of Colonel Mundy in London on the 14th May, 1863, the command of the battalion devolved on Colonel Robert Onesiphorus Bright, who had been appointed to the Green Howards as an Ensign on the 9th June, 1843. He had passed all his regimental service with them, and as we have already shown, had seen good service in the Crimea, at the Alma, Inkerman and before Sebastopol.

The regiment now had detachments at Umritsar, Kangra and Kussowlie. Early in 1864 these marched to Jullundur, where they were joined by the headquarters on the 13th March, together with the remaining companies which had been at Ferozepore.

During every half-yearly inspection since its arrival in India the Nineteenth had won unstinted praise from the various general officers under whom it had served. Its high state of efficiency and discipline; the steadiness of the men under arms and their soldier-like bearing; the interior economy as well as the institutions of the battalion were all most favourably commented on, and must have been extremely gratifying to Colonel Bright as well as to the officers and men.

After being nearly two years at Jullundur the regiment marched for Peshawar on the 1st November, and arrived there on the 8th December, 1865.

Early in February, 1867, the Green Howards moved to Nowshera (26 miles 4 furlongs), sending out a detachment of two companies to Attock, a few days after arrival.

When stationed at Jullundur the officers acquired a pack of English foxhounds from the 51st Regiment, which showed excellent sport. They were kennelled at Murree in the hot weather. Colonel Bright was the first master of "the Green Howards' Hunt," and was succeeded by Major Chippindall. When the regiment left the district the hounds were handed over, and a station pack established at Peshawar, the origin of the present "Peshawar Vale Hounds."

In 1866 the handsome regimental lace on the drummers' tunics of white, red and green was ordered to be discontinued, and a universal pattern was adopted for the drummers of the army, viz., white with small red crowns.

The officers' blue frock coat was given up in April, 1867, and replaced by a blue patrol jacket; steel scabbards took the place of the black leather ones so long in use. Field officers, however, retained their brass scabbards.

When the regiment had been at Nowshera for close on a year it was moved to Rawul Pindi, where it arrived on the 4th February, 1868, under command of Colonel R. O. Bright, C.B.

The very flattering reports which continued to be received at the various inspections caused the Commanding Officer to publish the following order shortly after reaching Nowshera:

"It is with great pleasure that the Commanding Officer congratulates the regiment on the brilliant inspection they have made before His Excellency the Commander-in-Chief. To be pronounced perfect on all points by so high an authority cannot but afford the greatest satisfaction to every member of the regiment.

"The same high opinion was expressed on a former occasion by Lord Strathnairn; to be so favourably reported on by two successive Commanders-in-Chief is a distinction of which the regiment may well be proud."

On the 1st May a detachment, consisting of seven companies, under command of Bt.-Lieutenant-Colonel Chippindall, proceeded as a working party on the Murree roads, and rejoined headquarters at Abbotabad on the 13th August. The latter had marched in the same day, accomplishing a

Ens. J. F. Miller. Lieut. J. C. Humfrey. Ens. W. McClintock. Ens. J. Jameson. Capt. S. W. Bent, Capt. M. Williams. Col.-Sergt. E. Pennington. Q.-M. J. Keatinge. Ens. T. B. Emerson. Col.-Sergt. Keyford. Pay-Mr. T. Palmer. Capt. W. H. Moffatt. Ens. E. S. Mott. Lieut. J. G. Moir. Capt. E. Kindersley. Lieut. J. F. Fraser. Lieut. C. Skinner. Ens. H. Baddeley. Lieut. E. Dickinson. Lieut. W. Bennett.

Ens H. Boulcott. Asst.-Surgn. F. P. Staples. Ens. Lorn Campbell. Lt. & Adj. J. Knox. Lt.-Col. E. Chippindall. Surgn.-Maj. H. B. Hassard. Capt. A. B. Morgan. Lt. C. C. Tribe. Lt. A. M. Handley.

OFFICERS, PESHAWUR. 1866.

distance of sixty-five miles in fifty-nine hours in the middle of the hot weather without any sick or a single casualty.

The reason for this move being that an expedition was shortly to be undertaken against the hill tribes on the Black Mountain.

These hill tribes had committed sundry depredations in the Agror Valley, and it was as a punitive measure that the force was proceeding against them. Heavy fighting was anticipated, as little was known of the Black Mountain itself or of its inhabitants; and with the experience of the fighting with the hill tribes at Umbeyla a few years before, and a narrowly escaped disaster, a larger force had been collected than might have appeared necessary. There were 10,000 men all told, the European troops being—

"D" and "F" Batteries Royal Horse Artillery, under Colonel Renny, V.C.

The 2/24 Battery Royal Artillery, under Captain Jackson.

1st Battalion 6th Foot, under Lieutenant-Colonel Osborne.

1st Battalion 19th Foot, under Lieutenant-Colonel Chippindall.

Major-General Wilde, C.B., C.S.I., was in command of the whole force.

The regiment remained at Abbotabad for two or three days while the Hazara Field Force was being organised. On the 15th August Colonel Bright arrived from Murree and took over command of the battalion, but only for a short time, as he was appointed Brigadier-General to command No. 1 Brigade, and Colonel Chippindall assumed command of the regiment.

We cannot now do better than follow Colonel G. E. Langford's account of the expedition:—

"On the night of the 17th August we were to have left Abbotabad for Khakee, in the Agror Valley. Tents were struck at 6 p.m., and the camels and baggage were sent on, the regiment being ordered to parade at 8 p.m., when it was cooler. While on parade a violent storm of rain and lightning, which continued almost unabated throughout the night, caused us to seek shelter in the native lines, where we passed the night in the verandahs, wet through in our khaki clothing, as all our baggage and great coats had been sent on. Fortunately some of the Commissariat remained, and the Commanding Officer ordered a ration of rum to be served out to all in the morning. We left Abbotabad for Khakee on the evening of the 18th August.

"There we remained a whole month, in a most vexing state of uncertainty as to whether we were to proceed or not. A lull seemed to have occurred in the political storm, and at one time we almost thought that we should have to return to Rawul Pindi without seeing any fighting at all.

"However, we were by no means idle. We had long parades daily to accustom us to hill climbing and the various tactics of hill fighting—sixteen and seventeen miles a day being nothing unusual. I may also mention that in the afternoons there was any amount of sport for those who had guns, as quail, duck and snipe were to be found close by in numbers.

"At the end of September the regiment at last got orders to move on through the Soosul Pass to Oghee, where the whole force (about 10,000 men) had assembled, and this was in reality the basis of operations of the expedition.

"There were two forces acting against the Black Mountain tribes at the same time—one *viâ* Durband on the Indus; the other from Oghee.

"On the 3rd October No. 1 Brigade was ordered to commence the advance up the Black Mountain, 10,200 feet high. This brigade consisted of one European regiment (the Nineteenth), the 20th Native Infantry, a Ghoorka regiment, and a Mountain Battery of Artillery. No tents were taken, and one great coat, one blanket, and a change of clothes was the only baggage allowed.

"The brigade reached Munna-ka-Dunna, the first halting place, about 4 p.m., after a stiff clamber. Several villages on the way were found deserted, and little sign of the inhabitants did we see. We were not, however, to remain long without hearing of them.

"There were no tents to pitch on arrival at our halting place, so arms were piled and various guards and sentries posted.

"Munna-ka-Dunna formed a ridge about half-way up the Black Mountain, and lying almost at right angles to it. Its height is about 5,000 feet, whereas the Muchai Peak, which forms the summit of the Black Mountain, is 10,200 feet. On the side of the latter were dense forests of blue pine and oak, the sides of the ridge itself being indented with ravines, and covered with low bushes and rocks. The brigade occupied the whole summit of the ridge.

"It was about 6 o'clock when we first became aware that some of the tribesmen were not far distant. We were taking our evening meal when shots (commencing with a sort of volley) began to be fired from one of the ravines into our bivouac.

"The order to stand to arms was given, and a deployment took place, the brigade entirely encircling the crest of the ridge. An entrenchment of earth and stones was made as far as circumstances admitted, but shelter trench tools were then unknown, and the natural cover afforded by the ground had mostly to be resorted to.

"It was curious how the enemy managed to get within firing distance without our being able to see them. From behind rocks and bushes their firing seemed incessant, and yet we could see nothing of them. Their shooting, however, was without judgment, and occasioned but small harm to our side. A volley directed at any particular spot where the enemy seemed to be in force generally succeeded in effecting a dislodgment; but as night advanced they appeared to increase in numbers. On several occasions they broke cover and made rushes at different parts of our entrenchments, but were stopped by well directed volleys ere they could reach us, and they had to retreat to the shelter of their rocks again.

"So passed the night of the 3rd of October, during which none of us slept or left our entrenchments. As day dawned on the 4th the enemy seemed to retire, and there was a cessation of firing for a time.

"Our casualties were two men killed and a dozen wounded (all native troops). We never discovered how much damage we had inflicted, but heard afterwards that the enemy suffered severely, and as we commenced our onward movement during the morning thirty bodies were discovered in one place, hidden away under the brushwood.

"More troops from Oghee had been ordered to occupy our position on Munna-ka-Dunna, and by 10 a.m. on the 4th October we received orders to push forward as far as Chuttabut, an intermediate rise between where we then were and the Muchai Peak.

"The guns of the horse artillery were brought into position at the end of the spur, and under cover of their fire we commenced the advance. The Ghoorkas led, then came the two mountain batteries, after them the 20th Native Infantry, the

Nineteenth and the 2nd Punjab Native Infantry bringing up the rear.

"There was a great deal of beating of drums and desultory firing on the part of the enemy before the advance began, but the guns soon dispersed them, and there was little opposition.

"By 4 p.m. Chuttabut was reached, and there was ample time to erect 'sangars'—entrenchments capable of holding about twenty men each—at intervals round our camp, and to post outlying and inlying picquets.

"During this march there were a few casualties amongst the native regiments, who were in advance, as the enemy would lie in ambush for them in parties of from two to a dozen till the skirmishers were almost on them, then fire rapidly and bolt down the ravines or behind rocks or trees, and be lost to sight.

"They also played sad havoc among the *bheesties* (water carriers) and such camp followers as had occasion to diverge in the least from the line on which we were moving. Many of these were killed, and in some cases horribly mutilated, so that the *bheesties* who were obliged to descend into the ravines to draw water had to be protected by strong guards, and our allowance of water was limited to about one pint a man.

"On the morning of the 5th the regiment was under arms by 6 o'clock, and a busy day it was for all.

"The enemy had assembled in large numbers on the slope of the Muchai, and we could hear their drums beating, and watch them assembling with their flags even at that early hour. The 20th Native Infantry were left to take care of Chuttabut and keep up touch with Munna-ka-Dunna, whilst the Ghoorkas and Nineteenth pushed on.

"The first part of our march was through pine forests, and we had skirmishers on both flanks, with the Ghoorkas in advance. About 10 a.m. we emerged from this forest, and found ourselves at the foot of a lightly-covered slope, which culminated in the famous Muchai Peak.

"No sooner were we sighted than the enemy's drums began to beat afresh. The hill-top seemed alive with men, and several parties, evidently fanatics, rushed forward, waving their flags and *tulwars* (swords), urging on the remainder.

"We expected a fight in earnest; the Ghoorkas were ordered to advance, skirmishing on both flanks, while the Nineteenth deployed and advanced up the slope direct. When within range, a few shots from the Ghoorkas and ourselves seemed to change the determination of the tribesmen, and they gradually disappeared. A few fanatics remained, shouting and brandishing their arms at us, but they also were soon dispersed by a few well-directed shots, and we were allowed to gain the summit without further opposition.

"It was curious how they continued to carry away with them their dead and wounded; we saw many apparently killed outright and others fall, but not a body did we find on our advance up the mountain.

"With this terminated practically the resistance during the expedition, although we all had much exposure and discomfort still to go through, and continuous outlying and inlying picquet duties day and night. Occasionally a shot or two directed at some *sangar* or picquet would remind us that the enemy were still hovering about; it was impossible to make the camp followers comprehend this, and several of them lost their lives, and were mutilated, through wandering away in search of wood or water.

"An amusing episode happened to a party of officers on the second day of our occupation of the camp on Muchai Peak. We had none of us washed or tubbed since we left Oghee, and on the morning of this day a tempting pool of water was discovered about fifty yards down one of the ravines. The temptation to bathe was strong, and some half-dozen of us resolved to accomplish this. The spot was quickly reached, and ten minutes later we were enjoying a refreshing dip, when our ablutions were suddenly arrested by a volley of some dozen bullets, which struck the rocks all round us. How we escaped was a miracle; clothes were hurriedly snatched up, and the shelter of a conveniently adjacent rock provided us with a dressing-room. We were not long in getting back to camp again, nor did we ever say much during the remainder of the expedition about our bath on Muchai Peak.

"That night we were greeted with a snowstorm, and a wretched time we had. The men were entirely without shelter; the cold—as may be imagined at 10,000 feet—was intense, and a ration of rum was served out in the morning to avert evil consequences.

"The following day the regiment was ordered to improvise wooden shelters:[1] wood was present in abundance, and the shelters were made company by company, each captain vying with the other in his ideas of architecture. It was understood that these were to be completed before night, and this was carried out successfully. This was the first bit of shelter the men had had since leaving Oghee, and half the regiment enjoyed a good night's rest, the other half as usual being on picquet.

1 These huts were erected in the course of twelve hours, so well designed and built that at the particular request of the Chief Commissioner they were left standing on the Muchai Peak, being abandoned in view of their having a moral effect on the enemy.—Regimental Records.

"Now ensued for some days another period of stagnation. It had been expected that our force would push on and retaliate in some measure against the hill tribes by burning their villages and destroying their crops. But the political element stepped in, and, thinking that the tribes seemed inclined to give in, issued a mandate to the effect that if the chiefs of the rebellious tribes (Afridis, Akagai, Keyl, and others that I do not now remember) would give themselves up a peace would be concluded.

"This was very bad news for us; for we had been put to an immense amount of hardship and discomfort, and there had been but little fighting.

"On the 9th October peace was proclaimed, and we were ordered to retire from the Black Mountain.

"In order to make a demonstration the Nineteenth was ordered to march through a part of the disaffected district (the Koonsh Valley) and then return to Oghee.

"To show how little the tribes cared for the punishment that had been inflicted on them, the rear-guard and camp followers were continually being harassed and fired upon during our evacuation of the Black Mountain. So aggressive were they at times as to provoke retaliation, and on one occasion a small party of Ghoorkas (a portion of our rear-guard), headed by a sergeant, made a demonstration on their own account, from which they slowly emerged with five bleeding heads, which they had lopped off the enemy with their *kookries*.

"I need hardly say that this incident was not chronicled in general orders, and the little Ghoorka sergeant and his mates were severely reprimanded for their gallantry; but I have no doubt that their own satisfaction at having had a really good ten minutes with their foes must have been great in the extreme.

"After we left the Black Mountain little more of interest happened. The battalion had a pleasant enough march through the valley of the Koonsh. The same precautions as to the protection of front and flanks by skirmishers during the marches, and outlying and inlying picquets at night, were taken; and we reached Oghee again on the 22nd of October, having been without shelter or bedding for twenty nights.

"On the 24th October the Hazara Field Force paraded before His Honour the Lieutenant-Governor of the Punjab, and the Nineteenth had the honour of furnishing the guard at the afternoon Durbar.

"The following day the force was disbanded, and by easy stages the regiment returned to Rawul Pindi, which was reached on November the 4th."

Her Majesty was pleased to grant the Indian General Service Medal to all those who took part in the operations, with the clasp "North West Frontier."

Towards the end of March, 1869, a detachment, numbering 225 of all ranks, under command of Major Griffiths, proceeded to Allahabad, and arrived there on the 3rd April. The men were employed in making roads in the district, and did not rejoin headquarters till the 2nd November.

The regiment now got orders for Saugor, and a farewell inspection was made of it on the 9th of November by Major-General W. O'G. Haly, C.B., commanding the Jhelum division, who afterwards issued the following divisional order:

"Rawul Pindi,

"25th November, 1869.

"The 1st Battalion 19th Regiment, after a service of five years in the Peshawar and Jhelum divisions, is now leaving the division. Major-General Haly desires to record his very high opinion of the most excellent state of efficiency, discipline and good conduct of this very fine battalion during the long period it has now served under his command. In every respect the battalion has been deserving of approbation, the interior economy and several institutions of this battalion leaving nothing to be desired.

"This high state of discipline and efficiency reflects the highest credit on Colonel Bright, C.B., and all the officers, non-commissioned officers and soldiers in their respective positions. Major-General Haly begs to assure this battalion of his best wishes and regards."

On the 22nd January, 1870, the headquarters and ten companies, under command of Colonel R. O. Bright, C.B., arrived at Saugor, having left Rawul Pindi on the 26th November.

During its stay at Saugor the battalion furnished a detachment of two companies at Jubbulpore, increased to three in June, 1871, by a company which had formed part of a working party at Puch-Murree.

The shooting of the regiment was so good this year as to draw forth the following favourable remarks from the Commander-in-Chief in India:—

"The shooting is excellent, and evinces a degree of care in the performance of the annual course of musketry which is very creditable to the battalion.

"The exertions of the officer instructors, Captain Bennett and Lieutenant Emerson,[1] have been reported to His Excellency as being worthy of special commendation, an expression of which Lord Napier of Magdala accordingly desires may be communicated to those officers."

In May, 1871, Colonel Bright was appointed to the command of the Saugor district, and Major Henry de Renzy Pigott was gazetted Lieutenant-Colonel to fill the vacancy in August. This officer had been commissioned to the 83rd Foot in 1845, and had been transferred as Major to the Nineteenth in 1863. He had served in the Indian Mutiny, for which he held the medal and clasp.

Orders now came for the return of the regiment to England after a service in India of fourteen years almost to a day. It accordingly marched from Saugor on the 1st November, and picking up the detachment at Jubbulpore, embarked in H.M.S. "Crocodile" at Bombay on the 24th November. A total of 184 non-commissioned officers and men had volunteered for service with other regiments, so that altogether the battalion only mustered 561 of all ranks on board the transport.

---

1 Theodore Bosville Emerson. Retired as hon. Lieutenant-Colonel 17th December, 1881. Died at Bath 15th November, 1883.

THE HAZARA ROLL.—1868.

Rank.	Name.	Remarks.
Lieut.-Col.	Robert O. Bright, c.	See text.
Major	Edward Chippindall, l.c.	See text.
,,	Edward St. J. Griffiths	Retired as Lieut.-Col., half pay 31st March, 1877.
Captain	William F. T. Marshall	Served in the Indian Mutiny. Retired 1st April, 1870. Died in 1872.
,,	Henry Edward Jerome Bt.-Major	Awarded the Victoria Cross for the Indian Mutiny. Retired as Major-Gen. 15th Sep., 1870. Died at Bath 25th February, 1901.
,,	Montgomery Williams	Served in the Indian Mutiny. To 100th Foot 13th March, 1878.
,,	William Henry Moffatt	Retired as hon. Lieut.-Col. 19th Mar., 1878. Military Knight of Windsor.
,,	Godfrey Baldwin ...	Retired as hon. Lieut.-Col. 1st Oct., 1878. Died at Limpsfield, Surrey, 3rd February, 1880.
,,	Alexander B. Morgan	To 9th Foot as Major 7th Aug., 1878. Later Sir A. B. Morgan, K.C.B. Died at Ilkley, Yorkshire, 13th Aug., 1911.
,,	Edward William Evans	Brigade-Major to 1st Brigade. Mentioned in despatches. Retired as hon. Lieut.-Col. 7th June, 1879. Died in London 16th Jan., 1910.
Lieutenant	William Bennett ...	See Nile Roll.
,,	John C. Taylor Humfrey	To Army Pay Department 8th August, 1880. Died at Blackheath 28th Jan., 1905.
,,	Augustus M. Handley	Retired as Colonel 29th Sep., 1888. Died at Clifton 27th Feb., 1906.
,,	Robert Gayer Traill...	Retired as hon. Major 20th March, 1880. Died at Belfast 5th March, 1908.
,,	James Francis Fraser	Retired with hon. rank of Major 27th April, 1881. Died at Richmond, Yorkshire, 27th April, 1910.
,,	Constantine C. B. Tribe	To Army Pay Department 16th May, 1879. Died May, 1879.
,,	Frederick S. S. Brind	To Captain, half pay 1st April, 1870. Served in the Indian Mutiny. Orderly Officer to Gen. Bright in the Hazara (mentioned in despatches). Later 17th Foot.
Lieut. and Adjutant	James Gordon Moir...	Retired 28th October, 1871. Died in London 11th July, 1903.

RANK.	NAME.	REMARKS.
Lieutenant	John Jameson ...	Retired with hon. rank of Lieut.-Col. 19th July, 1882. Died at Strathblane 4th January, 1899.
,,	Herbert Chas. Boulcott	Retired 10th May, 1870. Died at Upcross, West End, Hants, 18th February, 1909.
,,	Lorn R. H. D. Campbell	To Indian Army 28th July, 1869. Now Major-General and Colonel 38th Dogras.
Ensign	John Francis J. Miller	To Indian Army 15th April, 1869.
,,	William G. McClintock	See Nile Roll.
,,	George E. Langford...	To Army Pay Department 4th October, 1878.
,,	William Alex. Curtis	Retired in 1872.
,,	John Henry Barnard	Promoted Captain 101st Foot 7th March, 1877. Afterwards Colonel and A.D.C. to Queen Victoria. Died in London 11th May, 1891.
,,	William Augs. Burnett	Transferred to 103rd Foot 12th April, 1879. Died at Mannamead, Plymouth, 29th December, 1897.
,,	Charles Archd. Mercer	To Indian Army 18th January, 1870.
Qr'master	John Keating ...	To 62nd Foot 19th September, 1871. Served in the Indian Mutiny. Died at Hazareebaugh, India, 21st Mar., 1878.
Paymaster	F. Sargent-Openshaw..	Retired 12th September, 1896.
Surgeon	Henry Bolton Hassard	Took part in the Kaffir War 1851-3. Surg.-General 20th November, 1884. Died at Portsmouth 2nd July, 1892.
Asst.-Surg.	Francis Patrick Staples	Retired as Bde.-Surgeon-Lieut.-Col. 1st February, 1888.
,,	George Atkinson, M.B.	Served in the Bhootan Expedition 1865. Died at Hafaizai, near Ghuzni, 25th April, 1880.

CHAPTER XII.

Home Service—Changes in Army Organization—Portsmouth and Aldershot—Best Shooting Regiment—Chester and Sheffield—Presentation of new Colours by H.R.H. The Princess of Wales, 1875—Aldershot, Bermuda and Halifax, N.S.—The Territorial System—Malta and Egypt—The Nile Expedition—Assouan and Korosko—Engagement at Ginnis—Alexandria and Cyprus—The Nile Roll.

ON arrival at Portsmouth on the 27th December, 1871, the Nineteenth marched to the new barracks at Gosport, where it was quartered.

This was an important year in the history of the army, and many sweeping changes were introduced, one of which was the abolition of the purchase of commissions. Up to 1871 an Ensign's commission could be bought for £450. Every rank had its recognized money value, and if a promotion was to be secured, only the difference had to be made up between the amount already paid in and the sum then due. The only vacancies filled without charge by the promotion of the senior officers were those caused by death.

Owing to the increasing difficulty of attracting recruits for a long term of service as well as the desirability of forming a reserve, short service was introduced, by which no recruit could enlist for more than twelve years, six or seven of which were to be spent on the reserve.

In 1873 brigade depôts were formed, the 4th being at Richmond, Yorkshire, and allocated to the Nineteenth Foot, together with the two Militia battalions of the district.

NOTE.—By Royal Warrant of the 30th October, 1871, the ranks of Cornet and Ensign were abolished, and officers were gazetted as sub-Lieutenants, altered a few years later to 2nd-Lieutenants.

MAJOR GENERAL WILLIAM SPENCER COOPER.

Captain E. W. Evans, of the Nineteenth, was appointed the first Adjutant of the depôt under the new system.

From this date onwards the beautifully situated little town on the Swale has been associated with the depôt of the Green Howards.

Richmond was not altogether unknown to the Nineteenth, owing to the connection of the latter with the North Riding. The depôt companies of the regiment had been quartered there in the early part of the century, under command of Major Alexander Lawrence, and it was in Richmond on the 4th March, 1811, that his son, John Laird Mair, was born, the future statesman and Indian administrator, and first Lord Lawrence of the Punjab.

The Nineteenth remained stationed at Gosport till August, 1872, when it was split up and divided between Fort Grange, Fort Monckton and Fort Gomer, headquarters being at Fort Grange.

Just at this time Lieutenant-Colonel William Spencer Cooper, of the 70th Foot, exchanged with Lieutenant-Colonel Pigott.[1] He had been gazetted as Ensign to the 3rd West India Regiment in 1843, and had transferred to the 70th five years later. He had served in Burmah in 1852-1853 in the campaign there, also in the Indian Mutiny, and was in possession of both medals.

Early in March, 1873, "B" company, under command of Captain Bennett, proceeded from Fort Monckton to the Isle of Wight, and occupied Cliffend Fort.

On the 18th June the battalion reassembled at Aldershot, where it took over huts in the Guards' enclosure, Cove Common.

---

1 Promoted Colonel 25th August, 1876, and retired in 1882 with the honorary rank of Major-General. Died at Elkhorn, Manitoba, 14th Nov., 1889.

According to an army circular issued in May, 1874, the establishment of the battalion and depôt was fixed at 26 officers, 42 sergeants, 16 drummers, 40 corporals, and 480 privates.

Orders were now received for the white clothing hitherto worn by the band to be discontinued. A Glengarry cap was issued, instead of the small circular forage cap, and the regimental button worn by the men was replaced by one of a universal army pattern. Loose scarlet frocks took the place of the old shell jacket of the rank and file.

Whilst at Aldershot notification was received that in the annual musketry progress for 1872-1873 the 1st Battalion Nineteenth Foot was at the head of the infantry as the best shooting regiment in the army with a figure of merit of 97.00.

On the 3rd July, 1874, the battalion moved to Chester, headquarters and four companies being quartered at the Castle, the remainder being detached to the Isle of Man, Weedon and Liverpool. On the occasion of laying the foundation stone of the Seaman's Orphange in Newsham Park, Liverpool, by H.R.H. The Duke of Edinburgh, the detachment in the city formed the guard of honour, the band and regimental Colour being brought from Chester for the ceremony.

Colonel W. S. Cooper having been appointed to the command of a brigade depôt on the 27th February, 1875, Lieutenant-Colonel and Brevet-Colonel William Albert Stratton, from a brigade depôt, was gazetted to the vacancy He, however, retired almost at once, and Major and Brevet-Lieutenant-Colonel Bonar Millet Deane, from the supernumerary list, took over the command. This officer had

NOTE.—Regimental Surgeons and Assistant Surgeons were abolished by R.W. dated 1st March, 1873, which established an Army Medical Department.

Presentation of Colours by
H. R. H. The Princess of Wales, 17th August, 1875
By Permission of the Proprietors of the Illustrated London News.

joined the 22nd Foot in 1853, had been transferred as Major to the 18th Foot in 1864, and to the Nineteenth in the following year.

On the 22nd June, 1875, Their Royal Highnesses Prince Albert Victor and Prince George Frederick of Wales inspected the regiment.

It did not remain very long at Chester, as orders came for a move to Sheffield, where the headquarters and four companies, under Major E. St. J. Griffiths, arrived late in July, being joined at the same time by the detachments from the various out-stations.

The Green Howards had not been many days in Sheffield when an event took place which has always been looked back on with pride and pleasure by those who had the good fortune to be present on that memorable occasion. The Colours, which had long been in use, had been carried through the Crimean War, and were very much tattered and torn. Their Royal Highnesses the Prince and Princess of Wales, during August were on a visit to the Duke of Norfolk, and the Princess had graciously assented to giving new Colours to the battalion. The interesting ceremony took place on the 17th August in the grounds of the Duke of Norfolk. His Grace the Archbishop of York having first of all offered up the consecration prayer, H.R.H. the Princess of Wales then delivered the following address:—

"Colonel Deane, officers, non-commissioned officers and men of the Nineteenth Regiment,—It is a great pleasure to me to have the power afforded me of confiding in the name of Her Majesty the Queen these Colours to you to carry and protect. Your regiment, one of the oldest in the Service, is peculiarly connected with the county in which you are now assembled, being designated the First Yorkshire (North Riding) Regiment. It served under William III. at Steenkirke, and one of its earliest Colonels (Erle), at the battle of Landen, left his bed when ill with a fever. He led his men into action, and was wounded while doing his duty. Later, under the Duke of Cumberland, in the Netherlands, the 19th distinguished itself at Fontenoy and Val; and in 1795 it served in the campaign in the Netherlands under the Duke of York. It honourably shared in the victories of the British arms in the Crimea, and lastly it participated in the difficulties and successes which attended the suppression of the Indian Mutiny. With these annals of your regiment, I deliver you these Colours, confident of the gallantry of their defenders, and of their sense of duty to their Queen and country."

Lieutenant-Colonel B. M. Deane, in reply, said:—

"May it please your Royal Highness, on behalf of all ranks of the 1st Battalion of Her Majesty's 19th Regiment, which I have the honour to command, I beg you to accept our heartfelt gratitude for the honour you have bestowed on us. From the earliest days of chivalry no body of fighting men ever received their Colours from a more gentle and noble lady. Well may we exult in the great dignity conferred upon us, and I venture to hope that in the hour of trial we shall not prove unworthy of the great compliment you have paid us. If, however (which God forbid), our hearts should waver or our courage at the instant fail when hard pressed by the foe in time of hardship or distress, I can confidently assert that one glance at the Colours we have received and one thought of the much-beloved Princess who has presented them to us will nerve the youngest soldier in our ranks to struggle manfully to the last, and if needs be to die for the honour of his Queen and country."

The new Colours, which were received from Her Royal Highness by Lieutenant G. Reynolds[1] and Lieutenant J. Jameson,[2] were then raised in front of the regiment and acknowledged by a general salute, after which they were marched to their place in line, the band playing the National Anthem.

The regiment then formed column and marched past their Royal Highnesses to the music of "The Bonnie English Rose," after which it advanced in review order, when a royal salute was given.

At the conclusion of the ceremony the Princess was graciously pleased to accept the old Colours,[3] and to consent, at the request of Lieutenant-Colonel Deane, that the regiment should in future be designated—

"THE PRINCESS OF WALES'S OWN."

The Nineteenth remained at Sheffield till the 30th May, 1876, when it left by rail for Aldershot, under command of Lieutenant-Colonel Deane.

Early in July the battalion, with the Depôt 2nd Nineteenth attached, marched to Blackheath, where it was encamped for a fortnight, and then returned to Aldershot.

In October, 1877, a cork helmet, covered with blue cloth with a gilt spike and chin-strap, was issued to the battalion in place of the shako.

1 Retired as Major 18th November, 1882. Died at Ballyshannon, co. Donegal, 18th May, 1895.

2 See Hazara Roll.

3 These Colours are now at Sandringham.

NOTE.—Reading from left to right, those in the immediate foreground of our illustration are Lieutenant George Hewetson Reynolds, Lieutenant John Jameson, Lieutenant-Colonel Bonar Millet Deane, H.R.H. the Princess of Wales, H.R.H. the Prince of Wales, the Duke of Norfolk, Miss Knollys, H.R.H. the Princess Mary (Duchess of Teck), General Knollys, the Archbishop of York, and General Sir Henry de Bathe, Commanding the Northern District.

Orders were now received for foreign service, Bermuda being the station selected for the commencement of a colonial tour, and the Green Howards embarked on board H.M.S. "Himalaya" at Portsmouth on the 1st November, 1877, the strength being 708 of all ranks.

On arrival at the islands on the 17th November, the head-quarters and six companies were quartered at St. George's, under command of Lieutenant-Colonel Deane, two companies being sent to Ireland Island.

From time to time during the stay of the regiment in the colony companies were stationed at Boaz Island, Scour Hill and Ireland Island.

Early in 1879 Lieutenant-Colonel Deane retired on half-pay, the vacant command being given to Lieutenant-Colonel Lionel Hook, late of the Ceylon Rifles, who was gazetted on the 12th February. He, however, did not take up the command, as he retired, and Major and Bt.-Lieutenant Colonel Philip Doyne Vigors was promoted in his place.

Commissioned originally to the 11th Foot in 1846, he had been promoted Captain in the Nineteenth on the formation of the 2nd Battalion in 1858. He was in possession of a Staff College Certificate, but had seen no war service.

Colonel Deane went out to South Africa, and served on the staff of Major-General Colley in the first Boer War, when he was killed at the battle of Laing's Nek. The following inscription is on his tombstone at Mount Prospect, South Africa:—

> "In memory of Bonar Millet Deane, Colonel late 19th Regiment, who on the 28th January, 1881, whilst commanding the Natal Field Force, fell in action at Laing's Nek, at the head of a storming party, 10 yards in front of the foremost man."

There is also a cross to his memory in the South Camp Churchyard, Aldershot—"Erected by his Brother Officers and Friends in remembrance of a good and brave officer."

Major-General Colley, in reporting his death, wrote as follows:—

"I have especially to deplore the death of Colonel Deane, commanding Natal Field Force, in whom this force has suffered an irreparable loss. His experience and knowledge of all staff and regimental work, and his unremitting attention to every detail, having for its object the comfort, the efficiency and the security of the men under his command, coupled with his charm of manner, had made him alike beloved and looked up to by all serving with him; and his death was in keeping with his character as a chivalrous gentleman and officer."

After being about a year at St. George's the battalion moved by march route to Mount Prospect, in Hamilton Island, in relief of the 46th Foot, and returned to St. George's in February, 1880, its place being taken at Prospect by the 99th Foot.

The next move of the Green Howards was to Halifax, Nova Scotia, where they arrived in H.M.S. "Crocodile" on the 23rd November, 1880, having left Bermuda four days previously. This was a very sudden change in climate for all. When the vessel left Bermuda the temperature was 78° in the shade, and on entering Halifax harbour the water was freezing on her bows, and the thermometer registered 12° of frost.

Lieutenant-Colonel Vigors retired in January, 1881, and was succeeded in the command by Major Charles Hereford,

NOTE.—Regimental Paymasters were abolished by R.W. dated 22nd October, 1877, when the Army Pay Department was established.

whose promotion was dated the 12th January, 1881. Posted as an Ensign to the 39th Foot in 1855, he had been transferred as a Lieutenant to the Nineteenth in 1858 on its augmentation.

On the 1st July, 1881, the new territorial system came into being, and the title of the regiment was altered to—

THE PRINCESS OF WALES'S OWN (YORKSHIRE REGIMENT),

which was composed as follows:—

1st Battalion 19th Foot.		
2nd	"	"
3rd	"	5th West Yorkshire Militia.
4th	"	The North York Rifles.

The historic grass-green facings were changed to white, and "The Rose" was ordered to be borne on the Colours as the badge of the regiment, subsequently altered to "The White Rose" to show the connection with Yorkshire.

At the same time the Cypher of H.R.H. Alexandra Princess of Wales, interlaced with the Dannebrog, and the whole surmounted by the Coronet of the Princess, was ordered to be worn on the buttons and appointments of the officers.

Marches for the different regiments were also approved of, "The Bonnie English Rose"[1] being sanctioned as the "quickstep" or "march past" of the Princess of Wales's Own. This march had been in constant use since 1868, but till now was not officially recognised.

The Green Howards, in common with other regiments, greatly regretted the loss of their facings and number. Both had been associated with their history for such a long period

1. The music of "The Bonnie English Rose" was composed by Sydney Nelson and the words by Charles Jefferys. The song was first published in 1858. Ten years later the air was arranged as a march by Mr. Antcliffe, Bandmaster of the 1st Battalion.—M.L.F.

—the former for nearly two hundred and the latter for one hundred and thirty years—consequently, there was much sentiment connected with each. Repeated application had to be made ere the grass-green facings were restored twenty years later, and it is still hoped that the XIX., which is by no means forgotten or in disuse, will become once again the official designation of the regiment, and obviate the confusion which already exists amongst so many Yorkshire regiments.

The establishment at this time was altered to—

Lieut.-Cols.	Majors	Captains	Lieuts.	Staff
2	4	4	16	2
Sergts.	Corpls.	Drs.	Ptes.	Total
50	40	16	760	904

During the long Canadian winter the officers and men when on duty wore knee boots with fur caps and mitts as a protection against the cold. The officers had a specially made braided great coat trimmed with unplucked otter fur, and this with the white sword belt worn outside made a very picturesque uniform.

The battalion moved from Wellington Barracks to the citadel of Halifax on the 1st May, 1883, in relief of the 101st Royal Bengal Fusiliers.

On being relieved by the 86th Foot (Royal Irish Rifles), the Nineteenth embarked at Halifax on board H.M.S. "Himalaya" for conveyance to Malta, and sailed on the 28th February, 1884.

On disembarking at Malta on the 17th March the regiment took up its quarters in the lower St. Elmo Barracks, the Auberge de Bavière being handed over to the officers.

NOTE.—On the 1st July, 1881, all 2nd-Lieutenants were ordered to be styled Lieutenants, in conformity with the provisions of the R.W. of 25th June, 1881, by which the rank of 2nd-Lieutenant was abolished. The establishment of Lieutenant-Colonels in all battalions was increased from one to two, and the Majors from two to four.

During the summer preparations were being made by the War Office for an expedition on a large scale to relieve General Gordon, who was besieged in Khartoum, and orders soon came for the Nineteenth to reinforce the garrison in Egypt. Previous to its embarkation the battalion was inspected by Major-General Daniel, commanding the Infantry Brigade, who expressed himself as highly satisfied with the general appearance of the regiment and with its conduct during the time it had been under his command.

On the 4th August the battalion sailed for Alexandria in the hired transport "Hankow," where it arrived four days later. The headquarters and five companies proceeded to Ramleh by train, the other three, under Lieutenant-Colonel Burland, marching to Ras-el-tin Barracks, but rejoined at Ramleh a few days later on relief by the 3rd Battalion 60th Rifles.

Ramleh did not prove a very healthy station, and there was a good deal of sickness amongst the men, chiefly enteric fever. The barracks had formerly been used by the Egyptian cavalry, and as there were no drains in any shape or form,

---

NOTE.—The following are some of the changes of conditions of service, 1847-84.

1847 Enlistment for ten years introduced.

1867 Enlistment for twelve years introduced.

1867 Rates of Pay:—Sergeant-Major, 2s. 11$\frac{1}{2}$d.; Quartermaster-Sergeant, 2s. 3$\frac{1}{2}$d.; Colour-Sergeant, 2s. 1$\frac{1}{2}$d.; Sergeant, 1s. 7$\frac{1}{2}$d.; Corporal, 11$\frac{1}{2}$d.; Drummer, 8$\frac{1}{2}$d.; Private, 7$\frac{1}{2}$d., after deductions for rations.

1870 Short service introduced.

1871 Abolition of purchase.

1872 Rates of Pay (no deductions):—Sergeant Major, 3s. 6d.; Quartermaster Sergeant, 2s. 10d.; Colour-Sergeant, 2s. 8d.; Sergeant, 2s. 2d.; Corporal, 1s. 6d.; Drummer, 1s. 3d.; Private, 1s.

1881 Punishment by flogging abolished.

1881-84 Rates of Pay (exclusive of deferred pay):—Sergeant-Major, 5s.; Quartermaster-Sergeant, 4s.; Colour-Sergeant, 3s.; Sergeant, 2s. 4d.; Corporal, 1s. 8d.; Drummer, 1s. 1d.; Private, 1s.—"The Army Book for the British Empire."

By Royal Warrant dated 30th October, 1876, the rank of Sub-Lieutenant was abolished, and that of 2nd-Lieutenant substituted.

this may have accounted in some degree for its extreme unhealthiness. Consequently, the regiment was split up into several detachments, and was quartered all over Alexandria, one company, under Lieutenant Ferrar, going to the Red Barracks at Gabarri; another, under Captain Mathison, to Mex, where it was encamped; a third, under Captain Buckle, to the Ravelin Barracks, whilst a fourth went to Port Said, with Major McClintock in command. The remainder of the regiment was moved from the cavalry barracks to the disused palace of the Khedive close by.

On the tragic fate of General Gordon becoming known, Lord Wolseley's plan was to prepare for an autumn campaign on the Nile, and with this object in view a strong force was ordered to Suakin to hold that place and to co-operate with him in keeping the road to Berber open.

On the 9th February the battalion received orders for Suakin, and the next day Captain F. B. Briggs and Lieutenant J. T. Cotesworth, with sixty men, went to Cairo to join the mounted infantry battalion which was forming there.

One of the companies was under command of Captain Briggs, and consisted of two divisions of twenty-five men each from the Nineteenth, one division of Marines and one of Royal Engineers. They afterwards took part in the campaign under General Graham, and were present at the engagements of Hasheen, Tamai and Takdoul. They were granted the Egyptian medal with clasp, "Suakin, 1885," and the Khedive's bronze star for their services.[1] Captain Briggs was mentioned in despatches and awarded a brevet majority. To the regret of all ranks, he died at Cairo on the 15th September, 1885. He was an exceptionally smart officer, had been Adjutant to

1 A graphic account of the services of this detachment appeared in "The Green Howards' Gazette," vol. iii., written by Lieutenant J. T. Cotesworth.

the 2nd Battalion from 1878 to 1880, and his death was a great loss to the regiment.

The headquarters were to have left for Suakin on the 7th March, and already the advanced party of thirty men, under Lieutenant Orr, with all the camp equipment, had reached Suez, when counter orders came at the last moment for the battalion to proceed to Assouan to reinforce the troops on the river. So it entrained at Ramleh and proceeded to Assiout, where it embarked on four barges and two steamers for passage to Assouan.

Lieutenant-Colonel Hereford had been invalided and gazetted out on half-pay on the 4th March, and Lieutenant-Colonel William Harris Burland was now in command. He was first commissioned to the 9th Foot in 1854, and had been present at the siege of Sebastopol with that regiment. In 1878 he was transferred as a Major to the Nineteenth.

The voyage up the Nile commenced on the 11th March, and two days later Souhag was reached, where the steamers stopped to coal. The opportunity was taken to have a bathing parade, when two men were unfortunately drowned—Lance-Corporal Bushby and Private Kelly. Their bodies were recovered and buried next morning in a Coptic graveyard, about seven miles inland, on the left bank of the river.

Owing to the rapid falling of the Nile much delay was caused by running on the various sandbanks, which became more numerous daily, and it was not till the 18th March that Keneh was reached. Luxor was the next stopping place, and most of the officers were given leave to ramble amongst the famous ruins of Karnak, or to go across the river and see the still more interesting tombs at Thebes.

After passing by Esneh, Edfou and Komombo, the regiment arrived at Assouan on the 23rd March, and disembarked the day after. The headquarters and six companies marched up to Tagoog Heights, about 500 feet above the river, where they encamped, whilst two companies remained below on the river bank north of Assouan.

For the first few days the men at Tagoog were kept busy clearing the ground for a hut encampment, the temperature 99° in the shade, and the Khamseen wind at its worst. The bricks used in building were made of Nile mud by the natives, and carried up the hill by camels and donkeys. In a few weeks the men were all under huts, and as comfortable as possible under such circumstances.

About this time, owing to complications with the Russian Government, the withdrawal of the Nile expedition was definitely settled on by the Gladstone Ministry, in opposition to the wishes of Lord Wolseley, who was most anxious for an autumn campaign to shatter once for all the power of the Mahdi, and secure lasting peace for the frontier of Egypt. Had his advice been taken many valuable lives and much treasure would have been saved England. The Gladstone policy was, however, confirmed by the Conservative Government, who came into office in June, and who telegraphed that they were not prepared to reverse the orders given by their predecessors by countermanding the retreat of the troops from Dongola.

In June also it was decided that medals for the expedition would be granted up to March 7th, but only to the troops south of Shellal, which was a great disappointment to all ranks of the regiment.

Up to this time the intense heat of a Nile summer seemed to affect the men but little, and there were only a few deaths, one of them being that of Paymaster-Sergeant Park, a most deserving non-commissioned officer.

It was arranged that the 38th South Staffordshire Regiment and the 20th Hussars were to remain at Assouan for garrison duty, and the battalion was ordered to build huts for the former, which was shortly expected from the front. This gave the men something to do, and anything was better for them than to lie sweltering in their huts during the hot weather. The thermometer in May averaged 98° in the shade and in June 107·6°.

On the 15th July the 38th arrived, and a week later the 20th Hussars. All the other corps of the Expedition had in the meantime passed down the river to Cairo, with the exception of the 79th Cameron Highlanders, who remained at Korosko.

Rumours were at this time being circulated as to the renewed activity of the dervishes, and at last intelligence was received that their outposts had actually occupied Dongola. Accordingly on the 4th October the Camerons were ordered to Wady Halfa from Korosko, and the Nineteenth at the same time got orders to relieve them. So on the 10th October the battalion, 550 strong, having marched to Shellal, four miles distant at the head of the cataract, embarked on three steamers, towing daahabeyahs, under command of Lieutenant-Colonel W. Bennett, and arrived at Korosko three days later.

At this station the battalion furnished a section of the Frontier Force Camel Corps, consisting of twenty-nine men and a bugler, under command of Lieutenant Hadow.

On the 10th November a draft of 200 men, under Captain Chauncy and Lieutenant King, arrived, bringing up the strength to 700 effectives.

NOTE.—The regiment was paid the Nile gratuity, a Lieutenant-Colonel's share being £120, Major's £80, Captain's £60, Subaltern's £37 10s., Sergeant's £12, and a Private's £5.

It was now reported that the dervishes had come still further down the river, that the 79th were in fact besieged in Kosheh Fort, about one hundred miles south of Wady Halfa, and that it was decided to send up a force to relieve them.

Consequently the right wing, under command of Major McClintock, left for Halfa on the 2nd December, and another company, under Lieutenant Ferrar, was ordered up a few days later owing to a report that the enemy were in the vicinity of Halfa. However, there was no truth in this, and the company was ordered to return to Korosko. Finally, the headquarters and remaining companies left for the front on the 28th December, being relieved by the 56th Essex Regiment. The band and Camel section remained at Korosko.

The steamer, the s.s. "Fayoum," arrived at Abu Simbel late the next afternoon, and tied up at the river bank to wait for the moon to rise, so that everyone had time to inspect this wonderful temple hewn out of the solid rock. The whole face of the cliff rising 200 feet from the water's edge, has been hewn away and smoothed in order that at the summit four colossal figures of the great King Rameses II. might sit for ever as guardians of his temple overlooking the Nile.

Wady Halfa was reached on Christmas Day, and the regiment at once proceeded to Akasheh, at the head of the second cataract, ninety miles by rail, and occupied the various forts in the vicinity of the station.

The next day the right wing embarked in whalers and nuggers for Dal, followed on the 28th by the left wing. The boats took about three hours to reach Dal, helped by a good breeze. Previous to leaving Halfa the men's valises were placed in store, and each man had a shirt, pair of boots and a blanket carried for him by the regimental transport. They were ordered to discard their khaki and put on their red serges, blue trousers and putties, so as to look more formidable in the eyes of the dervishes. This was probably the last time red was worn by a British force in action with the enemy.

From Dal the battalion marched to Firket, ten miles further south, where the 49th Berkshire, 50th Royal West Kent and 106th Durham Light Infantry were already in bivouac. "C" company, under command of Captain Buckle, was left to garrison Firket, and the remainder of the force marched on to Mograkeh, another five miles.

The Nineteenth was told off to the 2nd Brigade, which was under the command of Colonel Huyshe, C.B., 49th Regiment, and was composed of the 79th Cameron Highlanders, 1st Battalion Egyptian Army, 9th Battalion Egyptian Army (Soudanese), Cameron Section of the Camel Corps, Gardner Gun Battery, manned by a detachment of twenty-seven non-commissioned officers and men of the Nineteenth, under the command of Lieutenant W. L. Mercer.

The 1st Brigade, under Major-General W. Butler, C.B., was composed of the 49th, 50th and 106th. The remainder of the force comprised the 20th Hussars, Mounted Infantry, three sections Camel Corps and three divisions Royal Artillery. Lieutenant-General Sir Frederick Stephenson, K.C.B., commanded the entire force, with Major-General Grenfell, C.B., commanding the division.

The position of the enemy was in the bend of the river, about three miles long, running from west to east. At the eastern end was a loop-holed village, which had for thirty-five days engaged the serious attention of the Cameron Highlanders in Kosheh Fort. At the western end were the headquarters of the Emirs. In the intervening three miles

THE ENGAGEMENT AT GINNIS, 30TH DECEMBER, 1885.

By permission of the Proprietors of "The Illustrated London News."

were groups of mud houses, with the usual palm trees along the river bank. The plan of attack was as follows:—

A portion of the 2nd Brigade, consisting of a mule battery, the Nineteenth Regiment, and the Camel Corps, was to appear at daylight on the ridge above the loop-holed houses near Kosheh, and after that position had been well shelled, the 79th and the blacks marching along the river bank were to take it at the point of the bayonet.

General Butler's Brigade was to appear on the ridges overlooking the west of Ginnis and engage the enemy. The rest of the 2nd Brigade was then to co-operate on General Butler's right, and converging on Ginnis, clear the groups of houses as they advanced.

At about 4 a.m. on the 30th of December the Nineteenth marched off to its position, escorting the battery, with which were also the machine guns. After making a detour in the desert over very rough ground with deep khors or dongas between the ridges, the battalion was halted in quarter column just before dawn, and was deployed on either side of the guns. The enemy opened a brisk fire almost immediately, and as it was not quite daylight the flash of their rifles could be clearly seen, whilst the bullets whistled harmlessly overhead.

The guns at once replied, and as soon as the men could see their sights the companies were ordered to fire long range volleys at the dervishes, who were moving about in large numbers in the vicinity of the houses on the river bank.

It was not long before the regiment was ordered to advance in conjunction with the 79th, and reached the village about 9 o'clock with practically no opposition. In the meantime the 1st Brigade had been engaged, and after some sharp fighting had seized the enemy's camp. The dervishes retreated along the river bank, more or less under cover, in the greatest disorder, leaving quantities of ammunition and all encumbrances behind them. Their total loss was estimated at 600. The Field Force had forty-five officers and men killed and wounded, including Egyptians. The only casualty in the regiment was one man wounded.

The following officers were present at this engagement:—

Lieut.-Colonel W. Bennett (in command).
Major W. G. McClintock.
,, J. O. C. Möller.
,, J. H. Eden.
Captain G. H. Mathison.
Lieut. A. G. Cartwright.
,, M. L. Ferrar.
Lieut. W. L. Mercer.
,, C. A. King.
,, A. B. Boyd-Wilson.
,, A. F. Paxton.
,, E. Somervell.
Capt. & Acting.-Adjt. W. A. Chauncy.
Quartermaster C. Organ.

Captain E. J. Buckle was at Firket in command of "C" company, and Lieutenant M. H. Orr was on train guard between Halfa and Akasheh during the action. Captain G. C. Handcock arrived the day after.

The 1st Brigade advanced to Abri, a few miles up the river, but the 2nd remainded at Ginnis till the 4th January, when the force being ordered to return, the Nineteenth marched off to Dal, and eventually arrived at Korosko on the 9th, where it encamped until the departure of the 56th Foot on the 17th January.

The Egyptian medal and Khedive's bronze star was awarded to all who served at and south of Wady Halfa between the 30th November, 1885, and the 11th January, 1886.

Three days later the strength was reinforced by a draft under command of Captain J. W. R. Parker, with Lieutenants

NOTE.—It was after this battle that the curious Dervish map of our fort at Kosheh was found, with its rather superfluous heading: "This is the fort of the infidels, the enemies of God, the liars, God curse them!"

Cotesworth and Connop, consisting of 153 non-commissioned officers and men.

Early in February the battalion was inspected by Colonel E. P. Leach, V.C., C.B., commanding at Korosko, who expressed himself highly satisfied with the efficiency of the men and their appearance on parade.

In April, when the weather began to get hot again, several deaths occurred from enteric fever and heat apoplexy, amongst them being that of Lieutenant W. M. Douglas, a young officer who had just joined.

On the 1st May the Green Howards left Korosko by companies in stern wheel steamers, and arrived at Shellal on the 3rd, when they went by train to Assouan, and were quartered once more in their home-made huts on the heights of Tagoog.

The 106th Durham Light Infantry were at this time stationed at Assouan, together with the 20th Hussars, who were at the north end of the town, and the 39th Dorsetshire Regiment, who had only recently arrived, were at Shellal. There were a great many deaths in the garrison, chiefly from enteric fever and heat apoplexy, in the months of May and June, and the temperature in the huts often stood at 122°. Batches of men were sent down the river to Alexandria, and others went to Cyprus, where they recruited their health on Mount Troödos.

The station hospital at Assouan was on Tagoog Heights, and the Green Howards had not only to bury their own dead, but those of the other corps as well. Funerals took place regularly, both morning and evening, and the crêpe was never off the drums. In order that the spirits of the men might not be affected by the "Dead March," all music at funerals was forbidden, and the bodies were buried in the desert without even the customary three volleys fired over their last lonely resting place.

Lieutenant-Colonel W. H. Burland retired on half-pay on the 28th July, and Lieutenant-Colonel William Bennett was shortly afterwards gazetted to the command. He had joined the Nineteenth in 1858, had served with it in the Hazara campaign of 1868 and on the staff during the Nile Expedition. For his services during the recent frontier operations culminating at Ginnis he was one of the first recipients of the Distinguished Service Order, a new decoration instituted by Queen Victoria on the 6th September, 1886.

The Colonelcy of the regiment also became vacant this year by the death in London on the 26th of October of General Sir Abraham Cloeté. It was now conferred on Lieutenant-General Sir Robert Onesiphorus Bright, K.C.B., and the Green Howards were glad to welcome as their head one who had such a distinguished regimental career. He had taken part, as we have shown, in the Crimean War with the Nineteenth, including all the principal actions, and in 1868 as Brigadier-General was in command of the 1st Brigade of the Hazara Field Force, when he was thanked for his services by the Government of India. He also served in the Afghan War of 1879-80, in command of the Khyber Line Field Force, including the operations in the Hissarik Valley and the expedition against the Wuzaree Khagianis (mentioned in despatches, received the thanks of both Houses of Parliament, medal and K.C.B.). On Queen Victoria's birthday, 1894, he was created a Knight Grand Cross of the Bath.

"He was in his prime one of the smartest officers in Her Majesty's service, and in every respect the beau ideal of a

GENERAL SIR ROBERT ONESIPHORUS BRIGHT, G.C.B.

commanding officer. No one who ever came under his command was ever treated unjustly, and he was absolutely worshipped by both officers and men.

"The excellent state of the 1st Battalion, to which all who were in authority at headquarters in India from 1857 to 1870 could testify, is not perhaps generally known, and Sir Hugh Rose, the Commander-in-Chief, frequently expressed his high appreciation of his eminent qualities as a commanding officer. During the Crimean War he was one of the very few officers who was never absent, either from sickness or other causes—always cheery, always ready for duty—during the dreadful winter of 1854. For his intelligence and soldier-like qualities he was selected to command the sharpshooters of the 2nd Brigade of the Light Division, which was no small compliment to him."[1]

After nearly two years' service on the banks of the Nile, the Green Howards at last received orders to move down the river, and the right wing, under command of Lieutenant-Colonel A. J. Paterson, arrived at Ramleh on the 9th February, 1887, followed by the headquarters and remaining companies a fortnight later. The regiment then marched to Ras-el-tin Barracks, Alexandria, in relief of the 56th Foot.

Here it remained for nearly a year, and previous to its departure for Cyprus the annual inspection took place by Major-General the Hon. R. H. de Montmorency, commanding the garrison. It lasted for three days, and at its close the General complimented the battalion in high terms on its fine appearance and general efficiency. To be told that it was as near perfection as possible was very high praise.

1 Extract from a letter to "The Green Howards' Gazette," on the death of General Bright in 1896.

In 1887 the rank of 2nd-Lieutenant was re-established and the Second-Lieutenant-Colonel abolished in all regiments.

During the stay of the regiment in Egypt and the Soudan the following return shows the number of deaths caused by fever, sunstroke and accidents (chiefly by drowning in the Nile):—

	Officers	Sergts.	Corporals	Privates.
Fever, etc.	2	2	3	77
Accidents	—	1	2	10
Total ...	2	3	5	87 *

The Nineteenth arrived at Limassol, Cyprus, in H.M.S. "Himalaya" on the 2nd February, having left Alexandria the day before. Strength—25 officers and 823 non-commissioned officers and men, under the command of Lieutenant-Colonel Bennett, D.S.O.

The headquarters and seven companies marched to Polymedia Camp, three miles inland, and took over camp and quarters from the 49th Berkshire Regiment. The remaining company, made up to 119 strong and four officers, under command of Captain W. A. Chauncy, disembarked at Larnaka on February 3rd, and marched to Nicosia, the capital of the island, there to be stationed.

On the 16th February the battalion paraded for the purpose of being presented with the Khedive of Egypt's bronze star, by Colonel Simpson Hackett, commanding the troops on the island.

In order to escape the heat of the plains in the summer the battalion, on the 25th May, commenced its march to

* The names of all will be found in the Appendix.

NOTE.—A memorial brass on a carved Purbeck marble mounting was erected in the south choir aisle of York Minster in memory of those who died in Egypt and the Soudan. The inscription in embossed letters reads as follows:—

"To the Glory of God and in memory of the officers and men of the 19th Princess of Wales's Own (Yorkshire Regiment) who died in Egypt and in the Soudan campaign, 1884-1887. This brass is placed by their comrades past and present." Here follow all the names.

Mount Troödos (6,000 feet above the sea) by companies, and by the end of June all were encamped on the top of the mountain, where they remained during the hot weather.

A pack of fourteen couple of harriers had been sent for from England, and arrived when the regiment was on Troödos, where kennels had to be built for them. They showed excellent sport in the winter with the hares, which were very plentiful in the island, and on the close of the season in February, 1889, attention was then turned to the foxes, which gave many a good run. Lieutenant F. Connop was master, with Captain Handcock, Captain Cotesworth and Lieutenant Ferrar acting as whips. An account of the various runs will be found in the "Field."

The return march to Limassol commenced on the 26th September, and by the last week in November all the companies had reassembled at Polymedia.

A somewhat similar move was gone through in May, 1889, and the battalion was once more encamped on Mount Troödos by the middle of June.

On the 11th July a telegram was received, ordering the Nineteenth to hold itself in readiness to proceed to Egypt immediately, for the purpose of reinforcing the army of occupation, as some difficulty was anticipated in dealing with the dervishes, who were advancing from Upper Egypt in considerable force. On the 19th July Major-General S. Hackett inspected the regiment prior to its embarkation, and addressed it in the most flattering terms as to its efficiency, fine appearance and the good conduct of the men.

On the 16th July Colonel W. Bennett, D.S.O., retired on half-pay, and in the next day's "Gazette" Lieutenant-Colonel Alfred James Paterson was appointed to the command. Joining the Nineteenth in 1863, he had been Adjutant of the 1st Battalion from 1870 to 1878. He held the Staff College certificate and had been for five years a Brigade Major at Aldershot.

On the 22nd July the battalion commenced its march down to Limassol so as to be ready to embark on H.M.S. "Himalaya." By the 28th the move was completed, and the troopship arrived on that date with the headquarters and half a battalion of the 56th Foot in relief. They disembarked that night, and by 12 noon the next day the whole regiment was on board with the exception of a few details and the women and children.

The Green Howards disembarked at Alexandria on the 31st July, and took over quarters at Ras-el-tin Barracks till their services should be required on the Nile. Unfortunately they were not considered necessary, and the battle of Toski, putting an end to the hopes of all ranks, was fought on the 4th August by the Egyptian Army troops, in addition to a squadron of the 20th Hussars.

Orders soon came for the Nineteenth to say farewell once more to the land of the Sphinx and the Obelisk, and it sailed for Portsmouth in H.M.S. "Himalaya" on the 7th September, where it arrived a fortnight later. On disembarking the regiment marched to the Anglesey Barracks, which it occupied for the next three years.

THE NILE ROLL.

Officers of the Regiment who served on the Nile between the 23rd March, 1885, and the 14th February, 1887.

RANK.	NAME.	REMARKS.
Lieut.-Col.	William Harris Burland, Bt.-Colonel	In command. Retired as Major-General 1st April, 1887. Died 28th July, 1890.
,,	William Bennett, D.S.O.	On the Staff and with regiment. Retired on half-pay 17th July, 1889.
,,	Alfred James Paterson, p.s.c.	Retired as Colonel 29th March, 1893.
Major	William G. McClintock	Retired as Colonel 28th July, 1896.
,,	John Olaus C. Möller	Retired as Lieutenant-Colonel, half-pay 2nd May, 1890.
,,	John Henry Eden ...	Retired as Lieutenant-Colonel, half-pay 4th March, 1892.
Captain	John William R. Parker	Retired as Major 27th May, 1896. Commanded 3rd East Lancashire Regiment in Boer War. Best shot in the army 1883. C.B. 1910.
,,	Charles J. Spottiswoode	Retired as Colonel 20th April, 1904.
,,	Henry Bowles, Bt.-Major	On the Staff. Mentioned in despatches. Bt. of Major. See Tirah Roll and Boer War.
,,	Gilbert H. F. Mathison	Retired as Major 18th November, 1896.
,,	Edward John B. Buckle	Retired as Major 12th January, 1898.
Captain and Adjutant	James Ahmuty Fearon	See Tirah Roll and Boer War.
Captain	Gerald C. S. Handcock	See Boer War.
,,	William A. Chauncy	Retired as Major 16th December, 1896.
,,	Arthur de Salis Hadow	With Korosko Section Camel Corps. Retired as Colonel 29th Sept., 1910.
,,	Alfred G. Cartwright	Boat Officer at Korosko. See Tirah Roll.
Lieutenant	James T. Cotesworth	Served at Suakin with the M.I. See Tirah Roll and Boer War.
,,	Michael Harrison Orr	See Boer War.
,,	Harry O. D. Hickman	On Staff and Egyptian Army. To Inniskilling Fusiliers 8th January, 1890.
,,	Michael Lloyd Ferrar	Acting District Officer R.E. Korosko. See Boer War.
,,	George Pearson ...	With the Egyptian Army. See Boer War.
,,	William Lindsay Mercer	Retired 27th April, 1898.
,,	Edward Malcolm Esson	See Boer War.
,,	Charles A. C. King...	See Boer War.
,,	Francis Connop ...	With Korosko Section of Camel Corps. Killed by a fall from his horse at Fifehead Neville, Dorset, 5th May, 1896.

RANK.	NAME.	REMARKS.
Lieutenant	Adam B. Boyd-Wilson	To Inniskilling Dragoons 7th April, 1886.
,,	Archibald F. P. Paxton	To Indian Army 14th January, 1887.
,,	Ernest Somervell ...	See Boer War.
,,	William M. Douglas...	Died at Korosko 13th April, 1886.
,,	Edward Snell Wallis	Died at Portsmouth 25th March, 1890.
,,	Shelley Alexander Seale	Retired on half-pay 16th March, 1889.
,,	Berkeley C. W. Williams	See Tirah Roll and Boer War.
,,	Charles William Gale	Retired on half-pay as Captain 1st May, 1899.
,,	Leslie Michael Farrell	Retired 17th August, 1898.
,,	Morton Henry Eden...	Died in London 5th November, 1889.
Lieut. and Qr'master	Charles Organ ...	See Boer War.

CHAPTER XIII.

Flattering report by the G.O.C. Cyprus—Meeting at Portsmouth of the two battalions—Ordered to London—Jersey—"The Green Howards' Gazette"—The Curragh—The Evelyn Wood Competition—Dublin—Gibraltar—Bradford and Strensall—Restoration of Facings—The Boer War.

SHORTLY after the Green Howards arrived home orders were received to send to the Army Reserve all men of five years' service. Consequently, when the battalion embarked for Jersey three years later it was only 400 strong, as very few recruits were received in the interval.

Portsmouth was never a hot-bed of guards like Gibraltar, but the battalion had to find an officers' guard for some time after its arrival, and during the whole of its stay in the garrison a guard of an officer and twenty men was mounted daily over the convicts at work on the dockyard, which was being extended.

A few days after landing the following letter was received by the Governor, General the Hon. Sir Leicester Smyth, K.C.B.:—

"Sir,—I have the honour by the desire of the Commander-in-Chief to transmit for your information a copy of the report which has been received in this department from the G.O.C. Troops, Cyprus, wherein General Hackett communicates the result of his inspection of the 1st Battalion Yorkshire Regiment prior to the embarkation of the battalion for Egypt, and at the same time to request that you will be so good as to cause H.R.H.'s appreciation of this very satisfactory report to be conveyed to the Officer Commanding the 1st Battalion Yorkshire Regiment and those serving under his command.

"I have, etc.,

"(Sd.) GODFREY CLERK,

"D.A.G."

"The G.O.C., Alexandria.

"Enclosure.

"My Lord,—I have the honour to report that the 1st Yorkshire Regiment embarked for Egypt yesterday in H.M.S. 'Himalaya,' and in so doing I desire to bring to the notice of H.R.H. the Commander-in-Chief the satisfactory and efficient state of the battalion.

"The men are physically in excellent condition and good health. There have been but few serious cases, and they have suffered but little from fever or climatic disease for the last six months. On medical inspection none were found unfit for service.

"Only those fit for duty actually embarked; of the few (8) left in hospital, none are serious cases and will shortly be able to join.

"In the recent monthly returns I have been able to report a marked improvement in the general conduct of the battalion, and the fact that there are no court-martial prisoners speaks for itself.

"The officers generally know and take an interest in their work: they have supported sports and healthy recreations and amusements for the men. All can ride, many very well, and their gentlemanly conduct and bearing while in the island have earned universal respect and goodwill.

"I have, etc.,

"(Sd.) S. HACKETT, M.-Genl.,

"Commdg. Troops, Cyprus."

On the 1st November the establishment of the battalion was fixed at 24 officers and 777 non-commissioned officers and men.

On the 31st December the 1st and 2nd Battalions of the regiment met for the first time in their history. The 2nd was sailing for India from Portsmouth the next day in H.M.S. "Euphrates," and the 1st Battalion took advantage of the occasion to entertain their brother officers at dinner, and many other Green Howards were invited to meet them. The band of the 1st Battalion being on furlough, the officers of the Royal Marine Light Infantry were kind enough to allow their string band to perform during the evening.

The following were present:—

Lieut.-General Sir R. O. Bright, K.C.B.
(Colonel of the Regiment).
Major-General E. Chippindall, C.B.
,, W. S. Cooper.
,, C. Hereford.
Colonel Sir Lumley Graham, Bart.
,, W. Bennett, D.S.O.
,, A. M. Handley.
Lieut.-Colonel J. H. Kirke.
,, T. H. Kirby.
,, G. W. Hughes.
Major J. H. Eden.
,, W. E. Franklyn.
,, A. W. Money Kyrle.
Captain H. J. Charrington.
,, T. D. Kirkpatrick.
,, D. Grant.

2ND BATTALION.

Lieut.-Colonel G. Oakes (commanding).

Major E. A. Bruce.
Brevet-Major H. Bowles.
Captain J. A. Fearon.
,, A de S. Hadow.
Captain A. G. Cartwright.
,, J. T. Cotesworth.
,, A. B. Elton.
Lieut. A. L. Napier.

Lieut. B. C. Hood.	Lieut. H. R. S. Maitland.
,, D. L. Hartley.	,, B. C. W. Williams.
,, W. F. T. Robinson.	2nd-Lieut. R. D'A. Fife.
Lieut. H. G. Holmes.	,, E. L. Vans Agnew.
,, W. S. Brooksbank.	,, H. F. Lea.
,, C. L. Nicholson (Adjutant).	,, W. H. Dent.
	,, H. E. Raymond.

1ST BATTALION.

Lieut.-Colonel A. J. Paterson (commanding).

Major W. G. W. McClintock.	Lieut. W. J. Todd.
,, J. O. C. Möller.	,, P. Lees.
,, E. S. N. Dickenson.	,, F. Connop (Adjutant).
Captain J. W. R. Parker.	,, E. Somervell.
,, E. J. B. Buckle.	,, C. W. Gale.
,, G. C. S. Handcock.	,, L. M. Farrell.
,, M. H. Orr.	,, C. Organ (Quartermaster).
,, W. A. Chauncy.	
,, G. Pearson.	2nd-Lieut. M. H. Tomlin.
,, W. L. Mercer.	,, F. H. Hodge.

Shortly after the arrival home of the regiment the use of the officers' blue-frogged patrol jacket was discontinued, and was replaced by an undress coat of blue serge with brass buttons. The tunic was ordered to be only used for ceremonial and specially defined duties, a red serge jacket with regimental facings being adopted for all ranks for route marching, manœuvres and guard duties.

On the 6th June, 1890, the battalion, 460 strong, left for Aldershot to form part of a flying column, and returned on the 14th June.

The following highly favourable remarks were received by the Officer Commanding as to the fire discipline of the regiment during the manœuvres.

"I am greatly pleased to hear from the Colonel Commanding the Flying Column that the discipline of your battalion when in action in the firing line was admirable, that the section commanders gave excellent fire commands, and that Lieutenant-General Sir Evelyn Wood observed that he was much struck thereat, and that the fire discipline of the battalion was an example to the whole force.

"(Sd.) L. SMYTH, General,

"Commandg. Southern District."

Early in July, 1890, half the battalion was at musketry on the Browndown ranges, and was quartered at Fort Gomer. On the 8th it got orders to rejoin headquarters, as the regiment was to be despatched to London the same night. The reason for this sudden and unexpected summons was that a battalion in the London garrison had refused to turn out on parade owing to some alleged grievance, and that the police also were disaffected.

In one and a half hours from the first receipt of the order 614 officers, non-commissioned officers and men were entrained at the Portsmouth town station, and proceeded to London. They arrived at 1 a.m. the following day, and marched to Wellington Barracks.

When the services of the regiment were no longer required, it was ordered back to Portsmouth on the 12th July, and entrained at Waterloo at 3 p.m. Previous to this it was inspected in the station by H.R.H. the Princess of Wales, as she had expressed a wish to see her regiment before its departure. Her Royal Highness was accompanied by H.R.H. the Prince of Wales, the Princesses Victoria and Maud, the Duke of Cambridge, Lord Wolseley, and the whole of the Horse Guards headquarter staff. The Princess expressed her great satisfaction at the appearance and bearing of the men to Lieutenant-Colonel A. J. Paterson.

A detachment of 200 non-commissioned officers and men, with a due complement of officers, remained in London, under command of Major C. J. Spottiswoode, and were quartered at Kensington Barracks till the 20th July, when they rejoined headquarters at Portsmouth.

Relative to this visit of the battalion to London the following letter was received by Lieutenant-Colonel Paterson from Lord Wolseley:—

"War Office,
"11th July, 1890.

"Dear Colonel Paterson,

"His Royal Highness the Duke of Cambridge desires me to express to you his satisfaction at the manner in which your men have done their duty and behaved themselves during their stay in London.

"The battalion, leaving 200 men in Kensington for some time longer, will return to-morrow to Portsmouth, carrying with them a good record as to their soldier-like conduct and appearance, both on and off parade.

"His Royal Highness regrets that he has not been able to see this battalion of the Yorkshire Regiment during their stay here.

"I have to request that you will express to all ranks under your command His Royal Highness's appreciation of their soldier-like qualities.

"Believe me to be,
"Very truly yours,
"WOLSELEY."

General Sir Leicester Smyth, K.C.B., was succeeded in the Portsmouth command by H.R.H. the Duke of Connaught, who arrived on the 25th August, and was received by a guard of honour from the Green Howards.

On the 1st January, 1891, a detachment of one company was quartered at Tipnor Magazine, and remained there till the move came for Jersey, being relieved from time to time.

On February 26th two large men-of-war, the "Royal Sovereign" and the "Royal Arthur," were launched by Her Majesty Queen Victoria in the dockyard. The battalion, in company with the other regiments of the garrison, was drawn up inside the dockyard, and presented arms as Her Majesty drove by.

On the 4th January, 1902, "D" company, under command of Captain M. H. Orr, proceeded to Fort Gomer to assist in making a new rifle range there, and returned to Portsmouth by the end of February.

The Martini-Henry rifles were now withdrawn from the battalion and the ·303 (Lee-Metford) issued in their stead.

The next move of the Green Howards was to the Channel Islands, and they embarked for Jersey in H.M.S. "Assistance" on the 8th September, 1892, arriving at St. Helier's the next day. Strength—22 officers, 2 warrant officers, and 398 non-commissioned officers and men.

On landing, a detachment of seven officers and sixty men proceeded to St. Peter's Barracks, and another of three officers and seventy men to Grève de Lecq, the remainder of the battalion being stationed at Fort Regent.

The following letter was received by the Commanding Officer shortly after the arrival of the regiment, from the Chief Staff Officer Southern District: —

"H.R.H. the Duke of Connaught wishes me to express to you his satisfaction at hearing how well 'The Princess of Wales's Own' marched out of Portsmouth, and his great regret that he was unable to be present to wish the battalion 'good-bye.'"

In April, 1893, the regiment started a monthly journal, called "The Green Howards' Gazette." Major J. W. R.

Parker was the first Editor, and the magazine has been published regularly every month since this date.

Four years was now the regulated period for an officer to hold the command of a battalion, but Colonel Paterson, whose time expired in September, 1892, was granted an extension of six months, when on the 29th March, 1893, he was succeeded by Major Edward Archibald Bruce, who was promoted from the 2nd Battalion. The latter had joined the Nineteenth in 1869, and had been Adjutant of the 2nd Battalion from 1873 to 1878. He did not take over the command till September, and up to that time Major A. C. Fryer acted temporarily.

On the 6th May, 1895, the battalion left Jersey in H.M.S. "Jumna" on relief by the 58th Regiment, and landed at Kingstown on the 8th, when headquarters and five companies were at once entrained for the Curragh, the remainder coming on with the baggage the following day. The battalion was quartered in the huts of "K" lines, which were decidedly the worse for wear.

On the 29th July, 500 men and 15 officers from each of the five regiments in camp marched to Maryborough Heath (twenty-one miles) in order to take part in the autumn manœuvres. Next day the march was continued to Ralish Camp, near Abbeyleix (fifteen miles), and on the 31st to the permanent camp at Glashaire, in the county Kilkenny, ten miles from Ballyragget station.

On the 2nd August the manœuvres commenced under the personal direction of Lord Wolseley, Commanding the

---

NOTE.—In the winter of 1892 the Jersey Drag Hunt was reconstructed, and the management taken over by the regiment. New kennels were built at Fort Regent and ten couples of hounds brought over from England. Lieutenant F. Connop was elected master with Captain Ferrar and Lieutenant Gale as whips. The drag was hunted for two seasons with great success.

Forces in Ireland, and finished on the 10th, when the battalion returned to the Curragh by much the same route, and on arrival on the 14th August was quartered in "H" lines, vacated by the 20th Fusiliers, who went to Athlone.

From the 27th January to the 1st February, 1896, the battalion was engaged in six days' continuous route marching, the total distance covered being 103 miles. The men marched well, and only one fell out during this training.

General Sir R. O. Bright, G.C.B., Colonel of the regiment, died on the 15th November at his residence, Normandy Park, near Guildford, when he was succeeded by Lieutenant-General Edward Chippindall, C.B., an officer who had been identified with the Nineteenth since 1849. He had joined the 32nd Foot as Ensign in 1847, and was transferred to the Nineteenth two years later. With the 32nd he was present during the latter part of the siege operations before Mooltan, the surrender of the fort and garrison of Cheniote on the 9th and battle of Goojerat on the 21st February, 1849, under Lord Gough. We have already mentioned his services in connection with the Crimean and Hazara campaigns, for the latter of which, when in command of the Nineteenth, he was mentioned for his "untiring energy and zeal." He was gazetted an Aide-de-Camp to Queen Victoria on the 14th March, 1872.

The regiment suffered a great loss this year by the death of Captain Francis Connop on the 5th May, from the effects of a fall from his horse. He had joined the battalion in 1884, and was Adjutant of it from 1889 to 1893. A good soldier and sportsman, it was mainly due to him that the battalion started its pack of hounds in Cyprus, which he hunted so successfully. His death was much regretted by his brother officers, with all of whom he was most popular.

The manœuvres of this year, 1896, took place again in the county Kilkenny, and on the 4th August the Curragh Brigade set out on its march to the training area, Tubrid. The standing camp was reached on the 8th, and there the brigade remained for ten days. Colonel Bruce was in command, with Captain Ferrar acting as brigade-major and Lieutenant H. A. Stansfeld as galloper. Lord Roberts, Commander of the Forces in Ireland, was the chief Umpire.

In June, 1897, the annual inspection of the battalion by Major-General Boyce Combe took place. At this time it was under orders for Dublin, and at the close of the inspection the General addressed the men as follows:—

"Colonel Bruce, Officers, Non-Commissioned Officers and Men:—

"I have great pleasure in making a very satisfactory report of this battalion. You have turned out this morning for my inspection very well indeed, considering the short time you have had this new equipment. You drill remarkably well, and are very steady in the ranks and under arms. I have been glad to notice a considerable decrease in the number of summary awards, which shows how well you conduct yourselves. Your signalling is, as I know, excellent, and you attained in it an exceptionally high figure of merit, and as regards musketry, though I am sorry there has been a slight falling off from the high percentage of last year, still, this can be accounted for easily by the bad weather you experienced this year as compared with last. You have given me little or no trouble since you have been under my command, and when you go I shall be sorry to lose you."

NOTE.—In 1896 the manual exercise was assimilated for all regiments. This involved the abolition of the "long shoulder."

LIEUT. GENERAL EDWARD CHIPPINDALL, C.B.

In addition to this complimentary speech, the following remarks by the Major-General on his inspection were published in regimental orders of the 12th June:—

"The Major-General Commanding the District desires to convey to the battalion his satisfaction at its general condition.

"The turn out of the whole regiment has been good.

"The men are smart, well set up, and remarkably well drilled.

"The regiment is conspicuous by its steadiness under arms.

"The conduct of the men has been very good during the last year.

"The interior economy and system existing in the regiment may be considered as especially having his satisfaction and approval.

"It is a great pleasure to the Major-General Commanding to be able to place on record so satisfactory a result of his annual inspection.

"By order,

"E. T. H. HUTTON, Colonel,

"Chief Staff Officer."

The All-Ireland "Evelyn Wood" Competition took place at the Curragh on the 16th and 17th July. The two days were exceptionally hot ones, and out of the seven teams sent in, comprising 203 men, not one fell out, and they all marched up to the firing point in capital style. The competition included a 10½-mile march, finishing up with the attack 800 yards from the targets. Of the twenty-three teams which competed the battalion took five places out of the first nine, including "G" company second and "B" company third. Still greater success awaited the regiment in this competition a few years later at Aldershot.

On the 9th August the Green Howards set out for Dublin, *viâ* Dunlavin, Ballymore Eustace and Rathcoole. The march was combined with the annual manœuvres, the battalion belonging to the force which was trying to capture Dublin. In this they were deemed by the umpires to have been unsuccessful, and the regiment, after the result was made known, marched to the Phœnix Park, where it was encamped for a few days previous to relieving the 43rd Light Infantry at Beggar's Bush Barracks.

The battalion was only permitted to have a very short stay in the Irish capital, for when the 5th Fusiliers were sent from Gibraltar to Egypt in January, 1898, it received orders to relieve them. Consequently, it embarked for Holyhead, 550 strong, on the 11th January, and arrived by train early the next morning at the Albert Docks. The embarkation on board the P. and O. s.s. "Malta" took place at once, and the ship sailed the same afternoon for the Rock, where she arrived on the 17th January.

On landing, two companies proceeded to the North Front, whilst the remainder marched to the South Barracks, the identical block of buildings in which the Green Howards had been quartered in 1749.

After a sojourn of little more than a year at the Rock the battalion was ordered home, and a farewell inspection was made by the Governor, Sir Robert Biddulph, when he complimented the men on their successes and the good reputation they had left behind them.

NOTE.—The Green Howards won the Governor's Football Cup each season during their stay at Gibraltar.

NOTE.—Whilst the battalion was at Gibraltar Spain was at war with the United States, and in February, 1899, the U.S. transport "Grant" came into harbour with the 4th and 17th infantry regiments on board on their way to the Philippines. An invitation to dine at the officers' mess at the South Barracks was accepted for February 11th, probably the first occasion on which a body of officers representative of an American regiment had dined with a British one.

On the 16th March, 1899, the Green Howards embarked on the hired transport "Nubia," 520 all told, on being relieved by the 1st Battalion Coldstream Guards, and sailed for England the same day. On arrival at Southampton the headquarters and four companies entrained for Bradford, whilst the remainder were railed to Strensall Camp, under the command of Major T. D. Kirkpatrick.

At the end of March Colonel Bruce's period of command came to a close. He had held it for six years, a longer time than any previous commanding officer since the days of Colonel Bright. He shortly afterwards was posted to the 19th Regimental District, and was the last officer of his rank to hold the appointment.

He was succeeded in the regiment by Lieutenant-Colonel Henry Bowles, who was gazetted to the Nineteenth in 1876. He had served in the Nile Expedition of 1884-1885 on the staff, and had also taken part in the Tirah campaign. He had been mentioned in despatches, and awarded a brevet for both these expeditions.

In the middle of June the headquarter companies moved to Strensall Camp, so that the whole battalion should go through their annual field firing and any drill and manœuvres that might be ordered. On the 3rd July these companies returned to Bradford.

Some time previous to the battalion leaving Gibraltar application had been made for the restoration of the grass-green facings, which, together with the number, had been so wantonly removed on the introduction of the territorial title in 1881. Thanks mainly to the late Mr. Arnold Forster, M.P. for Belfast, who worked hard to regain some of the various regiments' lost distinctions, the request was favourably considered, on the understanding that all battalions of the regiment agreed to the change, and in August, 1899, permission was accorded to resume wearing the time-honoured green facings.

In October, on the outbreak of the Boer War, a party of thirty-five non-commissioned officers and men, under the command of Lieutenant Maurice Tomlin, were selected to join the 2nd Mounted Infantry Battalion for service in South Africa.[1]

Bradford was really only a temporary station, the accommodation in the barracks there, even for a few companies, being totally inadequate. It was not long therefore before orders were received to move to Aldershot, and two companies proceeded there early in November, under command of Captain Ferrar.

On the necessity of further reinforcements being sent to South Africa, the Green Howards were at once ordered to mobilize at Aldershot, where the remainder of the battalion arrived on the 17th November.

The scene when the headquarter companies marched down to the Midland Station at Bradford was one long to be remembered. As the men neared the centre of the town the crowd increased, and it was with the greatest difficulty that the "fours" were kept intact. Forster Square was one surging

1 They took part in the fighting round Colesberg under General French, and advanced to Pretoria with Lord Roberts, being present at the battles of Paardeberg and Driefontein. They also took part in the battle of Diamond Hill, and lost heavily in the engagement at Nooigedacht on the 7th of December, 1900, under Major-General Clements.

Captain Tomlin was mentioned in despatches on the 28th July, 1901, for having in a most daring manner ridden with twenty men through a deep kloop to head off a convoy, and under heavy fire stopped it.

Lieutenant W. B. Walker, who was also attached to the 2nd M.I., was mentioned on the 8th December, 1901, he having, when in command of a section retiring before superior force at Kaffir Spruit, on the 5th October, and a sergeant's horse being killed, returned under fire and helped him along to safety.

An account of the services of the 2nd M.I. is given in "With the Green Howards in South Africa."

mass of humanity, and there must have been 50,000 people within and around the station. Only privileged persons were allowed to pass the barriers, but every available hand-cart or vehicle was used as an improvised stand. The huge crowd cheered and roared again and again, and eventually burst into "God Save the Queen" as the train steamed slowly out of the station.

The experiences of the regiment in the South African War have already been fully related in "With the Green Howards in South Africa, 1899-1902," so that in the next chapter we shall only give a brief *resumé* of its services in that country.

CHAPTER XIV.

The 19th embarks for South Africa—On arrival entrains for De Aar—Takes part in the operations near Colesberg—Engagement at New Zealand Hill—Proceeds to the Modder River—Battle of Paardeberg—Attack on Kitchener's Kop—Battle of Driefontein—Occupation of Bloemfontein—Advance to Dewetsdorp—Occupation of Kroonstadt, Johannesburg, and Pretoria—Battle of Diamond Hill—The Advance Eastwards—Engagement at Belfast—The Komati Valley—Entrains for Pretoria and Ladysmith—Elandsfontein—Home—Roll of Officers—Honours and Awards.

WITHIN a week after the order for the Green Howards to proceed to the seat of war all the reservists had joined, and on the 23rd November the regiment paraded for inspection by Major-General Kelly-Kenny, commanding the Aldershot Division, previous to its departure next day from Southampton in the s.s. "Doune Castle."

The total strength on embarkation was 20 officers and 938 non-commissioned officers and men. Of this number 450 were reservists and 117 had previously served in the Tirah campaign with the 2nd Battalion.

The following officers embarked:—

Lieut.-Colonel H. Bowles, in command.
Major J. A. Fearon, second in command.
Captain G. Christian, Adjutant.
Captain C. Organ, Quartermaster.

COMPANY OFFICERS.

"A."—Captain M. L. Ferrar, Lieutenant R. H. Darwin, Second-Lieutenant W. G. Tarbet.

"B."—Captain E. Somervell, Second-Lieutenant M. M. H. Nevile.

"C."—Major G. C. S. Handcock, Lieutenant C. F. C. Jarvis, Second-Lieutenant E. S. Broun.

"D."—Captain M. H. Orr.

"E."—Major T. D. Kirkpatrick, Second-Lieutenant T. W. Stansfeld.

"F."—Captain G. Pearson, Second-Lieutenant E. V. L. Wardle.

"G."—Captain E. M. Esson, Second-Lieutenant C. H. de St. Pierre Bunbury.

"H."—Captain B. C. W. Williams.

MEDICAL OFFICER.

Major N. C. Ferguson, R.A.M.C.

The "Doune Castle" arrived at the Cape on the 15th December, and the battalion entrained next day for De Aar Junction, where it arrived two days later.

In company with the Royal Warwickshire and the Essex, the battalion took its share of picquetting the hills in the vicinity of the camp and in setting guards over the huge accumulation of stores near the station.

During the fortnight spent at De Aar the men were exercised daily for several hours up and down the kopjes near the camp, and were getting fit and ready for the work before them.

The useless valise had been placed in store at the base, and when the battalion took the field each man carried, in addition of course to his arms and equipment, one blanket rolled on his waist belt, a full haversack, ammunition and Wallace-spade. The great coats were carried by the regimental transport, and more than half of them were lost on the march to Bloemfontein, owing to the wretched condition of the mules through want of proper rations. This was partly due to the loss of a large convoy of over 200 waggons near Klip Drift, which when proceeding in rear of the army with insufficient escort was captured by the Boers.

Afterwards, when the winter set in, an extra blanket was issued, and sometimes both were carried by the men, but this was only on rare occasions, as, for instance, when the transport might have to go a long roundabout way or some difficult drifts were anticipated. So that with two blankets, arms, equipment, full haversack, and 150 rounds of ammunition, a man carried a total weight of some 50 lbs., and this in a tropical climate with bad or no roads and a scarcity of water, made marching anything but light work.

Whilst at De Aar three officers and 130 non-commissioned officers and men were selected to form a mounted infantry company,[1] which was from time to time supplemented by drafts from the battalion during the war.

On the 3rd January, 1900, the battalion entrained for Naaupoort, where three companies, under Major Fearon, were detrained, the headquarters and three companies going on to Arundel, whilst two under Captain Ferrar proceeded to the railhead at Rensberg. The remainder of the battalion joined up at Rensberg on January the 8th and 9th. A few

---

1. This company formed part of the 4th M.I. Battalion. It was at first under the command of Major G. C. S. Handcock, with 2nd-Lieutenants W. G. Tarbet and M. M. H. Nevile as subalterns. Other officers of the 19th attached to it during the course of the war were Captain H. G. Holmes, Captain D. L. Hartley, Lieutenants T. W. Stansfeld, M. R. Liddon, H. V. Bastow, J. C. Morgan and B. H. Leatham. The battalion took part in the march from Modder River station to Komati Poort with Lord Roberts, and fought in all the principal actions. It had twenty-one officers and men killed and wounded in the 19th company.

The following officers were mentioned in despatches by Lord Roberts:—Captain H. G. Holmes, Lieutenant W. G. Tarbet (Adjutant), Lieutenant T. W. Stansfeld, Lieutenant M. M. H. Nevile. Lieutenant Nevile was also mentioned by Lord Kitchener "for skill" and enterprise in surprising a laager in Bothasberg, July 13th, 1901, when he captured eight Boers and a considerable number of stock and waggons.

Captain (local Major) H. G. Holmes was mentioned by Lord Kitchener for good service in command of detached bodies of troops in Kroonstad district (8th December, 1901).

An account of the part played by the 4th Battalion M.I. is given in "With the Green Howards in South Africa."

days later two companies, under Captain Williams, went off to Maidar's Farm, twelve miles west of the railway, and two, under Captain Ferrar, were sent to Slingersfontein, a farm some fifteen miles north-east of Rensberg, which General French made his headquarters after a few days, transferring at the same time most of his force there.

The battalion had to furnish two picquets at Slingersfontein, one of them on a high hill on the west of the camp and overlooking a chain of kopjes running still further west. The company on New Zealand Hill, as it was called, had been reduced by one-half on the morning of the 15th, when the post was hotly attacked at about 10 a.m. The Boers were keeping up a heavy fire on the sangars on the crest, whilst others were climbing the hill, tactics very similar to what took place at Majuba. Some of them had actually gained the crest, but were driven back with loss by the Green Howards, helped by some men of the New Zealand Mounted Rifles, under Captain Madocks, R.A. In this affair Colour-Sergeant Roberts, Sergeant Jamieson and three privates were killed, whilst Captain Orr and five privates were wounded.

This hill was the key of General French's position. Had it been taken it would probably have spelt disaster to our troops.

Shortly after this General French moved most of his force nearer to Colesberg, and the headquarters of the regiment and two companies marched back to Rensberg, and there joined the two at Maidar's Farm. Thence they marched in company with the 99th Wiltshire Regiment to Hobkirk's Farm, distant about twenty miles from Rensberg. Lieutenant-Colonel Stephenson, of the Essex, was in command of the infantry, and had two companies of his own regiment with him.

The Green Howards had orders to hold a spur on the left flank of the Boer position, in conjunction with the two Essex companies, whilst the Wiltshire Regiment attacked in front, the idea being to cut the Boers' communication with Colesberg Bridge. Their position, however, was found to be too strong, and the troops were ordered to withdraw to Hobkirk's Farm, as there was no water obtainable nearer.

The 99th had eleven men wounded, but there were no casualties amongst the Green Howards, who did not come under fire.

The force now returned to Rensberg, and the regiment found awaiting it there a draft of 220 non-commissioned officers and men, under command of Lieutenant C. V. Edwards. These filled up the gap caused by those who had joined the mounted infantry.

The other outlying companies had come in, and the regiment was once more together, very ragged and weather-beaten it is true, their clothes and boots almost worn out, but their spirits high, and all cheerful and ready for the ensuing campaign.

Lord Roberts had now arrived at the Cape, and was busy making his plans for an advance into the Free State, to further which the Green Howards, together with other troops, were ordered to Modder River Station, whilst Major-General Clements, with a weak brigade, remained near Colesberg to hold the enemy in check there.

The regiment left Rensberg on the 31st January, and arrived at Modder River the next day, where it encamped on the south side of the River Riet. It now formed part of the 18th Brigade, which, with the 13th Brigade, composed the infantry of the 6th Division, under command of Lieutenant-

General Kelly-Kenny, C.B. The other battalions of the 18th Brigade were—

2nd Battalion Royal Warwickshire Regiment,
1st " The Welch Regiment,
1st " The Essex Regiment,

under the command of Brigadier-General T. Stephenson.

The Royal Warwickshire did not join the brigade till after its arrival in Bloemfontein.

On the 11th February the 18th Brigade received orders to entrain for Graspan. To deceive the enemy the men left all their tents standing, and with the exception of a few days at Bloemfontein and Springfield Kop, did not have canvas again till after their arrival in the Crocodile Valley, making a period of some ten months bivouacking.

The next day the division commenced its march with the object of turning Cronje's flank, who was strongly entrenched at Magersfontein.

Most of Lord Methuen's force lay encamped on the north side of the Modder. He, as well as Cronje, the Boer leader, had been practically inactive since the battle of Magersfontein, a few weeks before. The latter had refused to leave his entrenchments, believing that the English would never forsake the railway, but when he realised that a movement was in progress he hastily quitted his position and trekked off towards Bloemfontein with the English army on his heels.

The first march of the 18th Brigade was to Ramdam, on the Free State border, about ten miles distant. This was a new experience to the men, and they found what a difference it was marching with a division compared to what they had hitherto been accustomed, with the battalion only or by companies, and it took them seven hours to do the ten miles. The next march was to Waterval Drift, on the Riet River, another ten miles. A night march from Waterval brought the force to Wegdrai, about noon on the 14th February. The distance was only nine and a half miles as the crow flies, but it took the division eleven hours to complete the march. Starting again at 5 p.m. the same day, Klip Drift, on the Modder, was not reached till long after midnight. The darkness was intense during this march, and the men blundered on across the veld, tumbling continually over the many antbear and mere-cut holes as well as the loose stones and antheaps, which are so common in South Africa. Then down came the rain, not as it does at home, but in bucketsful, and in about a minute every man was wet to the skin; the slippery, muddy veld added further difficulties to the already worn-out men, and when the bivouac was reached a miserable night was spent, as no transport, with the long-wished-for blankets, arrived till daylight.

On the 16th February the 13th Brigade engaged the Boers' rear-guard, and had 130 casualties.

The next bivouac was at Klip Kraal, where the battalion was in company with Colonel Hannay's Mounted Infantry and the guns. Escorting the latter, the Green Howards marched to Brandvallei, a short six miles distant. Here there was a well-to-do farm, and the men revelled in grapes, peaches, apricots, to say nothing of goats, pigs, chickens and ducks, all of which fell victims to the hungry soldiers.

With a few hours' rest the pursuit was resumed, and after marching till 1 a.m. on the 18th a halt was called near Paardeberg Drift, where the men simply threw themselves down almost dead beat. On the move again at 3 a.m. they struggled forward with the day rapidly dawning as they advanced. The regiment was acting as escort to the guns, and was in column of half battalions.

Suddenly on topping a rise what appeared to be a small village on the river bank burst on the view of the expectant soldiers. In reality it was the laager of Cronje's force, now brought to bay in the river bed, and the guns at once opened fire, they and the regiment being at the same time subjected to a heavy rifle fire from the Boers.

The Green Howards were at once deployed facing the river, and lay down to await orders.

In the meantime the Welch and Essex marched straight on and finally took up a position on the east side of the laager.

After some delay the battalion was ordered to advance on the river and turn the Boers out of the entrenchments they were supposed to be occupying on the near side.

The veld sloped gently down towards the Modder, its surface rough and stoney, with absolutely no cover, save perhaps an odd boulder or not-to-be-trusted antheap, and the river bank quite 1,500 yards off. The advance, therefore, under a very heavy well-directed fire was as difficult as any that could well have been planned, and the men straight away found this to be the case. The cry for stretcher bearers became a frequent one, and they as well as the ammunition carriers had not an enviable time. But all the same they did their work bravely and well, and many a man owed his life to their bandaging. Three of them were killed, Corporal Kearns, Bandsman Davis and Drummer Coombs, Davis being hit as he was attending to Major Kirkpatrick, who was dangerously wounded in the throat. It was soon quite impossible to deal with all the wounded, and the majority of them had to lie out in the South African sun all day before their wounds could be attended to, or before they could be carried to the rear.

The men's fire was directed on the bushes fringing the river, where it was thought the Boers lay hidden in their rifle pits, but so excellent was their cover that it was quite impossible to locate them.

By about 9 a.m. the advance was carried to within 400 yards of the river. Just at this time Lieutenant-Colonel Bowles, who was with the leading sections, was severely wounded in the chest, and soon afterwards an order came to stop any further advance. Previously to this, however, a party of about sixty men, headed by Captain Ferrar, Captain Esson and other officers, had rushed down to the bank only to find that there were no Boers on the near side at this part, and that the river was quite impassable. In the final rush Second-Lieutenant Neave and many of the men were killed. Providentially most of this party happened to get cover near the bank from the enemy's rifles, which were not more than thirty yards from them. It was here that Sergeant Atkinson went several times for water for an officer of the 43rd Light Infantry and some wounded men, and was at last mortally wounded himself in performing these gallant acts. His relatives were given the Victoria Cross, which would have been his reward had he lived. Private Burns, who was also down at the river bank, volunteered to cross the bullet-swept veld to stop the firing of a machine gun in rear of this party. Many other Green Howards distinguished themselves, and several well-deserved medals were won.

It was not till dusk that the regiment could with safety make its way back up the slope, where the men spent the night in groups on the top. Stretcher parties were then sent down to the river to collect any dead or wounded men that they could find. On the roll being called the next morning

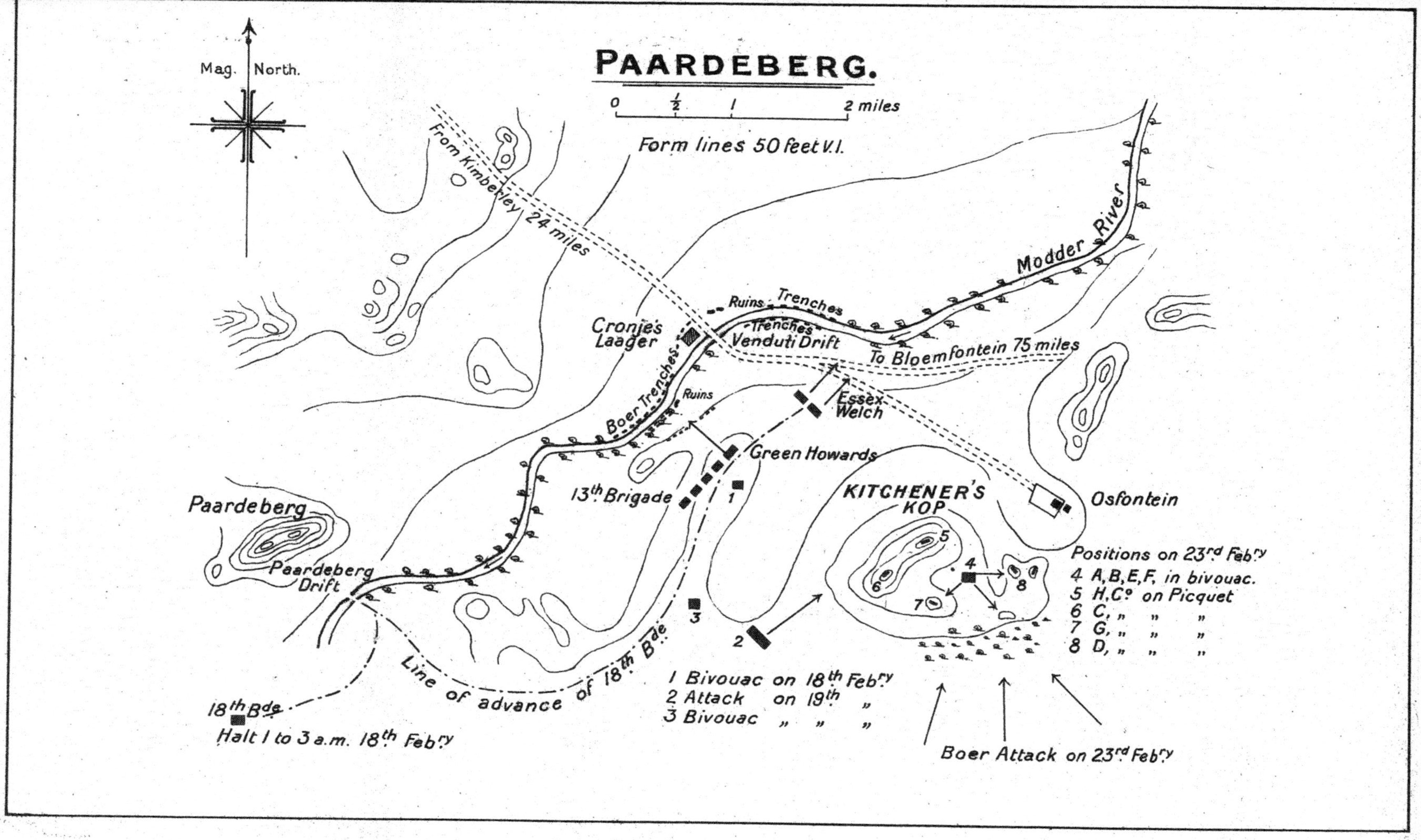
PAARDEBERG.
Mag. North.
0 ½ 1 2 miles
Form lines 50 feet V.I.
From Kimberley 24 miles
Modder River
Ruins
Trenches
Trenches
Cronje's Laager
Venduti Drift
To Bloemfontein 75 miles
Boer Trenches
Ruins
Essex
Welch
Green Howards
13th Brigade
KITCHENER'S KOP
Osfontein
Paardeberg
Paardeberg Drift
Line of advance of 18th Bde.
18th Bde.
Halt 1 to 3 a.m. 18th Febry
1 Bivouac on 18th Febry
2 Attack on 19th "
3 Bivouac " " "
Positions on 23rd Febry
4 A, B, E, F, in bivouac.
5 H, Co on Picquet
6 C, " " "
7 G, " " "
8 D, " " "
Boer Attack on 23rd Febry

the total loss of the regiment was found to be—Second-Lieutenant A. C. Neave, Colour-Sergeant Hughes, Sergeant Tate, and twenty-eight men killed. Wounded—Lieutenant-Colonel Bowles (severely), Major Kirkpatrick (dangerously), Captain Buckle[1] (dangerously), and Lieutenant C. V. Edwards (severely), besides ninety-five rank and file, ten of whom afterwards died of their wounds. Missing—three men, who were taken prisoners, and rejoined after the occupation of Pretoria.

The same day the battalion was ordered, in company with the 43rd and the 28th Gloucestershire Regiment, to attack and capture a kopje about two miles south of the laager. Just before dark the 28th obtained a footing on the eastern side, but found the enemy too strongly posted. The battalions retained their positions during the night, but were withdrawn at daybreak.

This hill, known as Kitchener's Kopje, was finally evacuated by the Boers on the 21st February, and the Green Howards were then ordered to occupy it. The kopje stands some 350 feet above the river, and about two miles from it. It has a series of under features running south for about half a mile, and on these, besides the higher ones on the north side, picquets were posted, four companies being detailed to carry this out, whilst the remainder of the battalion were in bivouac on a level space in the centre of the kopje.

Heavy rifle fire was heard on the 23rd at dawn, close to the southern defences. All the men rushed to their posts to find that the position was being strongly attacked by a reinforcement of the Boers under De Wet.

The main body, seeing that the battalion was prepared for them, galloped away to the number of 500, but their advanced

1 Captain A. C. Buckle, 38th South Staffordshire Regiment, was attached to the Battalion.

party had seized a position at the most southerly end of the kopje, from whence they kept up a well-directed fire. At about 8 a.m. another strong party of Boers were seen to retreat, and some men of "A" company rushed across the open to occupy the small under-feature evacuated by the enemy. They had no sooner reached it when they came under a hot fire, and two men were wounded.

Hearing the firing some of the Buffs had come up on the right, and "B" company, supported by "F" and "E," were ordered to turn out the enemy from their position. In doing so three officers were wounded, viz., Captain Pearson (severely), Lieutenants Gunthorpe and Wardle (slightly), whilst Sergeant Richardson and eight men were killed and seventeen rank and file wounded.

The Boers after this decided to surrender. All their horses were either killed or useless, and they gave themselves up to the number of eighty-five. Eight of their dead were found on the kopje, besides sixty dead horses, and amongst the prisoners were many wounded.

The battalion held Kitchener's Kop till the 7th March. This delay in a forward movement was necessitated by the capture of the convoy previously alluded to, by which ten days' supplies were lost. The rations in consequence were only two biscuits daily and half a pound of trek ox of the worst and toughest quality.

At this time there was a great deal of rain, and it was a common occurrence for the men's bivouacs to be rendered quite untenable during the night, which they had to spend either walking about or standing, waiting for the welcome rays of the rising sun.

On the 27th February the anniversary of Majuba, Cronje with all his force surrendered, and the 4,000 prisoners were very soon on their way to Modder River Station, *en route* to St. Helena.

On the 7th March Lord Roberts decided to continue his advance, as the Boers were reported to be occupying a range of hills in the direct line of march on Bloemfontein. In order to threaten their left flank the 6th Division moved from the bivouac at 2.30 a.m. to take part in this turning movement. The Boers, however, would not wait for it to be completed, and retreated precipitately without scarcely firing a shot. But though they did not come under fire, it was one of the hardest days the Green Howards had during the campaign, for the Modder, which meant the first drink of the day, was not reached till 4.30 p.m. Another three miles brought them to Poplar Grove, and the long-wished-for bivouac.

Bloemfontein was now the immediate objective, and the army started off again on the 8th March, and reached Wolvefontein just before dark.

The march was resumed on the 10th, and at first there was no signs of the enemy, but soon it was discovered that a series of kopjes near Driefontein Farm were strongly held. The 18th Brigade was in advance, the Essex on the left, the Welch in the centre, and the Green Howards on the right flank. The Buffs and Gloucestershire were in support, the remainder of the division forming the rear-guard with the transport.

The battalion was ordered to work wide to the right, and if possible come up on the Boer flank. But before this could be done the main position had been charged by the two leading battalions, and the Boers fled in all directions, leaving 103 dead on the field. The Green Howards in their final advance on the flank came under a heavy fire, when three men were killed and twenty-five wounded. The total losses of the division were six officers and fifty-two men killed and sixteen officers and 321 men wounded.

"C" company, under Lieutenant Jarvis, did not take part in this action, as it was left at Kitchener's Kop on picquet duty, and did not rejoin the battalion till after its arrival in Bloemfontein.

On the 11th March the advance was continued to Dam Vlei and the next day to Venter's Vlei, about twenty-six miles distant from Driefontein. Finally, on the 14th March, Bloemfontein was entered, the 18th Brigade leading, headed by the Essex, the Green Howards coming next, and then the Welch. Owing to so many casualties there were only ten officers actually with the battalion, viz.:—

Major J. A. Fearon (commanding).	Lieut. T. W. Stansfeld.
	,, R. H. Darwin.
Captain M. L. Ferrar.	2nd-Lieut. C. H. Bunbury.
,, E. M. Esson.	,, E. S. Broun.
,, B. C. Williams.	Major N. C. Ferguson (Medical Officer attached).
,, G. Christian (Adjt.).	
Captain C. Organ (Qr.-Mr.).	

On the 19th March the 6th Division was reviewed by Lord Roberts. Great praise was given to each regiment for their marching and for never losing touch with Cronje till he was safely trapped. To the Green Howards he attached great credit for the defence of Kitchener's Kop, thereby preventing Cronje from receiving his much needed reinforcements, and consequently hastening his surrender.

The battalion remained at Bloemfontein till the 4th April, when the 18th Brigade was ordered to proceed to Springfield Kop, seven miles east of the town, the increasing activity of the Boers rendering its occupation necessary.

The March from
GRASPAN TO BLOEMFONTEIN
12th Feb. to 14th Mar. 1900.

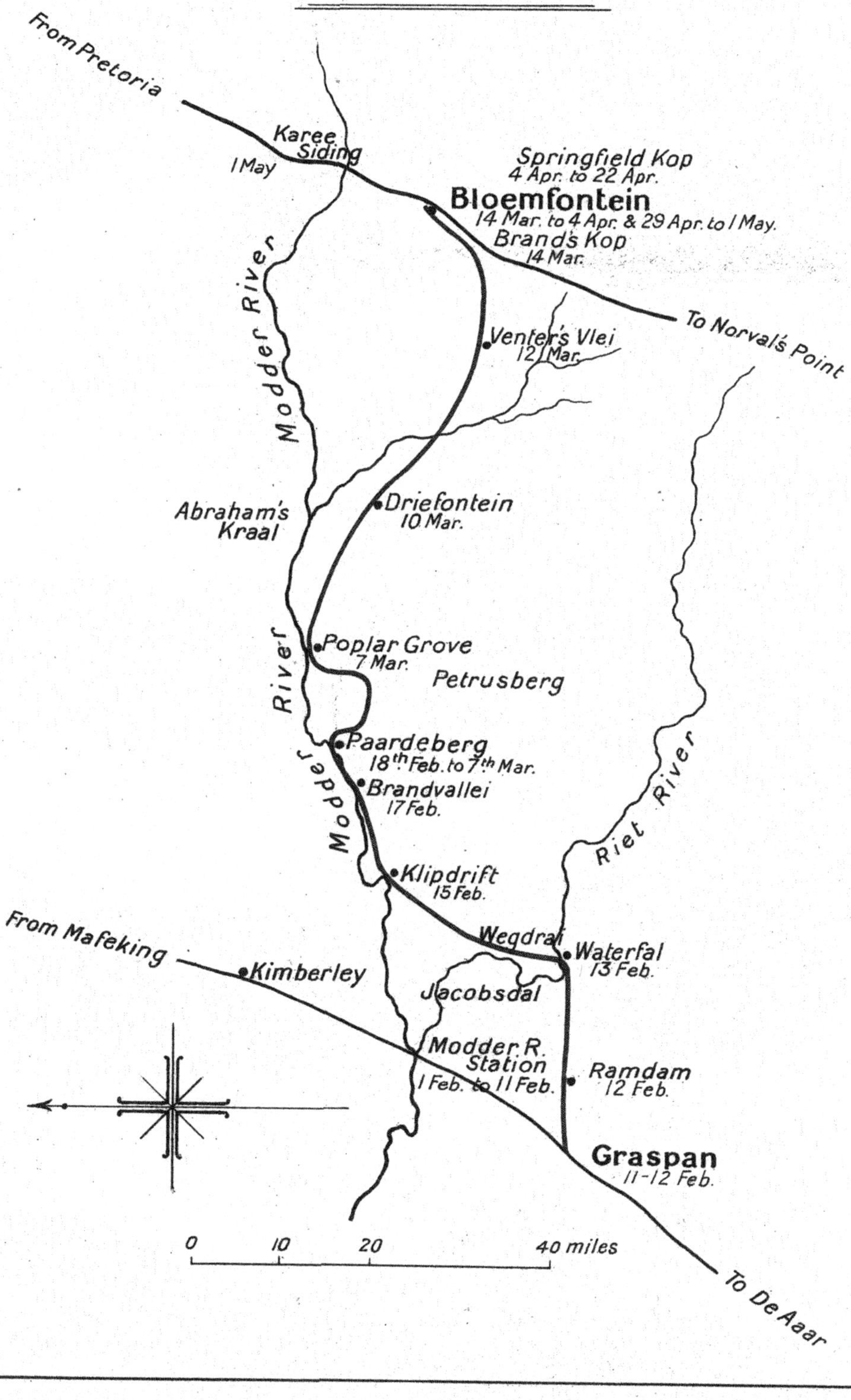

At this camp Lieutenant-Colonel Bowles rejoined, as well as many of the men who had been sick or wounded, and on the 14th April a company from the Volunteer Battalions arrived, consisting of three officers and 100 non-commissioned officers and men. It remained with the regiment till October, 1900.

On the reconstruction of the Field Force for the advance on Pretoria the 18th Brigade, in conjunction with a Brigade of Guards, became the new 11th Division, commanded by Lieutenant-General R. Pole-Carew, C.B. The Guards' Brigade was under Major-General Inigo Jones, and consisted of the—

3rd	Battalion	Grenadier Guards.	
1st	"	Coldstream Guards.	
2nd	"	"	"
1st	"	Scots Guards.	

The 18th Brigade left Springfield Kop on the 22nd April to take part with the Guards' Brigade in an expedition which had for its object the cutting off of the Boer force then threatening Dewetsdorp.

The same day there was some fighting at Leuukop, the Welch, Essex and Warwickshire having a few casualties amongst them. The latter corps had joined the 18th Brigade at Bloemfontein. On nearing Dewetsdorp communication was opened with General Rundle's column, when it was reported that the enemy had trekked off northwards. The 11th Division then returned to Bloemfontein on the 29th April, after having covered about seventy miles in its march.

Two busy days were spent in Bloemfontein drawing supplies and ammunition, besides issuing clothing, boots and other necessaries. Here the battalion was joined by ninety-nine reserve men, under Captain Orr, making up the strength to 700 effectives.

On the 1st May the general advance on Pretoria commenced, the 11th Division forming the centre column, which kept to the railway more or less all the way.

The first march was to Karee Siding, twenty-two miles distant, where 100 men, under Lieutenant Walker, were left as a guard to the camp, but at Brandfort, twelve miles further on, Captain Maitland arrived with a draft of ninety-four. On the advance being continued the battalion came under fire close to the Vet River. The Boer shells kept pitching right amongst the leading companies, but they did not burst effectively, and very little harm was done. Next morning, the 6th April, a fight was expected, but the enemy had trekked in the night, after demolishing the five-span bridge over the river. Smalldeel Junction was reached on the 7th, and Welgelegen, eighteen miles further, on the 9th. Again there was a prospect of a fight on the Zand River, where the Boers were located in force, but after hurriedly exchanging shots with the advanced troops they decamped, having previously blown up the bridge. With little or no opposition Kroonstadt was entered on the 12th, where Lord Roberts and Staff were drawn up near the market place to see the division march in. The following order was afterwards published:—

"The General Officer Commanding the 18th Brigade has been directed to convey to all ranks under his command the special gratification of the Field-Marshal Commanding-in-Chief at the splendid marching powers and plucky endurance shown by them during the past twelve days."

The division was obliged to remain at Kroonstadt for eight days, as the fine bridge over the Valsch River had been destroyed, and the various regiments were employed day and night by four-hour shifts in cutting a deviation for the line, which, in order to avoid the steepness of the banks, had to make a considerable detour. There was a good moon, and the work progressed by night almost as fast as it did by day.

Starting again on the 22nd May the division reached Honingspruit, twenty miles on, and the Rhenoster River the day following. Then came Vredefort, fourteen miles, Grootvlei another fourteen, and Taaibosch Spruit, the last bivouac in the Free State, six miles from the Vaal.

The Transvaal was entered on May 27th, and the Green Howards, with their brigade, bivouacked at Vereeniging, a little township consisting of a collection of corrugated iron huts, interspersed with coal shafts. The next halt was at Klip River, a long and weary march of twenty-two miles over the burnt veld.

The weather was getting very cold now, and the rations very short. There was 16° of frost in the early morning, and sometimes the allowance of flour was only a quarter of a pound per man instead of the five biscuits which they had been accustomed to since leaving Bloemfontein.

Johannesburg was seen in the distance on the morning of the 29th, and the division halted for the night at Elandsfontein, one of the suburbs. Two days later the troops marched into the town, which had surrendered peaceably, and the Union Jack was hoisted with some ceremony in one of the squares.

The 11th Division bivouacked at Orange Grove till the 3rd June, when it marched to Reitfontein, eight miles on, and the next day was at Six-mile-spruit, just outside Pretoria. Here there was some show of resistance, but there were very few casualties in the 18th Brigade, and only one man wounded in the battalion.

On the 5th June there was a formal parade through Pretoria, the Green Howards marching at the head of the troops in column of sections. Passing through the Grand Square, where Lord Roberts and Staff were drawn up, they reached their bivouac, two miles on the west of the town, and an historic march had come to an end.

On the 7th June the 11th Division marched out to Silverton, a little station ten miles off on the Delagoa Bay line. Four days later it moved off five miles to the south-east to take part in the battle of Diamond Hill, in which the 18th Brigade held the centre of the position, but was not actively engaged.

The brigade returned to Pretoria on the 15th June, and a week later the battalion, in company with the Royal Warwickshire, marched off to Frankpoort to picquet the hills there, whilst the Welch and Essex halted at Eerste Fabriken.

The battalion remained a month at Frankpoort, or Edendale, as it was sometimes called, where the men were issued with "British warm" coats, many of which were badly wanted, especially by those who had lost their great coats between Paardeberg and Bloemfontein owing to the break-down in the transport.

The welcome order for a further advance came at last, and on the 23rd July the division reached Eland's River Station, and the famous Bronkhorst Spruit the day following. The Welch were left here to make a deviation, whilst the remainder of the brigade marched on to the Wilge River. The bridge was again found to be destroyed, and the Essex remained to cut the deviation whilst the other two battalions continued their march to Balmoral, where they arrived on the 27th July.

A stay of three weeks was made at Balmoral, the place was put in a state of defence, and all commanding positions close to the camp picquetted. By the 15th August all the deviations in rear had been completed, and another forward movement took place. The Green Howards were then ordered to Brugspruit to relieve the Grenadier Guards, who had left for Middleburg.

On the 18th four companies, under Major Fearon, were ordered to Middleburg, and were followed the next day as far as Oliphant's River by the remainder of the battalion, on its relief by the Buffs. The rest of the 18th Brigade was following in rear, being relieved by other corps as it advanced along the line. Wonderfontein was reached by the 23rd, where the division was assembling prior to an advance on Belfast, as the Boers were said to be in position there.

About seven miles outside the town the battalion was detached to hold a hill opposite the extreme right of the enemy. The position was attacked the next day, the 24th August, the 11th Division being near the town, whilst General French made a turning movement on the left flank, and General Buller attacked the enemy's left and centre. Three days' fighting ensued, and on the 26th General Buller sucessfully pressed his attack, with the result that the Boers fled eastwards, leaving behind them a number of dead and wounded. The loss of the 11th Division was trifling, being only thirty-eight killed and wounded, and of these only seven belonged to the 18th Brigade.

On the 28th August the brigade left Belfast for Swartz Kopje, about ten miles north amongst the mountains, and then turned south to Helvetia. A few days were spent here, and on the 6th September the headquarters and five companies of the regiment were sent to Waterval-onder. In company with the Essex the battalion marched to Nooigedacht on the 12th September, and the next day reached the Godwaan River, three more marches over the De Kaap Hills bringing them to Nelspruit on the railway.

The Guards' Brigade was ordered to Avoca, and eventually reached Komati Poort, *viâ* Kaapmuiden, a few days later. Kruger had arrived at Lorenzo Marques on the 11th September, after resigning the so-called reins of Government to Schalk Burgher. General French had entered Barberton on the 13th September, and the war was supposed to be finished.

The march into Nelspruit was the last one the Green Howards made in South Africa as a battalion. From the start they had covered about 1,000 miles, and had now to settle down to eighteen weary months of blockhouse and picquet duty.

On the 29th September four companies, under Major Fearon, were sent by rail to Kaapmuiden to assist in cutting the deviation over the Kaap River, the three-span bridge there having been blown up by dynamite.

The departure of the Essex Regiment early in October, who were under orders for India, necessitated a change in the various garrisons, and the Green Howards were ordered to hold the line from Belmont Bridge to Crocodile Poort, headquarters as before remaining at Nelspruit.

Another change took place on the 23rd November, the headquarters and four companies going to Komati Poort, one to Hector Spruit, whilst the remainder, under Major Cotesworth, occupied Kaapmuiden. Shortly after their arrival at Komati the companies there went into huts on the Lebombo Hills, close to the Portuguese frontier, whilst at

Kaapmuiden tents were drawn from store and the long spell of bivouacking came to an end.

Barberton was the next move, where the whole of the regiment was assembled on the 17th January, 1901, the first time the companies were together since September.

Owing to the bad effects of malaria in the low veld the regiment was simply a wreck of its former self, for no less than ten officers and 300 men were in hospital.

On the 10th April an officer and twenty men were sent to Avoca, and two companies to Kaapmuiden under Captain Ferrar.

After being quartered for six months at Barberton the battalion was again moved to Komati Poort and Kaapmuiden, where it arrived between the 9th and 12th July. Five companies under Captain Orr were at Kaapmuiden, and the other three with headquarters on Lebombo Heights. Thirty-eight blockhouses had been built between Komati and Kaapmuiden. These were occupied by a non-commissioned officer and four men, so that 190 men were required for this duty, their relief taking place once a week.

On the 22nd July 400 Boers attacked and captured Bremersdorp, which was held by forty of Steinacker's Horse. Parties of the battalion were ordered into Swaziland to cover important points in the Boers' line of advance and one half-company made a very fine march of forty-seven miles in thirty-six hours with the temperature at 100° in the shade. The posts were, however, all withdrawn after a few days, as the Boers retired westwards.

On the 11th August four companies, under Captain Ferrar, left by rail for Pretoria, being joined a week later by the rest of the regiment. Here it took over the eastern defences of the town, having posts at Koodoespoort, East Fort, Silverton and Johnston's Redoubt, whilst one company under Captain Somervell marched to Commando Nek, in the Magaliesberg Range, about thirty miles distant.

Shortly after the arrival of the regiment in Pretoria the following farewell letter was received from Major-General Stephenson, commanding at Barberton:—

"General Stephenson cannot allow the 1st Battalion The Princess of Wales's Own Yorkshire Regiment to leave his command without expressing his very hearty appreciation of their splendid services throughout the campaign. The 1st Yorkshire Regiment joined the 18th Brigade on its formation at Modder River on the 6th February, 1900, and from the first showed fine endurance and the highest soldierly qualities.

"Whether on the march, in action, or on outpost duty, General Stephenson has always had the most perfect confidence in this grand old regiment, and he has never been disappointed. He knows well from long experience that any service that the 1st Yorkshire may have the honour of rendering for the British Empire in the future will be nobly and gallantly performed. It is with the keenest personal regret that General Stephenson has now to bid good-bye to his well-tried comrades of the last battalion to leave his command of the old 18th Brigade.

"May the highest honour and success attend them, one and all, wherever they go.

"T. E. STEPHENSON, Major-General,
"Commanding 18th Brigade."

The regiment only remained three weeks in Pretoria, for when the Natal border was threatened in September by Botha, the Green Howards were amongst the troops that were hurried down to guard the colony. Ladysmith was its

COLONEL HENRY BOWLES, C. B.

destination, and here the men were employed in building up the old defences, and outpost duty on the north side of the town. Posts were established as far as Colenso on the Durban line and at Bester's on that of the Orange River Colony. Besides these, 100 men, under Captain Esson, proceeded to Van Reenen's Pass on the 7th October to act as a covering party during the erection of some blockhouses in the Drakensberg.

Consequent on the defeat of the Boers at Itala and Prospect the battalion was ordered to Elandsfontein, where it arrived on the 23rd October, and took over the outposts there from the Yorkshire Light Infantry. Two days later detachments were sent to Boksburg and Springs, under Captain Esson and Captain Ferrar respectively. Another detachment was furnished in January at Vereeniging, consisting of 100 men, under Captain Orr. Other parties were at Viljoen's Drift and Klip River Station. The headquarters remained at Elandsfontein, and the above-named posts were held by the battalion up to the end of the war.

At this period orders were received changing the name of the regiment to—

Alexandra, Princess of Wales's Own
(Yorkshire Regiment),

in order that it might still further maintain its connection with H.M. Queen Alexandra, who had first identified herself with it in 1875.

On the 4th September, 1902, the Green Howards entrained for Cape Town, where they arrived on the 8th. They sailed in the "Lake Michigan" in company with the 3rd Battalion Grenadier Guards and the 1st Battalion Duke of Wellington's Regiment on the 12th September, and arrived at Southampton on the 6th October.

AUTHORIZATION OF BATTLE HONOURS.

His Majesty the King was graciously pleased to approve of the line battalions of the regiment bearing on their Colours and appointments the following honours for service in the South African War:—

SOUTH AFRICA, 1899-1902.
RELIEF OF KIMBERLEY.
PAARDEBERG.

HONOURS AND PROMOTIONS FOR THE WAR IN SOUTH AFRICA.

Non-commissioned Officers and Men.

THE VICTORIA CROSS.

Sergeant A. Atkinson.

During the battle of Paardeberg, February 18th, 1900, Sergeant A. Atkinson, 1st Battalion Yorkshire Regiment, went out seven times under heavy and close fire to obtain water for the wounded. At the seventh attempt he was wounded in the head, and died a few days afterwards.

AWARDED THE DISTINGUISHED CONDUCT MEDAL.

Sergt.-Major J. Walker.
Col.-Sergt. B. Williams.
Sergt. A. Couldrey (M.I.).
Corpl. G. F. Collings.
Lce.-Corpl. H. Beecroft (Vol. M.I.).
,, W. Philpot.
Private J. Hayes.
,, W. Usher.
Col.-Sergt. E. Pickard.
Col.-Sergt. H. Parkinson (M.I.).
Sergt. H. Wilson (M.I.).
Lce.-Corpl. J. M. Hyland (M.I.).
Private J. Edmonds (M.I.).
Lce.-Corpl. G. T. Atkinson (M.I.).
Private W. Pearson.
,, E. Carss (M.I.).
,, F. Eyre.

Sergt.-Major G. J. Smith, Qr.-Mr.-Sergt. J. Lowther — For service with the Militia (3rd Battalion).

NOTE.—The regiment received a war gratuity similar to that given for the Nile Expedition. Lieutenant-Colonel £120, Major £80, Captain £60, Lieutenant £37 10s., Sergeant £12, Private £5.

MENTIONED IN DESPATCHES.

Sergt.-Major A. J. Hughes.
Col.-Sergt. J. Walker (2).
,, W. Bryant.
,, E. Pickard.
,, B. Williams.
Sergt. H. S. Mackay.
,, B. Richardson.
Corpl. G. F. Collings.
Private J. Williams.
Col.-Sergt. W. Cushen.
Sergt. H. Wilson.
,, A. Couldrey.
Lce.-Corpl. J. M. Hyland.
,, G. T. Atkinson.
,, H. Beecroft (Vol.).
Sergt.-Major G. J. Smith (3rd Battalion).
Qr.-Mr.-Sergt. J. Lowther (2) (3rd Battalion).
Sergt.-Inst.-M. W. Moy (3rd Battalion).
Corpl. A. Hatton.
Lce.-Corpl. A. Philpot.
Private F. Eyre.
,, W. Burns (2).
,, J. Hayes.
,, W. Hewitt.
,, W. Usher.
,, W. Pearson.
,, T. Dunn.
,, F. Wright.
,, F. Bushby.
,, W. H. Hodge (2).
,, H. Edmonds.
,, T. W. Schofield.
,, E. Carss.
Col.-Sergt. P. Hall (3rd Battalion).
Sergt. H. Rushton (3rd Battalion).
Col.-Sergt. H. Parkinson.

RETURN OF CASUALTIES.

	Officers.	Sergeants.	Corporals.	Drummers.	Privates.	Total.	Remarks.
Killed in action or died of wounds	1	6	2	1	51*	61	*6 men 3rd or 4th Batt. attached to 1st.
Wounded	11	3	7	1	142*	164	*12 men 3rd or 4th Batt. attached to 1st.
Died from other causes	—	3	8	1	79*	91	*7 men 3rd or 4th Batt. attached to 1st.

The names of all the above are given in the South African War Record.

ROLL OF OFFICERS

Who served with the 1st Battalion during the War 1899—1902.

Rank and Names.	Queen's Medal and Clasps. Relief of Kimberley.	Paardeberg.	Driefontein.	Johannesburg.	Diamond Hill.	Belfast.	Cape Colony.	Orange Free State.	Transvaal.	South Africa 1901.	South Africa 1902.	King's Medal 2 Clasps.	Remarks.
Lieutenant-Colonel Henry Bowles, C.B.	1	1	—	1	1	1	—	—	—	—	—	1	In command of battalion during the war. Twice mentioned in despatches. Severely wounded at Paardeberg. Awarded the C.B. Retired on half-pay as Colonel 29 March, 1903.
Major and Bt.-Lieutenant-Colonel James Ahmuty Fearon...	1	1	1	1	1	1	—	—	—	—	—	1	Commandant Wynberg, also in command of Depôt battalion at Cape Town. Mentioned in despatches. Brevet of Lieut.-Colonel. Retired as Colonel 29 September, 1906.
Major Gerald Carlile Stratford Handcock	1	1	1	1	1	1	—	—	—	—	—	1	In command of 4th M.I. Afterwards in command of M.I. Depôt. Retired 10 February, 1905.
Major Thomas David Kirkpatrick	1	1	—	—	—	—	—	—	—	—	—	—	Severely wounded at Paardeberg. Invalided. Retired 25 December, 1905.
Major James Trevelyan Cotesworth	—	—	—	—	—	1	1	1	—	—	—	1	Joined at Edendale. Commandant Kaapmuiden. Retired 21 January, 1906.
Major Michael Harrison Orr, D.S.O.	—	—	—	1	1	1	1	1	—	—	—	1	Severely wounded at Slingersfontein, Cape Colony. Awarded the D.S.O. Twice mentioned in despatches. Retired 23 December, 1907.
Captain and Bt.-Major Michael Lloyd Ferrar	1	1	1	1	1	1	—	—	—	—	—	1	Commandant Springs, Transvaal, October, 1901, to end of war. Twice mentioned in despatches. Brevet of Major. Retired 18 February, 1905.
Captain George Pearson	1	1	—	—	—	—	—	—	—	—	—	—	Severely wounded at Kitchener's Kop. Mentioned in despatches. Retired 29 September, 1901. Captain of Invalids, Chelsea Hospital.
Captain Edward Malcolm Esson	1	1	1	1	1	1	—	—	—	—	—	1	Mentioned in despatches. Station Staff Officer, Komati Poort. Retired as Lieut.-Colonel 28 March, 1911.
Captain Ernest Somervell	—	—	—	—	—	—	1	1	1	—	—	1	Assistant Railway Staff Officer, Cape Town. Retired 19 May, 1905.

Captain and Bt.-Major Hardress Gilbert Holmes ...	—	—	—	1	1	1	1	1	—	—	—	1	Joined at Bloemfontein. Served with 4th M.I. In command of 15th M.I., and also of a mobile column. Twice mentioned in despatches. Brevet of Major. Retired 4 February, 1908.
Captain Berkeley Cole Wilmot Williams	1	1	1	1	1	1	—	—	—	—	—	1	Retired as Captain 22 November, 1902.
Captain Gerard Christian, D.S.O.	1	1	1	—	—	1	—	—	—	—	—	1	Adjutant 16 Dec., 1899 to 29 Nov., 1900. Served as Staff Officer Town Guards, Cape Town. Mentioned in despatches. Awarded the D.S.O. Promoted Lieut.-Colonel 28 March, 1911.
Captain Harold Futfoye Lea	—	—	—	—	—	—	1	1	1	1	1	—	Joined at Komati Poort. Station Staff Officer Elandsfontein. Afterwards on Staff as Asst.-Provost Marshal. Retired 28 October, 1908.
Captain Matthew Robert Liddon	—	—	—	1	1	1	1	1	—	—	—	1	Joined at Springfield Kop. Served with 4th M.I. Severely wounded in the Origstad Valley. Retired on half-pay 28 November, 1903.
Lieutenant Christopher Vaughan Edwards	1	1	—	—	—	—	—	—	—	—	—	—	Joined at Rensberg. Severely wounded at Paardeberg. Mentioned in despatches. Major 28 March, 1911.
Lieutenant Charles Francis Cracroft Jarvis	1	1	—	1	1	1	—	—	—	—	1	—	Invalided. Mentioned in despatches. Retired as Captain 12 May, 1909.
Lieutenant Maunton Edward Thomas Gunthorpe	1	1	—	1	1	1	—	—	—	—	—	1	Joined at de Aar. Present at Stormberg with Royal Irish Rifles. Served as Transport Officer and Adjutant to the battalion. Slightly wounded at Kitchener's Kop. Died at Rosieres, Sudan, of black water fever 7 Sept., 1907, when serving with the Egyptian Army.
Lieutenant Wilfred Beckett Walker	—	—	—	—	1	1	1	1	—	—	—	1	Joined at Bloemfontein. Served with 2nd M.I. Mentioned in despatches. Captain 25 June, 1904.
Lieutenant Robert Henry Darwin...	1	1	1	1	1	1	—	—	—	—	—	1	Regimental Transport Officer. Employed with Army Service Corps. Staff Officer for prisoners of war, Cape Town. Captain 1 November, 1904.
Lieutenant Ernest Vivian Livesey Wardle, D.S.O.	1	1	—	1	1	1	—	—	—	—	—	1	Slightly wounded at Kitchener's Kop. Awarded the D.S.O. Mentioned in despatches. Resigned 25 August, 1909.
Lieutenant Thomas Wolryche Stansfeld, D.S.O.	1	1	1	1	1	1	—	—	—	—	—	1	Acted as A.D.C. to Br.-General Stephenson at Paardeberg. Served with the 4th M.I. Mentioned in despatches. Awarded the D.S.O. Promoted to Royal Warwickshire Regiment 19 February, 1902. Transferred to 19th Foot January, 1908.
Lieutenant William Godfrey Tarbet, D.S.O.	1	1	1	—	—	1	—	—	—	—	—	1	Served as Adjutant to 4th M.I. Slightly wounded at Klipdrift and Belfast. Mentioned in despatches. Awarded the D.S.O. Promoted Captain Worcestershire Regiment 13 April, 1904. Died at Jhansi 20 April, 1911.
Lieutenant Malhert Martin Huyshe Nevile	1	1	1	1	1	1	—	—	—	—	—	1	Served with the 4th M.I. Twice mentioned in despatches. Promoted Captain 5th Fusiliers 19 April, 1902.

Rank and Names.	Queen's Medal and Clasps. Relief of Kimberley.	Paardeberg.	Driefontein.	Johannesburg.	Diamond Hill.	Belfast.	Cape Colony.	Orange Free State.	Transvaal.	South Africa 1901.	South Africa 1902.	King's Medal 2 Clasps.	Remarks.
Lieutenant Charles Hamilton de St. Pierre Bunbury ...	1	1	1	1	1	1	—	—	—	—	—	1	Employed with Army Service Corps, 1901. Captain 25 January, 1905.
Lieutenant Ernest Scott Broun	1	1	1	1	1	1	—	—	—	—	—	1	Promoted Captain 6 April, 1906.
Second-Lieutenant Arthur Cormack Neave	1	1	—	—	—	—	—	—	—	—	—	—	Joined at de Aar from New Zealand Mounted Rifles. Killed at Paardeberg 18 February, 1900.
Lieutenant Harold Vermuden Bastow	—	—	—	1	1	1	1	1	—	—	—	1	Joined at Bloemfontein. Employed with Remount Depôt. Afterwards in command of Convalescent Depôt, Germiston. Served with 4th M.I. Promoted Captain 4 March, 1908.
Lieutenant John Cecil Morgan	—	—	—	1	1	1	1	1	—	—	—	1	Joined at Brandfort. Railway Staff Officer, Nelspruit. Garrison Adjutant, Barberton. Served with 4th M.I. Resigned 8 September, 1909.
Lieutenant Francis Grey Oke Sanderson	—	—	—	—	1	1	1	1	—	1	—	—	Joined at Pretoria. Invalided. To Indian Army 10 Oct., 1903.
Lieutenant Charles Howard Marsden	—	—	—	—	—	—	1	1	1	1	1	—	Joined at Barberton. Captain 10 October, 1908.
Lieutenant Hugh William McCall	—	—	—	—	—	—	1	1	1	1	1	—	Joined at Barberton. Captain 29 June, 1906.
Lieutenant Edward George Clarkson Bagshawe	—	—	—	—	—	1	1	1	—	—	—	—	Joined at Wonderfontein. Resigned 6 March, 1907.
Lieutenant Bertram Henry Leatham	—	—	—	—	—	—	1	1	1	1	1	—	Joined at Barberton. Served with 4th M.I. Captain 21 July, 1906. Adjutant 1st Battalion 16 January, 1908—16 January, 1911.
Second-Lieutenant Brerton Charles Dalton Nash-Wortham	—	—	—	—	—	—	1	1	1	—	—	1	Joined at Barberton from 3rd Battalion. Station Staff Officer, Elandsfontein. Adjutant 16 January, 1905—1908. Captain 20 March, 1909.
Second-Lieutenant Seton Dunham Massy	—	—	—	—	—	—	1	1	1	1	1	—	Joined at Barberton. To Indian Army 5 September, 1903.
Second-Lieutenant Richard Fitzroy Howard	—	—	—	—	—	—	1	1	1	—	1	—	Joined at Elandsfontein. Resigned 21 November, 1903.
Second-Lieutenant Claud Gifford Jeffery	—	—	—	1	1	1	1	1	—	1	—	—	Promoted from ranks of special Service Company 14 Sept., 1901. Captain 25 August, 1909. Adjutant 2nd Battalion 1 March, 1910.
Second-Lieutenant Bertram Lionel Maddison	—	—	—	—	—	—	1	1	1	—	1	—	Joined at Elandsfontein. Captain 29 September, 1910.
Second-Lieutenant Stuart Lumley Whatford	—	—	—	—	—	—	1	1	1	1	1	—	Joined at Elandsfontein from 4th Battalion. Served with the M.I. Captain 9 October, 1910.
Second-Lieutenant Joseph Harold Stops Westley	—	—	—	—	—	—	1	1	1	—	1	—	Joined at Elandsfontein. Captain 20 January, 1911.

Second-Lieutenant Clarence Edward Moss-Blundell ...	—	—	—	—	—	—	1	1	1	—	1	—	Joined at Elandsfontein. Resigned 11 November, 1903.
Second-Lieutenant George Prendergast Stevens	—	—	—	—	—	—	1	1	1	—	1	—	Joined at Elandsfontein. Resigned 3 February, 1904.
Second-Lieutenant Archibald Frederick Bone	—	—	—	—	—	—	1	1	1	—	1	—	Joined at Elandsfontein. Served with Imperial Yeomanry and 4th Battalion. To Indian Army 1 December, 1903.
Hon. Major and Quartermaster Charles Organ	1	1	1	1	1	1	—	—	—	—	—	1	Quartermaster to 1st Battalion throughout the war. Mentioned in despatches. Granted hon. rank of Major. Retired 13 Oct., 1908. Quartermaster Chelsea Hospital.
Hon. Lieutenant and Quartermaster Arthur John Hughes	1	1	1	—	1	1	—	—	—	—	—	1	Sergeant-Major 1st Battalion. Promoted Quartermaster 3rd Battalion. Mentioned in despatches. Died at Marrow-in-Lincoln 9 February, 1909.

OFFICERS ATTACHED TO THE BATTALION.

Rank and Names.	Queen's Medal and Clasps. Relief of Kimberley.	Paardeberg.	Driefontein.	Johannesburg.	Diamond Hill.	Belfast.	Cape Colony.	Orange Free State.	Transvaal.	South Africa 1901.	South Africa 1902.	King's Medal 2 Clasps.	Remarks.
Captain H. R. S. Maitland... ... ... (late 19th Foot Reserve of Officers)	—	—	—	—	—	1	1	1	—	—	—	1	Joined at Brandfort. Granted rank of Major in Reserve of Officers. Died at Bath 16 May, 1911.
Captain M. H. L. Bell	—	—	—	1	1	1	1	1	—	—	—	—	
Lieutenant Houseman	—	—	—	1	1	1	1	1	—	—	—	—	Joined at Springfield Kop with 1st section of Volunteer Company.
Lieutenant Ewart	—	—	—	1	—	1	1	1	—	—	—	—	
Lieutenant H. C. Matthews	—	—	—	—	—	—	1	1	1	1	1	—	Joined at Kaapmuiden with 2nd section of the Volunteer Service Company.
Captain A. C. Buckle...	1	1	—	—	—	—	—	—	—	—	—	—	South Staffordshire Regiment. Joined at de Aar. Dangerously wounded at Paardeberg.
Lieutenant A. P. Strange	—	—	—	—	—	—	1	1	1	—	1	—	3rd Battalion Royal Berkshire Regiment. Joined at Elandsfontein.
Second-Lieutenant C. Newington	—	—	—	—	—	—	1	1	1	1	1	—	3rd Battalion Royal Lancaster Regiment. Joined at Elandsfontein.
Lieutenant W. A. S. de Gale	—	—	—	—	—	—	1	1	1	—	1	—	3rd Battalion Northamptonshire Regiment. Joined at Elandsfontein.

OFFICERS OF THE REGIMENT

Who served in South Africa, not included in the foregoing Rolls.

Rank and Names.	Remarks.
Captain Arthur Lenox Napier	Brought a draft of 150 men from India to Elandsfontein and returned with a similar number. Queen's medal and 2 clasps. Retired 2 March, 1904.
Captain and Bt.-Major Charles Arthur Cecil King	Served with 3rd Battalion as Adjutant. Twice mentioned in despatches. Brevet of Major. Queen's medal and 2 clasps. King's medal and 2 clasps. Promoted Lieutenant-Colonel 29 September, 1910.
Captain Desmond Lambert Hartley	Served with 4th M.I. Queen's medal and 5 clasps. Retired 6 July, 1904.
Captain and Bt.-Lieut.-Colonel Edward Stanislaus Bulfin	On Staff. Present at engagements of Belmont, Enslin, Modder River and Magersfontein. In command of a mobile column. Three times mentioned in despatches. Brevets of Major and Lieutenant-Colonel. Queen's medal with 4 clasps and King's medal. Manchester Regiment 28 November, 1903.
Captain and Bt.-Major Maurice Hilliard Tomlin	Served with the 2nd M.I., afterwards on the Staff. In command of 9th M.I. and also of a mobile column. Mentioned in despatches. Brevet of Major. Queen's medal with 3 clasps and King's medal. Retired 3 July, 1907.
Captain Harry Elliott Raymond	Asst.-Provost-Marshal to 9th Division. Queen's medal and 3 clasps. Major 3 July, 1907. Retired on half-pay 11 February, 1911.
Captain Arthur Francis Owen-Lewis, D.S.O.	Served as Adjutant to 6th Battalion Lancashire Fusiliers. District Commissioner in Cape Colony. Mentioned in despatches. Awarded the D.S.O. Queen's medal and 3 clasps. Retired 25 June, 1904. Governor of Mountjoy Prison, Dublin.
Captain Ernest Henry Lee Warner	Employed with Railway Pioneer Regiment. Queen's medal and 3 clasps. King's medal. Retired 10 October, 1908.
Lieutenant Ernest Gregorie Caffin	Attached to Devonshire Regiment in Natal. Present at Elandslaagte, Rietfontein, Lombard's Kop, and Defence of Ladysmith (severely wounded). Promoted Captain 5th Fusiliers 9 May, 1900, and Major 19th Foot 23 Dec., 1907. Queen's medal and 2 clasps.
Lieutenant Charles John Herbert Hay Noble	Employed as Transport Officer in Ladysmith. Promoted to Manchester Regiment 9 June, 1900. Killed near Bethlehem, O.R.C., 12 November, 1901.
Lieutenant Atwell Charles Dennet Pearson	At Cape Town. Queen's medal and 2 clasps. Resigned 27 April, 1904.
Second-Lieutenant George Clarke Denton	Served with 3rd Battalion. Queen's medal and 3 clasps. To Indian Army 25 June, 1904.
Second-Lieutenant Charles James Hookam Gardner	Promoted from ranks of 1st Royal Dragoons to 2nd Battalion. Served at Relief of Ladysmith, including operations on Tugela Heights. Action at Laing's Nek. Queen's medal and 6 clasps. King's medal and 2 clasps. Promoted Captain 22 July, 1909.
Second-Lieutenant Herbert James Kirkpatrick	Transferred from Durham L.I. Militia to 2nd Battalion. Queen's medal and clasp. Died at Benares 16 April, 1902.
Hon. Captain and Quartermaster George Croft	3rd Battalion. Queen's medal and 3 clasps. Mentioned in despatches. Retired 25 September, 1901.
Hon. Lieutenant and Quartermaster John Sheridan	4th Battalion. Queen's medal and 2 clasps. Promoted hon. Captain 13 December, 1909.

CHAPTER XV.

Sheffield—Presentation of South African Medals—Notes on Dress—Unveiling of South African War Memorial—Aldershot—The Evelyn Wood Competition—Farewell Order of Colonel Egerton—New Colours presented by Queen Alexandra—Egypt—New Badge adopted.

ON the afternoon of their arrival at Southampton the Green Howards entrained for Sheffield, and on arrival marched to Hillsborough Barracks.

Lieutenant-General Chippindall, C.B., died on the 12th September at Quendon Court, Newport, Essex, and the Colonelcy of the regiment was now conferred on Major-General William Spencer Cooper, who had formerly commanded the battalion.

The new Colonel was born at Greenock on the 22nd May, 1827, and his first commission was an Ensigncy in the 3rd West India Regiment in 1843. Two years later he was transferred as Lieutenant to the 70th Foot, in which he was promoted Lieutenant-Colonel in 1867. He exchanged to the Nineteenth in 1872, and was promoted Major-General ten years later.

He served throughout the campaign of 1852-53 in Burmah with the 18th Foot, including the capture of Rangoon, Prome and other minor affairs (medal and clasp). Also in the Indian Mutiny, 1857-58 (medal), and acted as Asst.-Quartermaster-General to the expeditionary force on the Eusufzai frontier, under Sir Stapleton Cotton, in 1858, and in the same capacity with the expeditionary force in the Waziri country under Brigadier-General Chamberlain in December, 1859 (clasp).

On the 27th February, the anniversary of the Paardeberg surrender, the South African medals were presented by

Colonel E. A. Bruce, commanding the 19th Regimental District. All officers, non-commissioned officers and men who served with the battalion from the start from Modder River Station on the 11th February, 1900, up to the engagement at Belfast in September of the same year, qualified for the Queen's medal and six clasps, of which we give an illustration. Altogether there were nine officers and about three hundred men. The King's medal with two clasps was only gained by those who were serving in South Africa in 1902, and who had eighteen months' previous service in the country.

The presentation took place in the Artillery Drill Hall, Sheffield, and was attended by representatives of the various corps in the district, as well as the guests of the officers, non-commissioned officers and men of the battalion.

For some years past the ordinary working dress of the soldier abroad had been of a khaki colour, the material being either of cotton drill or serge cloth. After the Boer War "khaki" became the universal dress of all arms of the service, the tunic being retained only for ceremonial and walking out purposes. A frock coat very similar to that worn in the sixties was reintroduced for all officers, the crimson silk sash being worn round the waist of both it and the tunic. The union locket with regimental badge, which had been in use since the Crimean War, was consequently discarded, and the sword was suspended by a web belt worn underneath the sash. When the service dress was worn the sword was suspended from a "Sam Browne" belt.

The officers' mess dress was of the usual pattern, green cloth forming the rolled collar, cuffs and shoulder straps, the waistcoat being green, with a specially approved pattern of Russian braiding running down the front and sides.

MEDALS & CLASPS AWARDED FOR THE SOUTH AFRICAN WAR.

The men were provided with bandoliers in which to carry their ammunition instead of the pouches, which had proved so unsatisfactory during the South African campaign.

With the introduction of khaki, leggings were no longer worn, their place being taken by putties, which were wound round the leg below the knee. A khaki great coat was also adopted for all ranks.

Lieutenant-Colonel H. Bowles' period of command came to a close on the 29th March, 1903, when he was succeeded by Bt.-Lieutenant-Colonel Granville George Algernon Egerton, specially promoted from the 72nd Highlanders, to which regiment he had been appointed in 1879. With them he had served in the Afghan War, and was dangerously wounded in the operations around Kabul in December, 1879. He took part in the march from Kabul to Kandahar, and was present at the battle of Kandahar (medal with two clasps and bronze decoration). As Adjutant to the 72nd he went through the Egyptian campaign of 1882, and was present at the battle of Tel-el-Kebir (medal with clasp and Khedive's bronze star). He had also served in the Soudan campaign of 1898, and was at the battles of the Atbara and Khartoum (mentioned in despatches, brevet of Lieutenant-Colonel, British medal and Khedive's medal). Lieutenant-Colonel Egerton joined on the 30th June, and took over command of the battalion.

After the annual training, the battalion having moved to Strensall for this purpose, proceeded by rail to Park House Camp, Salisbury Plain, on the 29th August, and took part in the manœuvres, returning to Sheffield on the 20th September.

On the 4th May, 1904, H.R.H. the Prince of Wales opened the exhibition at Bradford, when a guard of honour was furnished by the Strensall detachment, and another party of three officers and 120 men were sent from headquarters to assist in lining the streets.

On the 25th May the South African memorial, an obelisk of Aberdeen granite, subscribed for by past and present officers and men of the regiment, was unveiled by Lieutenant-General Sir Leslie Rundle, K.C.B., K.C.M.G., commanding the Northern District. At the base of the obelisk are engraved the names of all those belonging to the different battalions of the regiment who lost their lives in the war. Above these lists of names on three sides of the monument are bronze plates bearing the words Kimberley, Paardeberg, Driefontein, Johannesburg, Diamond Hill and Belfast, the names of the clasps on the Queen's medal. On the fourth side is the badge of the regiment.

After the unveiling ceremony the memorial was handed over to the Lord Mayor of York for safe custody. The obelisk graces the Tower Street approach to Skeldergate Bridge, and is the centre where three roads meet.

On the 20th June the battalion moved to Castleton in Derbyshire, the headquarters from Sheffield by march route and the detachment at Strensall by rail. Here the annual training took place, and there were constant field days with other corps in the same camp. On the 25th July the regiment struck its tents, and four companies, under Captain Esson, entrained for Strensall, the remainder going by road to Sheffield.

Prior to the departure of the battalion for its new station, Aldershot, it was inspected by Lieutenant-General Sir Leslie Rundle on the 14th February. In his farewell speech the General informed the men that he had sent in a most favourable report, particularly of what he had seen in July at Castleton, to the General in command of the Aldershot Army Corps.

The Green Howards arrived at Aldershot on the 18th February, 1905, and were quartered in Barrosa Barracks and attached to the 2nd Division.

A new method of training was in vogue at Aldershot, and the double company system had been introduced, which was adhered to by the battalion during its stay in the command.

In May the regimental team won the Douglas Shield, a competition open to the nine battalions of the 2nd Division. The team consisted of twenty-four files of "K" double company, and the attack practice had to be executed over a mile of country, the sections occupying a frontage of 400 yards.

On the 8th June the Aldershot troops were reviewed by H.M. King Edward and King Alphonso of Spain. Her Majesty Queen Alexandra was also present, and stood up as her own regiment marched past, an honour which was much appreciated.

On the King's birthday this year Colonel Egerton was appointed a Companion of the Bath, a compliment alike to the battalion as well as to its commanding officer.

In April, 1906, the Evelyn Wood Competition took place, in which the Green Howards were third, fourth and sixth in order of merit in the final tests. This competition is open to the army at home, and the teams, which numbered eighty-five, consisted of one officer, two non-commissioned officers and twenty men, who had to cover more than eleven miles in three hours, including halts, and execute an attack at the end of the march, commencing at 800 yards from the targets. Major King's team took the third place, and, in referring to the success of the various companies, Colonel Egerton very rightly remarked in regimental orders:—"Though we failed to carry off the cup, it is unlikely that any battalion at Aldershot will again have six out of seven teams in the final contest."

On the 1st October, 1906, Major-General W. S. Cooper, Colonel of the regiment, died in London, and was succeeded by Major-General William Edmund Franklyn, C.B., an officer who had passed all his regimental service in the Green Howards.

He was the eldest son of the late Rev. T. E. Franklyn, of Burton Grange, Cheshunt, and was born at Ventnor on the 14th May, 1856.

Appointed to a Sub-Lieutenancy in the 19th Foot in 1874, he was promoted Captain in 1881, Major in 1886, and Lieutenant-Colonel in command of the 2nd Battalion on the 28th July, 1896.

On passing out of the Staff College in 1887 he was appointed to the Staff at Aldershot, and whilst in command of the 2nd Battalion in 1898 was offered the post of Assistant Adjutant-General in Scotland, which he accepted. He was afterwards Assistant Military Secretary and Director of Personal Services at the War Office. He commanded the 10th Brigade at Shorncliffe 1902-1904, and the 3rd Division, quartered in the Southern Command, from June, 1906, to 1910. His advancement to Major-General took place in April, 1904, and to Lieutenant-General in August, 1910.

Under the history of the 2nd Battalion his services in the Tirah campaign are referred to, for which he was granted the brevet of Colonel.

He married in 1881 Helen, daughter of Edwin Williams, Esquire, of Craven Hill Gardens, London, and his son,

Lieutenant H. E. Franklyn, is at present serving in the regiment.

On the appointment of Colonel Egerton[1] to be Commandant of the School of Musketry, Hythe, in March, 1907, the Lieutenant-Colonelcy was given to Major Edward Malcolm Esson, who had joined the regiment in 1882, and had seen service with it on the Nile and also through the South African campaign.

Colonel Egerton, on taking farewell of the battalion, on parade addressed it in the following words:—

"The four years that it has been my honour and pride to command the Green Howards have been, I believe, not unsuccessful ones in the history of the regiment.

"We have played a very memorable part in all the Service competitions, such as the Douglas Shield, the Evelyn Wood Cup, etc.—contests, to my mind, far more fitting for soldiers to excel in than any football charity cups or cricket shields—and at manœuvres in the field we have more than held our own.

"All this would have been impossible had not everyone played the game and supported me as Colonel to the best of their power.

"There are many young soldiers who have recently joined in the ranks to-day.

"I would particularly ask them to remember that they have the honour to belong to a regiment that has always done well in Peace or War for nearly 220 years, and that on them will depend in the future whether this good name is to be kept up.

"*Esprit de corps* or pride in your regiment is a glorious thing, which on many a battlefield has been the saving of

1 Now commanding the Infantry Brigade at Malta with the rank of Brigadier-General.

our country, but let none here to-day forget that *esprit de corps* does not mean resting on your laurels.

"No regiment can remain prosperous and respected that is content to rely alone on the reputation made by those who have gone before.

"You must constantly strive to live up to that reputation, and to *add* to that honour and credit bequeathed to you by your predecessors.

"1st Battalion Nineteenth, I thank you one and all for your cheerful obedience and excellent discipline shown to me under all circumstances."

April, 1907, brought round the Evelyn Wood Competition once more. This time no less than ninety-six teams competed, and the Green Howards had the satisfaction of getting first place in their last try for this, the blue ribbon of the army. Captain Leatham commanded the team, which belonged to "B" company. He had taken eleventh place in 1905 and sixth in 1906, and now success had at last crowned his efforts.

In June a review was held on Laffan's Plain in honour of the King and Queen of Denmark, and later on in the year the battalion, in company with the other regiments of the 3rd Brigade, went to London to line the streets on the occasion of the visit of the Emperor and Empress of Germany.

On the 30th October the battalion had the honour of receiving a new stand of Colours from the hands of Her Majesty Queen Alexandra, in place of those presented by her in 1875. Major-General W. E. Franklyn, C.B., Lieutenant-Colonel Esson and Lieutenant and Adjutant Nash-

NOTE.—Captain M. E. Gunthorpe died on the 7th September, 1907, at Rosieres, Sudan, when attached to the Egyptian Army. He had been Adjutant to the battalion 1901-1905, and his death was much regretted by his brother officers and the regiment generally.—M.L.F.

Wortham accompanied the Colour party, which included Lieutenants C. H. Marsden and B. L. Maddison and four colour-sergeants.

The ceremony took place in the Bow drawing-room of Buckingham Palace at 1.15 p.m. When Her Majesty, who was accompanied by the Princess Victoria, and suite, entered the room, the Chaplain-General proceeded with the dedication service, after which Queen Alexandra, holding the Colours and crossing the poles, spoke as follows:—

"It is with the greatest pleasure that I confide these Colours to my own regiment, of which I am so justly proud, and for the second time within thirty-two years. May they ever lead you on to honour and glory."

Major-General Franklyn then made the following reply:—

"Although the Colours themselves will no longer be carried in the front line of battle, they will be treasured as emblems of your Majesty's most gracious favour, and your Majesty's name emblazoned in their centre will prove the highest incentive to a devotion worthy, I trust, of your Majesty's acceptance."

After the ceremony the officers were all presented to the Queen, who in course of conversation alluded to her former visit to Sheffield in 1875, when she presented the old Colours.

The new Colours are somewhat different in design from those presented in 1875. The white fimbriation around the broad red cross of St. George in the King's Colour has been reduced to a mere piping or edging, as well as the fimbriation on the outside of St. Patrick's Cross. In the regimental Colour the white rose appears for the first time, but the Union on the inside top corner has disappeared. The historic XIX. has also gone, its place being taken by the territorial title surmounted by the Tudor crown on both Colours.

After being stationed at Aldershot for nearly three years the Green Howards left the camp on the 16th of January, 1908, and embarking on the s.s. "Soudan" at Southampton, arrived at Alexandria the last day in January, where they took over quarters at Mustapha Barracks. On her way to Egypt the transport called at Limassol, Cyprus, and landed "C" company, under Captain W. B. Walker, to be stationed at Polymedia Camp.

Early in the year, in accordance with a new territorial scheme, instituted by the Minister for War, the 4th Battalion of the regiment was disbanded, and the present 3rd (Militia) Battalion became the 3rd or Reserve Battalion, the two Volunteer Battalions becoming the 4th and 5th (Territorial Force) Battalions. General regret was expressed at the disbandment of the 4th Battalion, whose historical record was probably unrivalled in the army, and every effort was made to retain it in the new force but without avail.[1]

In October, 1908, sanction was given for "The Cypher of H.R.H. Alexandra Princess of Wales in gold (thereon 'Alexandra') interlaced with the Dannebrog, inscribed with the date 1875, and the whole surmounted by the Coronet of the Princess" being adopted as the regimental badge in place of the White Rose. The latter badge had been authorized

1 An interesting history of this distinguished regiment, the old North York Rifles, has recently been written by Major R. B. Turton. Published by J. Whitehead & Son, Alfred Street, Boar Lane.

NOTE.—The following competitions were won by the battalion during its stay in the Aldershot Command:—

- The Douglas Shield.
- The Command Rifle Meeting Championship.
- The Brigade Judging Distance Competition.
- The Inter-battalion Field Firing Prize (won twice).
- The Brigade Inter-company Field Firing Prize (won twice).
- The Evelyn Wood Challenge Cup.
- The Aldershot Command Challenge Cup.
- The Marlborough Cup.
- The Individual Officers' Challenge Cup.

LIEUTENANT COLONEL GERARD CHRISTIAN, D. S. O.

COMMANDING 1ST. BATTALION.

on the introduction of the territorial system in 1881, but the new badge had always been worn on the buttons and appointments since that date.

On the 29th January, 1909, the battalion left Alexandria by train for Cairo, and on arrival there was quartered at Kasr-el-Nil Barracks in relief of the 60th Rifles. The Cyprus detachment arrived with the 4th Battalion Rifle Brigade, which relieved the regiment at Alexandria.

On the 18th October, 1909, the battalion moved to Abbassia Barracks, four miles from Kasr-el-Nil, in relief of the King's Own Scottish Borderers.

A sad accident in Cairo resulted in the death of a promising young officer, Second-Lieutenant H. L. Wheatley, who was killed by a tramcar on the 30th November, to the general regret of his brother officers.

Another move took place at the end of January, 1911, when the battalion embarked at Suez for Port Sudan *en route* to Khartoum, where it arrived on the 2nd February, having left two companies, under Major Noyes, at Alexandria as a detachment.

Lieutenant-Colonel E. M. Esson's period of command terminated on the 28th March, when he was succeeded by Major Gerard Christian, D.S.O. The latter had joined the regiment early in February, 1887, from Sandhurst, and had been Adjutant of the 1st Battalion from 1897 to 1901. With it he had served in the South African War, and for his services had been mentioned in despatches, receiving the King and Queen's medals with six clasps and the Distinguished Service Order.

The battalion is under orders for Sialkote, in the Punjab, and will probably leave Khartoum for India in January, 1912.

CHAPTER XVI.

Lieutenant-Colonel Warden raises the 2nd Battalion at Exeter—Stationed at Winchester, Shorncliffe, Aldershot, and Portsmouth—Roll of Officers, 1859—Service in Ireland—Quartered at Dublin, The Curragh and Newry—Embarks for Burmah—Report of Colonel Warden on the state of the battalion—Ordered to India—Bangalore, Calcutta, Allahabad, and Ranikhet—Returns to England—Parkhurst, Aldershot, and Sheffield—Embarks for Belfast—Moves to the Curragh—Wins the Queen's Cup, 1883—Buttevant—Embarks for England—Aldershot—Proceeds to India—Quartered at Bangalore—Moves to Burmah—Expeditions against the Katchins—Returns to India—Celebration of Queen Victoria's Diamond Jubilee.

AFTER the disbandment of Erle's 2nd Battalion in 1697, the regiment had only one battalion till 1756, when, as has been already related, it was again augmented. Two years later the new battalion was formed into the 66th Foot, now the 2nd Battalion Princess Charlotte of Wales's (Royal Berkshire Regiment). Exactly one hundred years were to elapse before the regiment for the third time in its history was composed of two battalions.

Owing to the difficulty in finding reinforcements for the Crimea and India during the Mutiny, it was decided to increase the army by twenty-eight new battalions, two being added to the 60th Rifles and Rifle Brigade, and one to each regiment from the 2nd to the 25th Foot.

Consequent on this order, on the 9th March, 1858, 350 men from the regimental depôt at Chatham were drafted to Exeter, which curiously enough had been the birthplace of the 1st Battalion, there to form the nucleus of the new one.

The command was given to Lieutenant-Colonel Robert Warden, C.B., an officer who had joined the Green Howards in 1841, and had served with them throughout the Crimean War, where, as we have already shown, he had greatly distinguished himself. A brother officer wrote of him as being "a soldier to the core—a Scot, honourable and true as steel, handsome, tall and spare, reserved and silent for the most part—one who by the mesmeric influence of his nature drew men to him as are steel filings to a magnet."

The establishment was fixed at forty-five officers and 1,081 of other ranks, and the battalion was armed at first with the Enfield pattern rifle of 1853.

The recruiting was general, there being under the system of the day no regimental control over the selection of the recruits, except in the case of those enlisted at the headquarters of regiments. Therefore the Nineteenth acquired a fair accession of Devonshire men, otherwise the greater number of the recruits were derived from parties stationed in Yorkshire, Leicestershire, Lancashire, Nottinghamshire, and one party in Ireland.

They came in nearly the following ratio:—Englishmen ten, Irishmen six, Scotch one, and the greater number were Protestants, about twelve Protestants to every three Roman Catholics.

The first move was to Plymouth, where two companies were detached early in November, under Captain Uniacke. They rejoined headquarters a month later, and on the 8th December the battalion was moved to Winchester.

In January, 1859, three companies were sent to Portsmouth, under command of Lieutenant-Colonel Lumley Graham. This was only a temporary move, as the whole battalion was transferred to Shorncliffe early in March.

LIST OF OFFICERS 2ND BATTALION—JANUARY, 1859.

Rank and Name.	Last Regiment.	Remarks.
Lieut.-Colonel Robert Warden	1/19th	Retired as Major-Gen. 18 April, 1868. Died in Edinburgh 15 June, 1890.
Major Lumley Graham l.c.	41st	Served in the Kaffir War and Crimea, Alma, Inkerman, and Sebastopol. Severely wounded, arm amputated. Afterwards Sir L. Graham, Bart. To 18th Foot 29 Sept., 1865. Died 25 October, 1890.
Major Robert Bates	45th	Served in the Kaffir War. To 83rd Foot 1 July, 1863. Died at Folkestone 12 July, 1894.
Captain Leonard D. H. Currie	1/19th	Retired 15 October, 1861. Served in the American Civil War (severely wounded). Died in London 3 Jan., 1907.
,, Henry Turner Uniacke	1/19th	Retired 2 Mar., 1860. Died at Sheringham, Norfolk, 19 August, 1907.
,, Edward N. Kindersley	1/19th	Retired 6 March, 1870. Died at Sherborne, Dorset, 11 November, 1907.
,, John Richardson Stuart	86th	Retired 30 December, 1859.
,, Philip Doyne Vigors	11th	Retired 12 January, 1881. Died at Holloden, Bagnalstown, co. Carlow, 30 December, 1903.
,, John Anderson... ...	51st	Retired 11 May, 1878. Died at Taunton 29 June, 1899. Served with the 51st Foot in the Burmese War, 1852-3.
,, Henry S. G. S. Knight	67th	Retired 29 Nov., 1876. Died at the Observatory, Littleton, Winchester, 13 October, 1904.
,, John Henry Kirke ...	1/19th	Retired half-pay 14 April, 1875. Died at Kempton Manor, Bedford, 24 February, 1892.
,, George Edward Bissett	55th	Served in the Crimea, Alma (severely wounded), and Sebastopol. Retired 31 May, 1859.
,, William F. T. Marshall	70th	Served in the Indian Mutiny. See Hazara Roll.
,, Thomas Madden ...	24th	Served in the Punjab Campaign 1848-9. Present at Chillianwallah and Goojerat. Retired 8 July, 1859.
,, Henry Edward Jerome	86th	Served in the Mutiny. See Hazara Roll.
,, William John Foster...	46th	Served in the Crimea (Sebastopol). Retired half-pay 5 July, 1872. Died at Hillsborough, Bath, 18 Nov., 1910.
Lieutenant Charles Hereford...	39th	Retired on half-pay 4 March, 1885. Died at Twickenham 14 July, 1891.

RANK AND NAME.	LAST REGIMENT.	REMARKS.
Lieutenant John Rinnie Mackenzie	30th	To 1st Foot 23 June, 1863. Retired 21 August, 1866.
,, James Knox ...	1/19th	Retired on half-pay 15 April, 1870.
,, Francis E. Biddulph	1/19th	To 9th Foot 5 November, 1870.
,, Conyngham J. Backas	1/19th	Died of cholera at Meerut 17 August, 1861.
,, Geo. Douglas Harris	1/19th	Retired 20 Sept., 1864. Died at Ryde 14 June, 1878.
,, Thomas Dennis Rew	1/19th	Retired as Bt.-Major 10 Nov., 1877.
,, William Robert Iles	1/19th	Retired on half-pay 12 July, 1878. Died 30 December, 1880.
,, Robert Hy. Hackett	1/19th	To 90th L.I. 20 Feb., 1865. Severely wounded at Battle of Kambula, 1878, losing the sight of both eyes. Died at Riverstown, Parsonstown, 30 December, 1893.
,, George Rogers ...	1/19th	Retired 19 August, 1862.
,, Beauchamp Colclough	62nd	Retired 28 October, 1873. Died at Southsea 30 March, 1900.
,, Charles J. F. Smith	32nd	Retired 17 April, 1866.
,, Arthur W. Burton...	1/19th	Retired in 1865.
Ensign William Bretton Kittson	1/19th	Superseded April, 1859.
,, Thomas Hartwell Kirby	1/19th	To 22nd Foot 1870. Served in the Burmese War, 1887-89. Died in Dublin 11 April, 1904.
,, Robert Biscoe	—	Retired on half-pay 14 June, 1876. Died at Quendon Court, Newport, Essex, 9 January, 1905.
,, Philip Downes Williams	—	Died at Penpont 2 April, 1869.
,, Brumhead Rogers ...	—	Retired 18 July, 1862. Died at Tooting, Surrey, 10 November, 1908.
,, William Read	R.E.	Retired on half-pay 29 Dec., 1865.
,, Duncan Campbell Affleck	—	To 42nd Foot 5 November, 1861.
,, Edgar Angelo Dickenson	—	Retired 3 March, 1868.
,, George H. Reynolds ...	—	Retired 18 Nov., 1882. Died at The Mullins, Ballyshannon, co. Donegal, 18 May, 1895.
Paymaster Frank W. Dundee	3rd Mdx. Mila.	To 3rd Foot 4 August, 1863. Died at Worthing 27 March, 1898.
Instr.-of-Musketry W. J. Foster	—	
Adjutant William Read ...	—	
Quartermaster J. J. Macdonald	2nd Jagers.	To 71st Foot 29 July, 1862. Died at Maidstone 14 July, 1867.
Asst.-Surgeon James C. Smith	21st	Served the Crimea with 21st Fusiliers (Sebastopol). Died at Ferozepore, Punjab, 13 February, 1864.
,, Donald Macpherson	—	Died at Gosport 12 May, 1873.

At this time the Nineteenth was 280 under establishment, and a year later it still required some thirty men to complete.

Little opportunity was given the new corps for settling down, for in August, 1859, it was again on the move—this time to Aldershot. Here its first Colours were presented on the 8th June by Major-General Knollys, who was then in command of the camp.

Changes of station followed each other in quick succession, and in September, 1860, the Nineteenth left the camp for Portsmouth, where it was quartered in the Clarence Barracks. Then on the 6th May, 1861, Portsmouth was vacated, the next move being to Ireland, where the battalion was stationed in Dublin at Beggar's Bush, a wing being detached, under Major Bates, to occupy Ship Street Barracks a few weeks after its arrival.

The next change was to the Curragh, where the regiment arrived on the 17th September, 1861, and was quartered in "B" lines.

On the 4th October, 1862, the headquarters arrived at Newry, four companies being detached to Enniskillen, under command of Major Bates.

A few months later, in April, 1863, the headquarters left Newry and arrived in Dublin on the 2nd May, where they occupied Richmond Barracks, and the Enniskillen detachment rejoined.

Orders now came for foreign service, and on the 24th August, 1863, the Nineteenth embarked at Cork, the headquarters on board the "Fanny Forsyth" with six companies, the remainder of the regiment, under Brevet-Lieutenant-Colonel Graham, on board the "Trafalgar." The latter arrived at Rangoon on the 15th December, and the "Fanny Forsyth" the next day.

The battalion did not remain at Rangoon, as on the 23rd December the headquarters and left wing, under command of Colonel Warden, embarked for the frontier, and steamed up the Irrawaddy. They arrived at Thayetmyo on the 2nd January, 1864.

The remaining companies, under Bt.-Lieutenant-Colonel Graham, proceeded to Tonghoo, where they were quartered.

As to the state of the 2nd Nineteenth at this time we cannot do better than to quote from the report of the Commanding Officer:—

"When it landed in Burmah the battalion was in a thoroughly efficient condition; it could be trusted to acquit itself well under any possible contingency. During its term of home service it had efficiently established its character.

"Its officers were a body of most honourable gentlemen, acting together with most cordial understanding. Its non-commissioned officers were of good promise, wanting only in more mature experience. The private soldiers were well contented and exceptionally well conducted, sober and intelligent.

"The *morale* of the battalion was high, and no base scandal had ever sullied its name.

"The training of the officers and men had not been limited to the prescribed exercises and movements. Every effort had been made to render them familiar with a more extended and liberal understanding of the real work of war. Their fighting drill was in advance of the requirements of the day. They possessed an undisputed superiority in all that concerned their action as 'light troops.' Their Lieutenant-Colonel had resolved from the first that the battalion, while preserving the allowed solidity of the British line, should in no whit yield to any foreign troops whatever in aplomb, or in *élan*, or in anything else on which they might choose to pique themselves.

"He succeeded in enjoying the interest and co-operation of those whom he had the honour to command, and he was quite satisfied with the result. The battalion acquired the utmost elasticity and facility of movement, ever ready on an instant to execute with extreme rapidity, and with safety and certainty whatever could possibly be required on the spur of the most sudden and unexpected emergency.

"The Captains could always lead their men, and their men would always follow their lead.

"The battalion was not an Adjutant's regiment. It was a sound, well-balanced weapon, well-tempered, and the ring of the metal was sterling and true."

The 2nd Nineteenth served in Burmah up to February, 1868, during which period the left wing was quartered at Thayetmyo and the right wing at Tonghoo. Being under orders for India, the headquarters and left wing, after steaming down the Irrawaddy to Rangoon, embarked there on the sailing transport "Alnwick Castle" on the 17th February.

The right wing arrived at Rangoon on the 12th February, where two companies embarked on the steamer "Cashmere," under the command of Captain Kirby, and after disembarking at Madras reached Bangalore on the 21st February.

The remaining three companies embarked on the steamer "Deccan" at Rangoon, under the command of Captain Biddulph. The "Deccan," then with the "Alnwick Castle" in tow, left Rangoon on the 18th February, and arrived in the Madras roads a week later. The companies on disembarking proceeded by rail to Bangalore, where they arrived

COLONEL ROBERT WARDEN, C. B.

on the 26th February, under the command of Major R. Barrett, there to be stationed.

The strength on arrival at Bangalore was as under:—1 field officer, 4 captains, 8 lieutenants, 7 ensigns, 47 sergeants, 40 corporals, 19 drummers, 680 privates.

On the 18th April, 1868, Colonel Robert Warden, C.B., retired on half-pay, the Lieutenant-Colonelcy being given to Major George Bingham Jennings, who joined on the 29th June, 1868, and assumed command. He had been appointed Ensign in the Nineteenth in 1843, and with it had served in the Crimea during the siege of Sebastopol. He had also served in India during the Mutiny, and had commanded the district of Dacca, receiving for his services the approbation of the Government.

Early in December, 1869, seven companies of the regiment, under command of Major R. Barrett, proceeded to Madras, and embarking there on the s.s. "Asia," disembarked at Calcutta. The remainder of the battalion, under Lieutenant-Colonel Jennings, waited for the return of the steamer, and did not arrive in Calcutta till the 21st December.

The battalion was quartered at Fort William, and occupied the Dalhousie and part of the South Barracks. On arrival the strength was as follows:—3 field officers, 5 captains, 7 lieutenants, 4 ensigns, 43 sergeants, 32 corporals, 17 drummers, 705 privates.

On the 6th March, 1870, Lieutenant-Colonel G. B. Jennings died of cholera at Fort William, and was succeeded by Lieutenant-Colonel Edward Chippindall, C.B., who joined on the 5th April, and assumed the command.

Originally gazetted to the 32nd Foot in 1847, he had been transferred to the Nineteenth in 1849. With the former corps he had taken part in the siege of Mooltan, 5th to 21st

January, 1849, and had also been present at the surrender of the fort and garrison of Cheniote on the 9th, and the battle of Goojerat on the 21st February, 1849, under General Lord Gough. With the Green Howards he had served in the Crimea from first to last, and had been present at the Alma and Inkerman, and throughout the whole of the siege of Sebastopol.

H.R.H. the Duke of Edinburgh landed at Calcutta on the 22nd December, 1869, when the battalion occupied the right of the line of infantry formed up for his reception.

The thanks of the Viceroy were conveyed to the Nineteenth for its general appearance, orderly conduct and efficient manner in which the several duties required from it during the visit of His Royal Highness were invariably performed.

In March, 1871, the battalion received Snider rifles at Fort William.[1]

---

1 This rifle took the place of the Enfield, a muzzle-loading rifle which had been in use up to this time by the battalion. It was the first breech-loading rifle issued to the English Army. Its dimensions were as follows:—

Weight	9 lb. 1 oz.
Length (without bayonet)	4 ft. 7 in.
Length of bayonet	17½ in.
Sights	100 to 1,000 yds.

The conversion of the Enfield to a breech-loading system was only a temporary measure, and after exhaustive trials a breech action invented by Mr. Martini, and a barrel and rifling invented by Mr. Henry, was chosen. This rifle, named the Martini-Henry, was introduced into the service in 1871, but the 2nd Battalion Nineteenth did not receive it until six years later. Its dimensions were as follows:—

Weight	9 lb.
Length (without bayonet)	4 ft. 1½ in.
Length of bayonet	22 in.
Diameter of bore	.45 in.
Muzzle velocity	1,350 f.s.

The Lee-Metford magazine-rifle with a calibre of 300 was introduced in 1888, superseded in 1895 by the Lee-Enfield, which is a breech-loader with a bolt lock, and detachable magazine for ten cartridges. Dimensions:—

Weight (without bayonet)	9 lb. 4 oz.
Length (without bayonet)	4 ft. 1.85 in.
Initial velocity	2,000 f.s.
Extreme range	3,500 yds.

There were now only eight service companies in the battalion, as two of the ten had been broken up in July, 1870, two companies still remaining with the depôt.

In November, 1871, the left wing, under Major R. Barrett, arrived at Allahabad, followed by the headquarters and remaining companies, under command of Lieutenant-Colonel Chippindall. Six companies were quartered in the new cantonments and two at Fort Allahabad.

In consequence of an outbreak of cholera, three companies, under command of Captain and local Lieutenant-Colonel H. S. Knight, proceeded into camp at Seorajpore early in August, 1872, and returned to Allahabad on the 15th October. The headquarters and three companies moved into camp at Burgurlo in the middle of August, and returned to Allahabad on the 18th October.

The two companies at the Fort were not attacked by cholera during this period, but suffered from a new disease called "dengue."

The battalion lost by cholera during the year 1872—4 sergeants, 1 corporal, 20 privates, 5 women and 7 children.

H.E. the Commander-in-Chief (Lord Napier of Magdala) was pleased to convey to the battalion the expression of his approbation of the admirable spirit which pervaded all ranks during the cholera epidemic.

In the middle of November, 1874, the battalion being under orders for Ranikhet, vacated the barracks for the relieving regiment, the 5th Fusiliers, and went under canvas near the rifle range.

Under command of Colonel Chippindall, C.B., A.D.C., it marched from Allahabad on the 26th November, and on arrival at Bareilly was detained there in camp from the 23rd December till the end of the cold weather. The route was resumed on the 4th March, 1875, and Ranikhet was reached on the 24th of the same month.

The marches averaged twelve miles, the longest being sixteen and a half and the shortest seven. Whilst at Ranikhet the barracks on Koompore Hill were occupied by the Green Howards.

It was not long before orders were received for home, and on the 20th November, 1876, the battalion marched to Moradabad and thence by rail to Bombay, where it embarked on the troopship "Serapis" on the 19th December.

It landed at Portsmouth Dockyard on the 23rd January, 1877, and was ordered to proceed to Parkhurst, Isle of Wight, but in consequence of an outbreak of smallpox on the troopship during the voyage, it was on landing segregated at the Portsdown Forts, and thus did not arrive at Parkhurst until the 15th March, 1877, when it joined the depôt companies.

The strength on landing was—3 field officers, 3 captains, 12 subalterns, 4 staff officers, 39 sergeants, 10 drummers and 510 privates.

Martini-Henry rifles were received on the 31st March, 1877, and the valise equipment on the 9th April. The Snider rifles had been handed into the arsenal at Allahabad.

During the stay of the Nineteenth at Parkhurst detachments were found by it at East Cowes in 1877 and at Bembridge, Marchwood and the Western Forts during 1878.

On the retirement of Colonel E. Chippindall, C.B., A.D.C., on the 23rd January, 1878, the vacant command was given to Major Richard Doyle Barrett. His father had served in the regiment for forty years, and the son had received his first commission in the 1st West India Regiment in 1846, to be transferred in 1852 to the Nineteenth. With it he had taken part in the Crimean War, and was present at the Alma and Inkerman, and the siege of Sebastopol up to February, 1855.

On the 1st October, 1878, the battalion was removed from Parkhurst to Aldershot, where it was quartered in the South Camp and attached to the 1st Brigade, under Major-General Pakenham.

Within two years the battalion was stationed at Sheffield, where it arrived on the 4th August, 1880, sending two companies on detachment to Weedon, under Captain Barton.[1]

On the 28th October, 1880, Lieutenant-Colonel Henry Cook, from the 100th Foot, exchanged with Lieutenant-Colonel Barrett. This officer had originally joined the 32nd Foot in 1855, and with it had served in the Indian Mutiny throughout the defence of the Lucknow Residency, where he had highly distinguished himself, being mentioned in despatches and receiving a year's service.

On the 10th April, 1881, the regiment embarked at Liverpool on the s.s. "Earl of Ulster" and s.s. "Galvanic" for conveyance to Belfast, where it disembarked the next morning in relief of the 1st Battalion 16th Foot.

A company was ordered to Carrickfergus on the 10th May, where it remained till relieved by the 69th Foot in January, 1883.

On the 29th September, 1882, Colonel H. Cook[2] retired on half-pay on completing five years' service as a regimental Lieutenant-Colonel. He was succeeded in the command by Lieutenant-Colonel Henry Edward Davidson, who had joined

---

1 Crosbie Barton. Ensign 19th Foot 29th August, 1863. Captain 23rd January, 1878. Retired as hon. Lieutenant-Colonel 13th October, 1886. Died at The Lodge, Frampton-on-Severn, September, 1902.

2 He was promoted a Major-General on the 5th August, 1888, and appointed Colonel of the Dorsetshire Regiment 17th June, 1910. Created a C.B. on the King's birthday, 1907.

the 100th Foot in 1858, and had been transferred to the Nineteenth twenty years later.

On the 14th August, 1883, the battalion moved by rail to the Curragh Camp, being relieved at Belfast by the 108th Fusiliers.

Early in December a detachment of 2 subalterns, 6 sergeants, 2 drummers, and 100 rank and file proceeded to Beggar's Bush Barracks, Dublin, for duty, under command of Captain W. E. Franklyn, and rejoined headquarters on the 8th February, 1884, owing to a draft being prepared for the 1st Battalion.

At the close of the musketry season of 1883 the 2nd Nineteenth had the satisfaction and distinction of winning the Inter-regimental Rifle match, which corresponded to the Queen's Cup of later years. The ranges were 200, 500 and 600 yards. Seven shots at each distance. Battalion prize, £60. The following composed the team:—

Private Parsons	94 points.
Lieutenant H. W. Boyce[1]	91 ,,
Sergeant G. Smith	91 ,,
Captain A. C. Fryer[2]	90 ,,
Colour-Sergeant Foley	89 ,,
Armourer-Sergeant Murray	87 ,,
Captain J. W. R. Parker[3]	85 ,,
Private R. Tong	81 ,,

A handsome silver tankard commemorating this event now decorates the officers' mess table.

1 Hugh Woolcombe Boyce. 2nd-Lieutenant 19th Foot 14th January, 1880. Transferred to 19th Hussars 11th March, 1884. Died in London 1st March, 1890, from the effects of a fall from his horse at Sandown Park the same day.

2 Alfred Charles Fryer. Ensign 61st Foot 16th March, 1867. Captain 19th Foot 30th March, 1881. Retired 28th March, 1894, as Lieutenant-Colonel on half-pay.

3 See the Nile Roll.

On the 27th January, 1885, three companies, under command of Major Levin,[1] were sent on detachment to Tralee, and another, under Captain Kirkpatrick, to Castle Island, co. Kerry. The next day the headquarters and remainder of the battalion went to Buttevant, relieving the 1st Battalion of the Buffs, ordered to Malta.

Late in July, 1885, the Castle Island detachment rejoined, and in August thirty men were ordered to Clare Castle. This latter detachment was increased in May, 1886, to fifty non-commissioned officers and men and moved to Killarney, under command of Captain T. D. Kirkpatrick.

By Army Circular 1st June, 1886, the establishment was fixed at 40 sergeants and 710 privates.

In August, 1886, the battalion embarked at Queenstown in H.M.S. "Himalaya," and disembarked at Portsmouth. It at once proceeded to Aldershot, and joined the 1st Infantry Brigade, under Major-General W. S. Cooper, a former commanding officer of the 1st Battalion, and was quartered in the South Camp.

Colonel H. E. Davidson's tenure of the command expired on the 29th September, 1886, and he consequently retired on half-pay.

He was succeeded by Lieutenant-Colonel Augustus Mourant Handley, who had been appointed to the regiment as an Ensign in 1858, and with the 1st Battalion had served in the Hazara campaign of 1868.

In December Captain C. T. Hennah[2] resigned his appointment as Adjutant, which he had held for the previous six

1 Lionel Henry Mocatta Levin. Ensign 19th Foot 20th February, 1870. Captain 28th July, 1879. Major 1st August, 1884. Died in London 31st March, 1886. Father of Lieutenant Hugh Levin, 1st Battalion.

2 Charles Tulin Hennah. 2nd-Lieutenant 19th Foot 2nd March, 1878. Adjutant 2nd Battalion 22nd January, 1881—22nd December, 1886. Retired as Captain 17th October, 1894.

years, on which the following tribute to his services appeared in regimental orders:—

"The Commanding Officer feels that he cannot allow Captain Hennah to resign the appointment which he has held for so long a period without placing on record his very high appreciation of the smart, soldier-like and highly conscientious manner in which this officer has carried out his duties for the last six years. The Colonel Commanding considers that the high character which the battalion has borne for the last six years and more is a proof of the very satisfactory way in which these duties have at all times been performed by Captain Hennah, who is now resigning at his own request."

The Green Howards took part in all the summer drills and manœuvres of the drill season of 1887, including the Royal Review on the 9th July in celebration of the Jubilee of Her Most Gracious Majesty Queen Victoria.

From the 26th July to the 3rd August, 1887, the battalion formed part of a field column equipped as for active service in the field, and received the thanks of the Major-General commanding for the efficient manner in which its duties were performed by all ranks, and the excellent conduct of the non-commissioned officers and men during these operations.

On the 13th July, 1888, the battalion received sudden orders to take the place in the 1st Field Column of the 2nd King's Own Scottish Borderers, ordered to Egypt, and within twenty-four hours of the receipt of the order was fully equipped as for active service in the field, and marched to Woolmer, fourteen miles distant. From the 9th to the 18th August this year the battalion formed part of the 3rd Field Column, and was on two occasions complimented by Lieutenant-General Sir Archibald Alison, commanding the division,

NOTE.—In the 1887-88 season the regimental Rugby football team played 26 matches, of which 19 were won, 6 drawn, and only 1 lost; goals for 104, against 18. They won the Surrey County Junior Cup and medals.

on the excellence of its marching, and gained distinction by having its camp pitched before the other regiments in the brigade throughout the operations.

On the 29th September, 1888, on the retirement of Colonel Handley, Lieutenant-Colonel George Oakes assumed the command. He had been appointed Ensign in the regiment in 1861, and had always served in the new battalion.

On the 31st May, 1889, the Nineteenth composed part of a division reviewed by Her Majesty the Queen in the Long Valley. It took part in all divisional field days during the drill season, and was attached to the 3rd Field Column, which left Aldershot on the 13th July and returned on the 20th to camp. The weather was good, and though there was a good deal of night work, only four men were reported sick during the week, all of whom except one rejoined the ranks before the battalion returned.

On the 7th August, 1889, the battalion formed part of the force exercised before the German Emperor on the Fox Hills, its strength being 17 officers, 28 sergeants and 680 rank and file.

In November, 1889, orders were received for the 2nd Nineteenth to hold itself in readiness for service in India, and it accordingly embarked at Portsmouth in H.M.S. "Euphrates" on the 31st December, having trained from Aldershot, and sailed the following day.

The troopship arrived in Bombay harbour on the morning of the 28th January, 1890, after a prosperous voyage.

After disembarkation the battalion left Bombay in two trains on the evening of the 28th, and arrived at Poona on the 29th. Thence it proceeded to Bangalore in three trains on the 31st January and 2nd February, and by the 6th Februaiy arrived at its destination, where it took up its quarters in the Parade Barracks. Its strength on arrival was—

25 officers, 2 warrant officers, 35 sergeants, 16 drummers, 37 corporals and 666 privates.

On the 2nd January, 1891, the battalion, leaving a sufficiency of non-commissioned officers and men to furnish garrison guards, etc., marched from Bangalore to take part in the field manœuvres of the district. During part of this time the division was divided into two forces, and manœuvred under service conditions for three days, when several long marches were performed and a considerable amount of work got through which was entirely new to the men. The annual field firing was also carried out during the field manœuvres, but as the battalion was brigaded with two native regiments it was impossible to tell its actual figure of merit. On returning to barracks a long march took place under service conditions. The baggage escorted by a guard marched twenty-eight miles, and the battalion covered twenty-three, besides fighting an engagement lasting for two hours with an opposing force from Bangalore.

On the 18th June, 1891, Lieutenant and Adjutant W. S. Brooksbank died from smallpox. The Officer Commanding, in referring to his death in orders, mentioned him as being "a good comrade, a zealous soldier, and an ornament to his profession."

In the report received from the Inspector of Army Signalling in Madras and Bombay, the signallers of the battalion examined by him proved themselves highly efficient, and their figure of merit was better by 8.95 points than that of the best regiment in India the previous year.

During the assault-at-arms in the Madras Presidency in October, 1891, the following events were won by the battalion:

Tug-of-war, escalading competition and gymnastic competition, first aid to the wounded competition, physical drill

Capt. J. T. Coatesworth. Capt. W. B. Silverthorne. Lieut. E. L. Vans-Agnew. Lieut. B. Williams. Lieut. H. R. Maitland. Lieut. F. W. Robinson. Lieut. H. F. Lea. Lieut. A. L. Napier. Bt.-Major H. Bowles. Lieut. H. E. Raymond. Lieut. W. S. Brooksbank. Capt. A. de S. Hadow. Lieut. R. D'A. Fife.

Capt. A. G. Cartwright. Lieut. G. Christian. Capt. J. A. Fearon. Lieut. W. H. Dent. Capt. A. B. Elton. Lieut. & Adjt. C. L. Nicholson. Major E. A. Bruce. Lt.-Col. G. Oakes. Lieut. A. F. Owen-Lewis. Capt. E. W. Mills. Lieut. E. S. Bulfin. Lieut. H. G. Holmes.

OFFICERS, BANGALORE, 1900.

competition. Sergeant Challice was awarded a medal for being the best man at arms in the Madras Presidency, and was also mentioned for having taken the prize at the Industrial Exhibition for having the greatest number of certificates in the Presidency.

On the 28th July, 1892, Colonel G. Oakes having completed six years as a regimental Lieutenant-Colonel, gave up the command of the battalion in accordance with the existing regulations.

Major William Graham Waugh McClintock was promoted to fill the vacancy. He was gazetted to the Nineteenth in August, 1865, and with the 1st Battalion had served in the Hazara campaign of 1868, and also in the Nile expedition of 1884-1885, including the operations on the Soudan frontier and engagement at Ginnis.

In October, 1892, the following parties were sent to Burmah for duty as mounted infantry:—

Lieutenant C. A. C. King and fifty n.c.o.'s and men to Bhamo.

Lieutenant D. L. Hartley and thirty-five n.c.o.'s and men to Shwebo.

Lieutenant H. R. S. Maitland and thirty-five n.c.o.'s and men to Bernardmyo.

On the 18th November the battalion left Bangalore in two trains for Madras, *en route* for Burmah, and went into camp on "The Glacis," Madras. Four days later it embarked on the "Canning" for Rangoon, where it arrived on the 27th November, 1892. Four companies, under command of Major H. Bowles, were at once sent off to Mandalay, which they reached on the 29th November. There they embarked on the s.s. "Sladen" with flats attached, and on arrival at Thabeitkyan two companies, under Major Bowles, disembarked and marched to Bernardmyo, sixty-two miles distant, over a hilly country.

The other two companies, under command of Major Mathison, went on to Bhamo in the same steamer, there to be stationed.

On the 30th November the headquarters and remaining four companies, under the command of Lieutenant-Colonel McClintock, proceeded by train to Tonghoo, where they were stationed in barracks till the 25th December, when they left by train for Shwebo, Upper Burmah, arriving there the next day.

On February 14th, 1893, a column, under command of Major Hammond, 32nd Regiment, with Major Mathison, Lieutenants King and Williams, 100 non-commissioned officers and men of the battalion, proceeded by march route to Namkham. They returned to Bhamo on the 31st March. Lieutenant King was in command of the mounted infantry. Ten years later the officers and men who served in this expedition were awarded the Indian General Service medal with clasp for the Katchin Hills.

A second column was sent out against hostile Katchins around Sima, under Captain Davis, 2nd Devonshire Regiment, Station Staff Officer at Bhamo. Lieutenants King, Williams, Fife (Staff Officer), Raymond, Second-Lieutenant Caffin, and Lieutenant Dent as intelligence officer accompanied it.

On April 7th the battalion received the Lee-Metford magazine rifle at Shwebo, the Martini-Henry rifles being given into the arsenal at Rangoon.

In May, 1893, a letter was received from the Commander-in-Chief, Madras, stating how gratified he was to hear of Lieutenant Dent's excellent services at a very trying moment during the attack on Sima, and when he himself was suffering from wounds previously received. His valuable services in the intelligence branch have been placed on record. He was also thanked by the Chief Commissioner of Burmah.

Lieutenant Dent was wounded in December, 1892, by a bullet which entered the left cheek and lodged in the back of the neck.

On the 7th April, 1894, Captain F. Pettitt, Quartermaster, died at Wellington, in the Nilgiris. He had been promoted to commissioned rank in December, 1881, after over thirteen years' service, and his death was a loss to the battalion.

In January, 1896, the battalion left Shwebo by rail for Rangoon on relief by the 87th Royal Irish Fusiliers, and embarked on the s.s. "Warren Hastings" for Calcutta, where it arrived on the 20th of the month.

A few days later Bareilly was reached, and after remaining there a few weeks the battalion proceeded by rail and march to Ranikhet, where it took over barracks vacated by the 37th Hampshire Regiment.

On the 27th July, 1896, Lieutenant-Colonel McClintock retired on half-pay, and was succeeded in the command by Major William Edmund Franklyn. He had joined the Green Howards in June, 1874, was in possession of a Staff College certificate, and had served on the Staff at Aldershot for five years as a Deputy-Assistant Adjutant-General.

On the 22nd July, 1897, the Diamond Jubilee of Her Majesty Queen Victoria was celebrated all over her dominions with great enthusiasm, and Ranikhet was in no way behind-hand in this respect. The officers of "The Princess of Wales's Own" gave a ball, followed by a gymkhana, and theatricals by the "Green Howards' Dramatic Club." Huge bonfires were lit on all the prominent hill-tops, and formed part of a complete chain which was fired simultaneously at every hill station from east to west of our great Indian Empire, thus forming a visible line of communication between many millions of Her Majesty's subjects. The whole week was one of pleasure and festivities.

CHAPTER XVII.

Battalion is ordered to Jullundur—Mobilizes there for active service on the North-West Frontier—Forms part of the 1st Division for operations in the Tirah—Its baptism of fire—Taking of the Sampagha and Arhanga Passes—Capture of Bagh—The Dwatoi reconnaissance—Death of Lieutenant Jones—The crossing of the Karu Darra Pass—Return to Bagh over the Chingaki Pass—The Tirah evacuated—Expedition to the Bazar Valley—The memorial brass in York Minster—Honours and rewards for the campaign—The Tirah Roll.

IN the summer of 1897 India was startled from a peaceful sense of security by the news of the treacherous attack on a civilian officer's escort at Maizar, which led to the formation of the Tochi Valley Field Force. Then before the end of June came the report of the outbreak on the Malakand, and more troops were hurriedly despatched to the Punjab frontier. Following this came intelligence of unrest among the great border tribes of the Orakzai and Afridi, and a force, under Major-General Yeatman-Biggs, was ordered to operate on the Samana range with the object of relieving the 36th Sikhs, who were disposed as garrisons of the frontier posts along that range.

On the 25th August, 1897, orders were received for the Green Howards to proceed at once to Jullundur, consequent on the above occurrences.

Accordingly the left wing, under Major Cartwright, started from Ranikhet a few days later, followed by the headquarters and right wing, under Lieutenant-Colonel Franklyn.

On arrival at Jullundur no further orders were received, and the battalion awaited anxiously the development of events. It was not till a fortnight afterwards that the summons came to mobilize for active service on the Punjab frontier. But this interval had not been wasted, for almost every day the men were exercised in route marching and extended order formations, so that when they left for the front they were hard and fit and in fine condition for the campaign before them.

Early in October the composition of the force operating in the Tirah was published, and the Green Howards were detailed to form part of the 2nd Brigade of the 1st Division. This brigade, which was under the command of Brigadier-General A. Gaselee, C.B., A.D.C., consisted of:—

1st Battalion The Queen's (Royal West Surrey Regiment).
2nd Battalion The Princess of Wales's Own (Yorkshire Regiment).
The 3rd Sikhs.
2nd Battalion The 4th Gurkha Rifles.

The general arrangement of forces was a main column, under Sir William Lockhart, to move direct to the heart of the Orakzai and Afridi country with an auxiliary force on either flank; on the right the Peshawar column, under Colonel Hammond, V.C., C.B., and on the left the Kurram movable column, under Colonel Hill. A reserve brigade, under Colonel Macgregor, was to be formed at Rawul Pindi, and the lines of communication were placed under Lieutenant-General Sir Power Palmer, K.C.B.

On the 4th October the battalion, 801 strong, left Jullundur in two trains for Khushalgarh, the remaining distance of twenty-seven miles to Kohat being completed in two night marches.

On the 13th October General Symons, commanding the division, inspected the battalion, after which he addressed the men. He alluded particularly to their healthy appearance, and to the excellent form they had displayed in the march from Khushalgarh.

For the coming campaign all ranks were issued at Kohat with field service warm clothing in the shape of thick khaki-coloured pea jackets, Cardigan waistcoats, Balaclava caps and warm mitts. The ammunition which was brought from Ranikhet was exchanged for the same quantity furnished with the Dum Dum bullet.

On the 18th October the 2nd Brigade left Kohat, and arrived at Usterzai the same day and Hangu the day following. For some distance now the march lay along the high road which leads to the Kurram Valley.

To the north lay the Samana range, on which could be distinguished Forts Lockhart and Gulistan and the intervening site and ruins of Saragarhi, where some men of the 36th Sikhs made a gallant but fruitless stand a few weeks before.

When about two-thirds of the march had been covered the column turned northwards from the high road on to a track leading to Shinawari. At Kai, where the brigade encamped, there could be seen from outside the tents the tiny forms of men on a cliff some six miles away silhouetted against the sky, and repeated volleys could be heard, as well as the boom of mountain guns.

Owing to various delays it was not till the 25th October that the battalion arrived at Khangarbur, on the Khanki River, where the whole Tirah Expeditionary Force was collected, under General Sir William Lockhart, K.C.B.

This was an important day in the history of the 2nd Nineteenth, as it was the one on which it was destined to receive its baptism of fire as a battalion.

Night Attack on Kharappa, 25th October, 1897.

On its arrival at Khangarbur the west or left of the camp was reserved for the 2nd Brigade, but owing to the crush of regiments at this point there was only room for four companies of the battalion along the perimeter, which was formed of a wall of loose stones built on the top of a razor-backed hill, with the ground sloping sharply to front and rear. The left half battalion was placed along this wall, and the right half bivouacked near a hill a short way in rear.

Some of the 1st Brigade who were foraging to the west were busily engaged with the enemy, and when the forage secured was brought in the covering troops fell back, followed persistently by the enemy. The bullets kept coming over the parapet and within the perimeter, and the men of the right half were consequently ordered to the reverse slope of their hill. But before they all could get clear six men were wounded, two of whom died, and Private Watson had his leg amputated afterwards. The left half opened fire on the enemy, but no persistent advance was attempted.

On October 28th the 2nd Brigade cleared and picquetted Ghundaki Hill. There was one casualty in the battalion, Sergeant Bromwich being badly wounded, and dying a few days later.

On the 29th the advance on the Sampagha Pass commenced, the 2nd Brigade in front. Very little resistance was met with, contrary to expectation. This was due principally to the rapidity of the march, whilst the continued occupation of the Ghundaki Heights by picquets apparently deceived

NOTE.—The accompanying illustration is from a photograph of the night attack at Kharuppa or Khangarbur, when the battalion was under fire for the first time. It was taken by a corporal of the Dorsetshire Regiment (which held the opposite side of the camp) by the light of a star shell, and shows a portion of the perimeter held by the Sherwood Foresters and some of the transport, with the enemy rushing to the attack. The Green Howards were posted a little further to the left, and an attack was directed on them somewhat earlier in the evening, but not pressed home to such an extent.—M.L.F.

the enemy as to the line of advance. The difficult pass was taken without a check and very few casualties, four men slightly wounded being the share of the battalion.

After crossing the Sampagha, the brigade bivouacked in the Mastura Valley, and passed a cold and miserable night, as only a small quantity of the baggage could be got over the pass before nightfall.

On the 31st October the Arhanga Pass was taken with even less opposition than the Sampagha. The 2nd Brigade made a flank attack, and it was a race between the two leading companies of the battalion ("E" and "F") and the scouts of the 5th Gurkhas as to which should reach the top of the pass first. It resulted in "E" company, under Lieutenant Edwards, getting in front, and he was the first man of the force to gaze on the Afridi Tirah. Major Cartwright, who was close on his heels with "F" company, at once opened volleys on to the opposite side of the pass, and the enemy were soon in full retreat. There were no casualties in the battalion, but Major Cartwright and Lieutenant Alexander were both grazed by bullets, and several others had narrow escapes.

The battalion remained on the heights from 10 a.m. to 3 p.m., while the 2nd Division and the baggage were streaming over the pass. It was then ordered to march to camp at Maidan through the Arhanga.

The division halted at Maidan from the 1st November to the 17th, the regiment being daily under fire. The men were employed in foraging, convoy duty, reconnaissances, and other work incidental to active service. Scarcely a night passed that the camp was not sniped, and the tents which arrived during the third week had to be pitched with their blue linings turned outwards, the floors being sunk and the earth piled up inside so as to ensure safety. The casualties during this period were one officer (Lieutenant Caffin), one sergeant and six men wounded.

On the 18th November the 2nd Brigade and 1st Divisional troops marched out of camp at Maidan, the object of the day's operations being the capture of Bagh, a collection of houses containing the principal mosque and meeting place, and for this reason might be called the capital of the country. It is situated almost opposite to the entrance of the Dwatoi defile and west of the main nullah which skirts the western end of Picquet Hill. As the operations up to the end of the month are full of interest, we cannot do better than quote Lieutenant-General Franklyn's account verbatim:—

"We assembled in the nullah south-west of the camp at 8 a.m., and with the banks picquetted marched to the west end of Picquet Hill, where we ascended the western bank, and found ourselves on an elevated plateau. The enemy were holding three conical hills due west of us, and about 2,000 yards distant. As we were the rear of the main body we remained lying down in reserve while the Queen's and 3rd Sikhs attacked these hills with the Derajat Mountain Battery supporting them with its fire.

"Shortly after our arrival I was ordered to despatch two companies due north to our right to protect the officers who were laying out the camp. After a long wait a staff officer came up to me at 11 a.m. with an order for the battalion to march into camp. We therefore moved to our right across a branch nullah on to another plateau, on which were dotted several substantial houses, among which the various camp-colour parties had planted several of their flags. On our arrival I was ordered to send another company, 'E,' to the western end of the plateau, as the enemy were opening an annoying fire from this direction. A message then arrived from the north-east from Captain Cotesworth (who was in command of the two companies, 'B' and 'C,' originally detached) for some support. I therefore sent Lieutenant Jones with 'D' company to act under his orders, and pushed Major Cartwright with two more companies, 'A' and 'F,' on to his left, with the object of endeavouring to connect 'E' company on the extreme left, but as this distance was $1\frac{1}{4}$ miles I had to move more men into the gap on Major Cotesworth's left until only one company remained with me.

"Later on the A.Q.M.G. 1st Division requested me to still further push the enemy back, as the continuous firing prevented the proper laying out of the camp. I therefore had to abandon any idea of supporting 'E' company, and moved in Major Cartwright's direction, upon which the enemy fell back.

"They clung on to the houses until actually driven out, and as Lieutenant Williams and a section of 'A' company rushed into the lower storey of one house the enemy dropped from the roof and continued firing from the nullah on the north.

"As this house was on the extreme north limit of the camp, I established headquarters there and brought up the maxim to still further disperse the enemy, now congregated across the nullah about 800 yards off. The fire of the maxim attained its object, and as Generals Symons and Gaselee now arrived I went forward with them to reconnoitre.

"It was decided to occupy some of the houses across the nullah, and Lieutenant Noyes, with 'G' company, went over and seized the nearest house. Meanwhile I ascertained that the three companies on our right, under Captain Cotesworth, were holding their own without difficulty, and as regards my left, General Symons told me that a party of the battalion on the west had with his sanction advanced against a village held by the enemy in this direction.

"To support this advance Lieutenant Chapman moved in a westerly direction with some men of 'F' company.

"Although the enemy on the north had withdrawn to a respectful distance, bullets continued to drop into camp, and two or three of our men were hit, but no one severely. As the mess mules had now come in some soup was made, which we partook of in rear of the house which Lieutenant Williams had captured. Whilst we were discussing this Lieutenant Fulton arrived in our midst, and told me that portions of 'E' and 'F' companies were in some houses west of the camp, under Lieutenants Chapman and Edwards, and that unless succoured they would be unable to withdraw; that his Colour-Sergeant, Wyatt, had been wounded and one or two men killed.

"I managed to withdraw some men of 'G' company, and sent them, under Major Fearon, to the west. I also obtained the help of two guns of the Mountain Battery, and with this assistance our men were able to retire without further loss. As this little fight on the west is interesting, I subjoin a description of it written by Lieutenant Fulton, who, in the absence of Captain Holmes with the maxim gun, commanded 'E' company:—

"'While occupying ground allotted to "E" company to guard the laying out of the camp, Generals Symons and Gaselee with staff rode up to my position, and interrogated me as to what was taking place. I explained that the enemy were harassing us somewhat from some houses about 800 yards distant. Whereupon General Symons ordered me to advance with my company and turn the enemy out of his position. I had about forty men with me, and advanced two sections in the firing line and two in support. The ground over which we had to traverse was almost entirely devoid of cover, and the advance was conducted by means of section rushes, until a point about 200 yards from the enemy's position was gained. The enemy during this time kept up a hot fire, and bullets kicked up the dust close to us, though fortunately no one was hit.

"'After a short consultation with Lieutenant Edwards I sent two sections under his command to make an attack on the enemy's right flank, the movement being covered by a hot fire from the men I retained with me. Soon after Edwards' departure with the two sections, my half company consisting of Colour-Sergeant Wyatt and about fifteen men, was reinforced by a small party, under Lieutenant Chapman, whom I at once sent round to assist Edwards in the village, as the latter officer was apparently heavily engaged with the enemy. About ten minutes later Sergeant House arrived with a message from Edwards reporting the village to be very strongly held by the enemy, and that three or four of his men had been hit. Soon after Sergeant House had rejoined the party in the village Edwards himself arrived at my position, and informed me of the death of Sergeant House and of Private Smithson, of "F" company. I also gathered from what he said that further reinforcements would be necessary before the enemy could be turned out. It was at this moment that Colour-Sergeant Wyatt was hit in the upper arm.

"'I now determined to go back for reinforcements from the camp, and having obtained them, after a short bombardment from the guns I withdrew my men, the enemy having also evacuated the village.'

"As the evening drew in, the companies under Captain Cotesworth were retired, leaving picquets in the nearer houses. Lieutenant Noyes was reinforced by some men of 'G' company, and remained for the night in the house he had seized across the nullah. The enemy retired, and the night passed off quietly.

"On the arrival of the 4th Brigade on the 19th they established their camp on our right. This enabled us to do away with the trench we had occupied facing east the preceding night, and as my battalion was therefore more restricted I was able to form two small reserves in case of attack. Our line was now facing north along the edge of a deep nullah.

"In the afternoon General Symons had the men of 'E,' 'F' and 'G' companies paraded, under Lieutenants Fulton, Noyes, Chapman and Edwards, and told them how pleased he was with their gallant conduct on the afternoon of the 18th.

"At night several volleys were fired by the enemy at a house on the south of the camp, which was divisional headquarters, and the battalion quartered near the spot opened a heavy fire about 8 p.m., but I did not hear with what success. The rest of the night passed off quietly.

"On the 20th November the 3rd Brigade, which had been left at Camp Maidan, marched into the new camp at Bagh, and the old camp which had been occupied for three weeks was thus abandoned. Our line of communication with the Arhanga was now slightly lengthened, and troops were told off daily to picquet it. Two of our companies were detailed for this duty, and three more were sent out foraging. The enemy seemed inclined to rest after his exertions of the preceding two days, and the foraging was carried out with very little opposition. I made a careful inspection of the perimeter allotted to us, and as I did not like the idea of the whole battalion being in the front line, I so rearranged the front as to allow of two companies being withdrawn and acting as reserves, one to each half battalion. The enemy woke up again at night, and amused themselves with more or less heavy firing into camp.

"As all the troops had now arrived at Bagh, the line of the camp was extended to very near the entrance to the Dwatoi defile, and this brought the troops on the north-east under possible fire from the peaked hill we had temporarily occupied on November 16th whilst out foraging with General Westmacott. As this hill, which was in reality the western end of the northern range, was on the side next the camp entirely devoid of trees or bushes, it received the name of Bare Hill. On the 19th it had been held by the 3rd Gurkhas, and on the 21st 'E,' 'F' and 'H' companies, under Major Cartwright, were sent up at 7 a.m. to take over this duty.

"Major Cartwright made the following distribution of his force:—On the low ridge blocking the defile was posted 'H' company, under Captain Todd and Lieutenant Alexander. Half-way up the hill was a picquet of thirty men, under Lieutenants Fulton and Chapman, and Major Cartwright held the summit with the remainder of his force, having with him Captain Holmes and Lieutenant Edwards.

"At 7 a.m. the rest of the battalion paraded with the 2nd Brigade for a punitive expedition, under General Gaselee.

"The western bank of the main nullah from the east was formed by a succession of spurs running from the western plateau, and on each of these spurs was a collection of well-built houses, which had been used as cover when our convoys to and from the Arhanga had been fired on. To prevent this and to further punish the tribes for the continual firing into camp, we were sent out to destroy the houses.

"The morning was bitterly cold, with 20° of frost, when we left camp. The two native regiments were fairly strong, but the Queen's and ourselves were much weakened by the heavy picquet duties; in fact, the former could only muster enough men to form a fairly strong company. We moved along the main nullah towards the Arhanga, and then were told off in parties for each spur. I was given charge of the three spurs nearest to camp, and moved up them with the Queen's and my own men, sending ahead up each spur a strong covering party with supports. General Gaselee took charge of the spurs further east with the Sikhs and Gurkhas.

"As soon as the covering parties had advanced and those in rear were in position, the brigade moved forward, setting fire to every house it passed. By the time the heads of the spurs where they joined the lower features of the southern range were reached, some fifty-seven houses had been set alight, and were burning furiously. We then took ground to our right to the conical hills, which the enemy had held on the 18th. There were several more houses here, but as enough damage had been done, we contented ourselves with carrying off the forage and returned to camp.

"Very few of the enemy were seen during the day, and only a few shots were fired at our advanced parties, who, however, heard a great deal of beating of tom-toms in the hills, showing that from their fastnesses the enemy must have looked down on our work with rage and dismay. The smoke rising from the burning houses spread all over the sky, and hung like a pall over the valley, quite obscuring the sun and rendering the atmosphere piercingly cold. The burning had a great effect on the jirgahs in camp, and they promised to use their influence to stop the nightly sniping.

"About 4 p.m. we heard rumours of our being included in an expedition planned for the next day. I could, however, obtain no authentic news before evening, when orders arrived that we were to take part in an important reconnaissance to be conducted by General Westmacott through the Dwatoi defile to the head of the Bara Valley. Further instructions showed that to us was allotted the duty of guarding the heights on the right of the advance.

"As the northern range of hills had shut out all view in the direction in which our march would lie, and as there were no reliable maps, I felt that this duty was rather a leap in the dark, so General Symons kindly gave me a note to the Chief of the Staff, asking for further details. General Nicholson naturally did not know much more than I did about the country lying north of the northern range, but he impressed upon me the importance of clinging to the highest points, as the Afridis never become venturesome with an enemy above them. Later on General Gaselee sent over Lieutenant the Hon. C. G. Bruce, an officer of the Gurkha Scouts, who had covered more ground than most of us during our stay at Maidan, and sitting in our mess-room after dinner he drew a rough sketch in my note-book of his idea of the country, which turned out to be wonderfully accurate. I found we were not to withdraw the picquets on Bare Hill, so that I had only five companies to carry out the duty, and from these had to be deducted baggage guards and men necessarily left in camp. As we were in communication wtih Major Cartwright by means of lamp signalling, I sent a message to him to send a party forward at daybreak to seize a lofty hill, which from Lieutenant Bruce's plan I gathered lay north-east of the outposts.

"Arrangements were made for our portion of the perimeter to be guarded by other battalions, and all our outlying picquets except those on Bare Hill were relieved. I also sent a non-commissioned officer to ascertain the best way out of camp, as we were to start before dawn.

"The next morning, the 22nd November, the battalion, consisting of 'A,' 'B,' 'C,' 'D' and 'G' companies, left camp at 5.30 a.m. It was very dark and bitterly cold. When we got down into the river bed in the dark we missed the way, but after much scrambling through icy-cold water, reached the southern slopes of Bare Hill. We had struck it a little too much to the east, but as the climb in front of us, though steep, seemed practicable, we commenced the ascent. Dawn broke whilst we were on our way up, and we found we were considerably east of the low ridge blocking the defile on which 'H' company was stationed. After a stiff climb of an hour or so we reached Major Cartwright's position at 7.30 a.m., and while the men were enjoying a temporary rest I went forward with him to reconnoitre.

"The position held by 'E' and 'F' companies consisted of the northern slope of Bare Hill as the main position. This slope was well wooded, and in a hollow close by was a loathsome looking pond whence their water supply was drawn. About 500 yards to the north rose another hill, which was now temporarily held by a detached party. Eastward from this latter hill a grassy ridge led to the predominant peak, about 12,000 yards north-east of the main position. This peak had been seized at daylight just in time to prevent its being occupied by the enemy, who were found congregated at the foot of its eastern slope. A few volleys had sufficed to disperse them, and the hill was now held by Captain Holmes with 'E' company. North of this grassy ridge and line of hills was a deep valley, but due north of the high peak a natural bridge or saddle led across the dip to a long flat-topped hill, the nearest part of whose summit was about 1,400 yards north of the high peak. The passage of the saddle entailed a descent of a few hundred feet from the ridge, and a corresponding ascent on to the flat-topped hill.

"As this hill flanked the line of advance of the Dwatoi reconnaissance, I decided to advance along its top. Captain Cotesworth, with Lieutenant Mactavish and 'B' company, was sent to the high peak with orders to keep Captain Holmes and his company there until we had possession of the northen hill, and after supporting our advance by the fire of the combined companies, to send 'E' company back to the outpost position.

"I kept 'C' and 'D' companies on the slope, south of and above the saddle, whilst 'A' and G,' under Major Bowles, advanced on to the northern hill. Captain Cotesworth opened fire over our heads on a few of the enemy who were seen to the north, but our advance was practically unopposed, save by a few desultory shots which did no harm, and by 9.30 a.m., after another severe climb, we found ourselves established on the flat-topped hill. The summit of this hill was shaped like the figure 8, and was about 800 yards long, lying north and south. At its southern end, overlooking the saddle, was a grove of trees and a small pond of water full of strange animals. In order to guard it and the saddle, and to act as a connecting link with 'B' company, I left 'D' company, under Lieutenant Jones, with orders to build a sangar or stone breastwork as quickly as possible in the grove. I then advanced with 'C' company to the northern end of the hill, where we found 'A' and 'G' companies, who reported that they had had a few shots fired at them whilst advancing, but that now the enemy had quite disappeared.

"The first problem was whether we could advance any further in a northerly direction. The hill, well wooded on the north and west, sloped fairly gradually into the Dwatoi defile, which skirted its base. On the east it fell more abruptly into another watercourse far below us. As there existed no means, therefore, of advancing further except by descending to the water level and making a subsequent ascent on the other side, and as my remaining force numbered only two and a half companies (actually 120 men), I decided to remain where we were, and signalled to this effect to the troops who were now entering the defile south-west of our position.

"From the foot of our hill the defile ran due north, and we could see the ground on which General Westmacott's force would probably encamp for the night. The watercourse on the east, which skirted our eastern slope, took its rise in a valley dotted with houses, lying just behind the north-eastern corner of Maidan, and probably forming the base whence the attack on the 48th Northamptons was delivered on the 9th November. The surface of the hill summit was turf, in which were more or less loosely embedded large stones, with which we began to build a sangar for our protection. Half the men were kept under arms, while the other half busied themselves in kicking up the stones and piling them up as a low wall enclosing a rectangle of about forty by twenty yards with a dividing wall joining the centres of the two longer sides.

"The work was progressing, when at 11.30 the enemy, who had crept up the precipice on the east, managed to open fire among us. Although no one was hit, the bullets began to fall inside our work, so 'C' company, under Captain Nicholson, was sent down a spur on the east which terminated in a natural rocky sangar. This had the effect of protecting the men working on the hill, though 'C' company itself was exposed to fire from the enemy, who had established themselves among the rocks half-way down the slope, due east of 'D' company's position. I therefore posted another party on the crest of the slope south-east of our sangar to keep them in check and to bring long-range fire to bear on the watercourse leading from the south-east.

"As I saw from our position that we should form the connecting link in heliographing between the camp at Bagh and the bivouac when established at Dwatoi, I had the heliograph set up on the north-western corner of our sangar, and told the signallers to establish communication with Bagh.

"Across the defile on the hills opposite to us on the west could be seen the line of the 2nd Gurkhas, to whom had been allotted the task of protecting the left flank of the advance. Below in the defile the long line of transport was observed, wading through the stream some 1,500 feet or more below us. We endeavoured also to get into communication with these in order to deflect our transport, which had been ordered to accompany that of the main column. I felt doubtful whether any mules could climb to where we stood, but I hoped to get them up to some point whence we could carry the bedding and rations up the hill by hand.

"Whilst I was busy here the Adjutant came up to report that Lieutenant Jones had been killed, that his subaltern, Lieutenant Watson, was dying, and that a lance-corporal had been severely wounded. It appears that Lieutenant Jones, while superintending the erection of 'D' company's sangar, had caught sight of some of the enemy who, ensconsed in some rocks below him, were firing on 'C' company. He pluckily determined to turn them out, and calling for a few volunteers, ran down the hill. After losing sight of the Afridis he suddenly again espied them, and calling out to Sergeant Griffiths, who had accompanied him, that he was going to outflank them, he ran along the rocky slope until he came within five or six yards of the enemy. He was then seen to jump on a rock and empty his revolver into the nest of Afridis, who bolted down the hill, followed by stones hurled at them by Lieutenant Jones. Unhappily his action had been observed by some of the enemy who were holding some houses about 600 yards to the south-east, and a shot from them cut short a promising life, and deprived the regiment of a loved comrade and the Service of a valuable officer. His companions reached him as he fell, and stood over the body to protect it from the enemy, who were now again advancing.

"Lance-Corporal Brunton was soon severely wounded, but Private Dangerfield stuck to his post, keeping up a rapid fire on the enemy, while Sergeant Griffiths went back to summon reinforcements. It was unfortunate that Lieutenant Jones' action had been seen and his call for volunteers heard by so few. Lieutenant Watson and Colour-Sergeant Simpson, who now at the summons of Sergeant Griffiths came hurriedly to the spot with half the company, had been in complete ignorance of what had occurred. As soon as Lieutenant Watson arrived he unstrapped his rolled warm coat and placed it under poor Jones' head, and took immediate steps to have the body and the wounded lance-corporal carried back to the sangar, while a fire was kept up on the enemy. The retirement having commenced, Lieutenant Watson turned—the last—to leave the spot, when he caught sight of a rifle left behind by one of the bearers. In reaching for it he exposed himself, and was shot through the body.

Colour-Sergeant Simpson arranged for the wounded officer to be carried back, and retired by alternate sections to the top of the hill without further loss.

"As soon as I learnt this sad occurrence I sent Major Bowles back to command the position held by 'D' company, which thus had been deprived of both its officers, and Surgeon-Captain Alexander also took up his post in 'D' company's sangar in order to attend the wounded."

The following extract from the "Standard" of the 21st December gives further particulars of the action at Dwatoi:—

"On the 22nd November a column under the command of Brigadier-General Westmacott, and accompanied by Sir William Lockhart and some of his staff, started off down the Shatoba Valley for a place called Datoi, or Diwatoi (two rivers), where the Shatoba and Rajgul rivers meet. The pass is a very narrow and difficult one, and to move down it was absolutely necessary that the heights on each side should be crowned. The Yorkshire Regiment took the heights on the right flank and moved along them, and the 2nd Goorkha Regiment performed a similar duty on the left flank. It was in performing this duty on the right flank that Lieutenant Jones, Yorkshire Regiment, was killed, and Second-Lieutenant Watson, of the same regiment, was dangerously wounded. Some of the enemy had taken up a position in some trees behind the crest of the hill, and Lieutenant Jones, quite a young officer, who had only lately given up his leave in England in order to rejoin his regiment on service at his own expense, went down the hill with only three men to turn the enemy out. The latter were so well concealed that though they kept on firing, their exact whereabouts could not be ascertained. Lieutenant Jones, regardless of all danger, approached nearer and nearer, until he finally perceived one of the enemy, and fired six shots from a revolver at him. In doing so he greatly exposed himself and received the shot which proved fatal. Second-Lieutenant Watson then brought a section of men down to his assistance, and in spite of the heavy fire which the enemy were keeping up, helped to get Lieutenant Jones into a stretcher, took off his coat to put under his comrade's head, and was himself almost immediately afterwards very dangerously wounded, as well as a Lance-Corporal who was with him. There are great hopes that Second-Lieutenant Watson may live to receive the reward of his courage which he has so gallantly earned, for the bullet has just missed the lung, and his fine constitution may carry him through. It is only a few days ago that Major-General Symons complimented highly two other young officers of this battalion in front of their companies for their dash and courage in leading their men. All this is treated as an every-day affair, but words cannot express the courage and devotion to the glorious duties of our service shown by our officers. Led as they are, nothing can stop our troops, who, British and Native, strain every nerve to follow closely their leaders."

The next important move of the battalion was on the 27th November, when it formed part of a small force which included the 2nd Gurkhas, a half battalion of the Royal Scots Fusiliers, and No. 2 Derajat Mountain Battery, with General Sir William Lockhart and Staff, for the march into the Khanki Valley through the Kahu Darra Pass. This little force was commanded by Colonel Spurgin, the battalion being detailed as baggage and rear-guard.

It took three days for the column to get through the pass, a very difficult one, and the half battalion on rear-guard fought an action every day, ably assisted by the battery. On the first day Sergeant Calvert and one private were mortally wounded, and the day following another man of the regiment was severely wounded.

The heaviest fighting, however, was on the third day. It lasted for twelve hours, and was often at very close range, the retirement being necessarily much retarded by the difficulty of getting the long transport train over the worst part of the pass. The rear-guard had to take up a series of defensive positions along the steep ridges on either side of the pass, and the enemy were wonderfully plucky, pushing up very close at times. The top of the pass was held by the 2nd Gurkhas, and the rear-guard did not reach it till after dark. By that time the casualties in the battalion had numbered eight wounded, including Lieutenant Williams, who was shot through the knee. He marched most of the way into the next camp, refusing to make use of stretchers or dhoolies, which were badly needed by others.

Once over the pass there was no fighting, but owing to the nature of the country and the severity of the winter, great hardships were endured by the battalion. The road down the other side of the pass was very steep and difficult, and lay along the bottom of a narrow gorge for some eight miles. In the dark and impeded as they were with the wounded, the last of the men did not reach the bivouac till 2 a.m.

On December 3rd the column marched to Maril Khan, about five miles to the south, and the day following arrived at Khanki Bazar at 9.30 a.m. The column consisted now of the Green Howards and half the Royal Scots Fusiliers, as the Gurkhas and the Derajat Mountain Battery had left on the 1st December to join General Gaselee in his operations against the Chamkanni Khels.

On the 5th December General Gaselee's force, consisting of his own and Colonel Spurgin's columns, now united, and with the addition of the 2nd Punjab Infantry, left Khanki Bazar at 7 a.m. to cross the Chingakh Kotal, a hitherto unexplored pass, and halted at Bagh till the 8th, previous to a return through the Arhanga Pass.

Mastura was reached the same day, and on the 9th the 2nd Brigade bivouacked at Haidar Khel, on the banks of the Mastura River. The 1st Brigade, which was in front of the 2nd, had arrived at the foot of the Sapri Pass by the 11th December, and as it required reconnoitring and improvement, the 2nd Brigade halted on the 12th and bivouacked at Bar And Khel.

On the 14th the long ascent of the Sapri Pass was commenced. The morning was bitterly cold, and the progress made was very slow. No European had ever crossed this pass before, and it was altogether new ground. The leading brigade had cleared away most of the difficulties, and the Kotal was reached in three hours where heliographic communication was opened with Peshawar, some twenty-five miles distant.

The next day the expedition arrived at Mamanai Camp, where the whole force, including both divisions and the Peshawar column, under General Hammond, was now assembled. On the 17th December the brigade arrived at Bara Fort, seven miles west of Peshawar.

The division moved out of camp on the 19th in two columns to cross the wide stoney plain which stretches from Peshawar to the neighbouring hills, and halted at Jamrud, where the camp was pitched only a mile or so from the entrance to the far-famed Khyber Pass, the scene of so much fighting on the part of British troops for the last seventy years. The battalion here received its tents again, which had been left at Bagh on the 26th November.

The Khyber Pass, with the forts therein, which had been successfully raided by the Afridis during the previous August, and which had remained closed ever since, had to be retaken, and the winter quarters of the Zakka Khels in the Bazar Valley had to be visited, and punishment inflicted in case of continued contumacy. Preparations for this expedition were made betwen the 20th and 23rd December, and on the 24th the 1st Division marched out of Jamrud, the 1st Brigade leading. The battalion was again in light marching order, and all tents were left behind. The Khyber was entered about noon, and the first night's bivouac was under the shadow of Ali Musjid Fort.

The Bazar Valley lay almost due west of the bivouac, and the plan of operations was briefly for the 2nd Brigade to enter it by the Chora Pass and form the left column; the 1st Brigade to enter by the Alachi Pass slightly to the north, while the Peshawar column was to move up the Khyber and along the heights bordering the valley on the north, and, if necessary, assist the retirement of the 1st Division over the

Bori Pass. Both brigades were to meet at the village of Walai, situated within easy striking distance of Cheena, the headquarters of the Afridis in the Bazar Valley.

On Christmas Day the 2nd Brigade started on its march, and bivouacked on the banks of the Chora stream, close to a sort of village of the same name, the total distance being about seven miles. Several towers, fortifications and houses were blown up, but the commonest forms of abode were caves scooped out of the hard shale in the banks of the nullahs or on the hillside, and there were hundreds of these.

The next day the march was continued along the bed of the Chora River. The men had to wade frequently, but after two or three miles the bed became a dry watercourse lying between steep banks. Any number of smoke-blackened caves were passed, and eventually the force reached a plateau about five miles in diameter. At the far end of it could be seen the towers and walls of Cheena.

The advance was not opposed, and the place was found deserted, so after occupying the heights commanding it the brigade bivouacked under the walls. The enemy, however, were not idle, and attempted to harass the rear-guard after their usual custom, when the battalion had one man killed and Sergeant Roche and another man severely wounded. Later on the picquets were fired on, when another man of the battalion was mortally wounded.

The next morning all the towers of Cheena were blown up by the sappers and miners, and with this the object of the expedition was accomplished, and the 2nd Brigade marched back to Jamrud, where it arrived on the 29th December.

A month later, on the 29th January, 1898, three companies of the battalion, under the command of Captain Grubbe, 2nd Battalion Connaught Rangers, took part in an extended operation with the object of capturing some herds of cattle on the Kajurai Plain, near Shin Kamar, which object was not attained. No resistance was offered by the enemy, and there were no casualties, but in another part there was severe fighting, resulting in heavy losses, particularly in the King's Own Yorkshire Light Infantry, who by their steadiness and good fighting qualities in the end cleverly extricated themselves from a very difficult situation.

This brought to an end the active work of the Green Howards in the Tirah expedition, one of the most memorable campaigns in Indian frontier warfare.

On the demobilization of the force the following complimentary order was issued by Major-General Sir W. Penn Symons, K.C.B., under whose command the battalion had served for over eight months:—

"I have nothing but praise to give the 2nd Battalion Yorkshire Regiment for their good work and good service during the Tirah campaign, and whilst under my command. I have formed a high opinion of the battalion, and have had the greatest pleasure in always reporting very favourably on it. Your patience and cheerfulness—both valuable soldierly qualities—lately displayed under very trying circumstances at Ali Musjid have earned my warm approval. I wish you all good fortune and active service as soon as may be."

General Sir William Lockhart issued the following farewell order to the force:—

"Tirah Expeditionary Force orders by General Sir W. S. A. Lockhart, K.C.B., K.C.S.I., commanding.

"Camp Peshawar,
"4th April, 1898.

"On relinquishing the command of the Tirah Expeditionary Force, which is about to be reduced to a single division, I thank all ranks for the work which through their bravery and devotion has been successfully accomplished in the past six months:

"From the beginning of October to the middle of January the Force was engaged in active operations, and seldom have troops been called upon to undergo greater fatigue, or to meet a more vigilant and enterprising enemy. After long marches in cold and wet, harassed by distant rifle fire and by assaults at close quarters, the columns bivouacked in positions which had to be protected by numerous strong picquets posted on commanding heights, and those picquets were always liable to determined attacks, and to molestation on withdrawal. There was, in fact, little or no rest for the force, the most carefully chosen camping-ground being generally open to long-range fire from scattered individual marksmen armed with the most accurate weapons.

"The boast of the tribes was that no foreign army, Moghul, Afghan, Persian or British, had ever penetrated, or could ever penetrate, their country, but, after carrying three strong positions, and being for weeks subsequently engaged in daily skirmishes, the troops succeeded in visiting every portion of Tirah, a fact which will be kept alive in the minds of future generations by ruined forts and towers in the remotest valleys.

"In this recognition of the gallantry and devotion of all ranks, British and Indian, I include the contingents sent by the Princes and Chiefs of India, corps which have fought side by side with the troops of the regular army, and have shared in the dangers and hardships of the campaign.

"For the past two and a half months the troops have been employed on the tedious duties of a blockade, and their discipline during this period is deserving of high commendation.

"I congratulate the soldiers under my command on the successful result of the operations. In no previous campaign on the North-West Frontier have the difficulties to be overcome been more formidable; in none has the punishment inflicted on the tribesmen been more exemplary, or their submission more complete.

"W. S. A. LOCKHART, General,
"Commanding Tirah Expeditionary Force."

The losses in the campaign amounted to one officer (Lieutenant D. E. O. Jones), three sergeants (J. E. House, T. Bromwich, and W. Calvert), and six rank and file killed.

Three officers (Lieutenants B. C. W. Williams, E. G. Caffin and O. C. S. Watson), three sergeants and twenty-six rank and file wounded.

On May 19th, 1900, a memorial brass which had been placed in York Minster to perpetuate the names of those who lost their lives whilst serving in the Tirah campaign was unveiled by Major-General Thynne, C.B., Commanding the North Eastern District. On it appears the names of Lieutenant Jones and forty-three non-commissioned officers and men, nine of whom were killed in action or died of wounds.

HONOURS AND REWARDS FOR THE TIRAH CAMPAIGN.

AUTHORIZATION OF BATTLE HONOURS.

North-West Frontier of India, 1897-8.

Her Majesty the Queen was graciously pleased to approve of the regiment being permitted, in recognition of the services rendered in the military operations on the North-West Frontier of India, 1897-98, to bear on its Colours the word

TIRAH.

Her Majesty the Queen was also graciously pleased to command that the India medal, 1895, with clasps inscribed "Punjab Frontier, 1897-8," and "Tirah, 1897-8," should be granted to the regiment on account of its services in the Tirah campaign.

MENTIONED IN DESPATCHES.

Lieutenant-Colonel W. E. Franklyn.
Major H. Bowles.
Lieutenant C. H. J. Noble.

AWARDED BREVET RANK.

To be Brevet-Colonel—Lieutenant-Colonel W. E. Franklyn (20 May, 1898).

To be Brevet-Lieutenant-Colonel—Major H. Bowles (20 May, 1898).

AWARDED THE MEDAL FOR DISTINGUISHED CONDUCT IN THE FIELD.

No. 930 Colour-Sergeant H. Himlan.
,, 3914 Corporal J. Matthews.
,, 4604 Lance-Corporal Dangerfield.
,, 4396 Corporal F. A. Brunton.
,, 3198 Private J. Purvis.

No. 3426 Sergeant C. J. Griffiths and No. 3412 Dr. Rose were recommended for this medal, but it was not granted.

A gratuity was awarded to all those who served in the campaign. The share was 24 rupees in the following scale:—Lieutenant-Colonel 32 shares, Major 16, Captain 12, Lieutenant 7½, Warrant Officer 4, Sergeant 3, Corporal 2, Private 1.

THE TIRAH ROLL.

Rank.	Name.	Remarks.
Lieut.-Col.	William E. Franklyn ...	In command. Mentioned in despatches. Brevet of Colonel. See text.
Major	Henry Bowles	Mentioned in despatches. Brevet of Lieut.-Col. See Boer War Roll.
Captain	James Ahmuty Fearon	See Boer War Roll.
,,	Alfred G. Cartwright ...	Retired 7 August, 1906.
,,	James T. Cotesworth ...	See Boer War.
,,	Arthur Bayard Elton ...	From 19 Dec., 1897. Invalided 15 Feb., 1898. Retired 2 August, 1902 as Major.
,,	James W. B. Silverthorne	Invalided 5 April, 1898. Retired 3 May, 1899.
,,	William Joseph Todd...	Retired 29 December, 1900.
,,	Hardress Gilbert Holmes	See Boer War.
,,	Cecil L. Nicholson ...	Invalided 1 Dec., 1897. Promoted Major Worcestershire Regt. 8 March, 1905.
,,	Henry R. S. Maitland...	From 5 Dec., 1897. See Boer War.
Lieutenant	Berkeley C. W. Williams	Took part with 18th Royal Irish in the operations on the Samana Range. Wounded 29 Nov., 1897. See Boer War Roll.
,,	Harry Elliott Raymond	Provost Marshall 2nd Brigade. See Boer War.
,,	David Edward O. Jones	Killed at Dwatoi 22 Nov., 1897.
,,	Ernest Gregorie Caffin	Wounded 1 Nov., 1897. See Boer War.
,,	Herbert Albrecht Fulton	Promoted Captain Worcestershire Regt. 20 June, 1900.
,,	Walter L. Alexander ...	Still serving.
,,	Earls Ainslie Hosford...	Still serving.
,,	Edward Bouverie Pusey	Invalided 11 Nov., 1897. Resigned 17 Nov., 1901.
,,	Ralph Elliot Noyes ...	Still serving.
,,	Charles J. H. H. Noble	Regimental Transport Officer. Mentioned in despatches. See Boer War.
,,	Edward Henry Chapman	Invalided 18 Dec., 1897. Still serving.
,,	Christopher V. Edwards	See Boer War Roll. Still serving.
2nd-Lieut.	Oliver Cyril S. Watson	Wounded 22 Nov., 1897. Retired 16 January, 1904.
,,	Charles Temple Morris	From 28 Jan., 1898. To Indian Army 12 April, 1900.
,,	William Joseph Roskell	From 28 Jan., 1898. Still serving
,,	Norman Edward Swan	From 11 March, 1898. Still serving.
Lieut. and Adjutant	Ronald D'Arcy Fife ...	Still serving.
Captain and Q'rmaster	James Greer	Retired 11 July, 1909.

Rank and Name.					Remarks.
2nd-Lieut.	A. E. Jewett	to 18/11/97			Officers of the unattached list attached to the battalion prior to mobilization.
,,	A. Shewell	,, 22/11/97			
,,	E. F. Holland	,, 18/11/97			
,,	F. G. Moore	,, 21/3/98			
,,	A. Mactavish	,, 21/3/98			
Captain	E. Grubbe	from 28/12/97	to	25/3/98	Officers of other corps temporarily attached to fill vacancies caused by wounds and sickness.
Lieut.	J. A. Strick	,, 28/12/97	,,	3/2/98	
,,	J. G. Chaplin	,, 24/12/97	,,	15/6/98	
,,	E. F. Rutter	,, 24/12/97	,,	1/2/98	
,,	A. Wood ...	,, 29/12/97	,,	12/5/98	

CHAPTER XVIII.

Battalion moves to Peshawar—Presentation of Medals for the Tirah campaign—Dagshai and Umballa—Sitapur and Benares—150 n.c.o's. and men leave for South Africa—The Delhi manœuvres and Durbar—M.I. detachment leaves for Somaliland—Bombay—South Africa—Unveiling of Paardeberg memorials—England—The Green Howards' Association—Presentation of new Colours by Queen Alexandra—Gift of a Challenge Shield by the people of Scarborough—King Edward's funeral—Irish manœuvres—York.

THE battalion remained in camp at Jamrud until the end of March, taking its share of picquet duty and in conducting convoys midway to Ali Musjid and Bara. A body of scouts at this time was organised and trained by Lieutenant Hosford.

The depôt, which had been left at Jullundur, was moved to Nowshera in February, 1898, under command of Major Fearon, where the women and children joined it from Ranikhet.

On the 1st April the Green Howards moved to Ali Musjid, and in the middle of June returned to Jamrud. After a short stay there the march was resumed to Peshawar, where the battalion occupied the Right Infantry Barracks vacated by the 2nd Royal Inniskilling Fusiliers.

During the hot weather several deaths occurred from heat-stroke, the men having become debilitated by the trying climate of Ali Musjid, and from the after-effects of the campaign.

On the appointment of Colonel Franklyn as Assistant Adjutant-General of the Scottish District on the 29th September, he was succeeded in the command by Major Charles John Spottiswoode. The latter had been gazetted to the Nineteenth in June, 1875, and all his service hitherto had been passed in the 1st Battalion.

On the 20th January, 1899, Lady Elles, wife of Brigadier-General Sir E. R. Elles, K.C.B., commanding the station, presented medals for the Tirah campaign to the battalion, which was drawn up in three sides of a square on the parade ground. Punctually at noon the General rode up, and was received with the usual salute, after which he addressed the officers and men. After briefly recapitulating the work of the battalion in the campaign, he concluded by saying:—

"Throughout your first campaign you have gained the reputation of being an excellent fighting regiment—a reputation which I am sure you will most jealously guard when again called out for service. The total losses of the regiment were one officer killed and three severely wounded, and nine men killed and twenty-nine wounded.

"In 1865 your first Colonel—Colonel Warden—wrote in the regimental records: 'The Captains could always lead their men, and the men could always follow their lead.' . .

"This was the spirit that animated the battalion in Tirah, and I refer to it now because in the present day, and more especially in hill fighting, it is on the company commanders that so much depends. It is for the commanding officer to train his company commanders so that they will act on their

NOTE.—During the season 1898-1899 the "Peshawar Vale Hounds," which originated as the Green Howards' Hunt in 1866, were hunted by Captain R. D'A. Fife, with Lieutenants E. G. Caffin and W. L. Alexander acting as whips.

Photo, Lafayette.

LIEUTENANT GENERAL WILLIAM EDMUND FRANKLYN, C. B.

own initiative; for cases may occur where not doing so may produce disaster.

"To quote again from your records: 'May the Princess of Wales's Own Yorkshire Regiment always be a sound and well balanced weapon and well tempered, and may the ring of its metal be always sterling and true.'"

When the General had finished his speech the officers were called to the front and their medals were pinned on by Lady Elles. She afterwards paid the same honour to the warrant officers and a number of selected men, while the officers pinned on those of the remaining men of their companies. When all had been decorated the battalion was formed up and marched past. It presented a fine sight, with some 700 men, each with a medal and two clasps on his breast. This terminated the proceedings, and the officers then adjourned to the mess, where they entertained their friends at luncheon.

On the 8th March the Nineteenth left Peshawar by march route, and arrived at Mean-Meer, 278 miles distant, a month later. From Mean-Meer they trained to Kalka, at the foot of the hills, and bivouacked there, finally reaching Dagshai on the 9th April, where they were stationed.

On the 3rd October the battalion, with the exception of "C" company, marched to Kalka, where it entrained for Umballa, arriving there the next day, and taking over the barracks vacated by the 92nd Highlanders.

On the outbreak of the war in South Africa, orders were issued from army headquarters for the retention in India of all time-expired men owing to the suspension of the trooping programme.

Early in April, 1900, a return to Dagshai took place, two companies being left at Umballa for duty.

Consequent on the 34th Pioneers' departure for China on active service, two more companies, made up to 200 strong, moved to Umballa in relief.

In November the battalion was again at Umballa, with two companies detached for the winter months at Dagshai.

February 2nd, 1901, being the day fixed for the funeral of Her Most Gracious Majesty Queen Victoria, a special memorial service was attended by all ranks of the battalion.

Later in the month the Green Howards entrained for Saharunpore, and on arrival there proceeded by march route for Sitapur, where they arrived on the 16th March. A few days later four companies, under command of Major Cartwright, left for Benares.

Orders were received in September authorising the issue of a bounty of £10 and a gratuity of £16 10s. to all men who were willing to extend their army service to complete twelve years with the Colours; 297 men extended their service under these conditions.

On the 2nd February, 1902, a draft of 150 non-commissioned officers and men, under command of Captain A. L. Napier, left Sitapur for Deolali, *en route* for South Africa to join the 1st Battalion. This draft was made up of men who had completed seven years' service and had not taken the bounty above alluded to. The men volunteered eagerly, and were as fine a body of soldiers as could be found anywhere. They were sent to the 1st Battalion in exchange for a similar number who arrived from South Africa in March.

"A" company, under Major Hadow, proceeded to Bhimtal from Benares on the 7th March to act as a guard over Boer prisoners. It was followed on the 1st April by another company, under Second-Lieutenant K. Simonet.

Colonel C. J. Spottiswoode completed his term of command on the 28th September, and was succeeded by Major and Brevet-Lieutenant-Colonel James Ahmuty Fearon. He had been appointed to the Nineteenth in January, 1878, when he joined the 1st Battalion in Bermuda, to which he was Adjutant for seven years, 1880-1887. He had served on the Nile and in the Tirah campaign, and had taken part in all the principal fighting in which the 1st Battalion had been engaged in South Africa.

All the outlying detachments having rejoined headquarters, the Green Howards left Sitapur by march route to attend the Delhi manœuvres and Durbar. A depôt was left at Sitapur, under command of Captain E. H. Chapman.

On arrival at Ghaziabad on the 20th November, the march was continued to Sagaar Jung's Tomb, seven miles outside Delhi, the concentration camp of the Southern army. Here the battalion was brigaded with the North Staffordshire Regiment and the 4th Infantry, Hyderabad contingent. This brigade moved out of camp on the 27th November, and reached Panipat on the 1st December.

After taking part in the manœuvres the Green Howards marched into the Durbar Camp at Delhi, and on the 29th December the Viceroy made his state entry into the city, the regiment being detailed as part of the troops lining the streets.

The great Durbar was held on New Year's Day, 1903, when the King Emperor was proclaimed, and a week later a

NOTE.—The battalion football team won the Murray Football Tournament, 1902, open to all corps in the Bengal Command.

review of all the troops took place, the total number on parade being 29,616.

On the 11th January, 1903, the battalion marched out of Delhi for Cawnpore where it arrived on the 6th February, having left at Muttra one company, under Captain W. L. Alexander, which proceeded to Fatehgarh. The total strength was thirteen officers and 590 non-commissioned officers and men.

In May, Colonel Sir Charles Leslie, Bart., commanding at Cawnpore, presented South African medals to the men of the battalion entitled to receive them, the total number being 205.

On it being found necessary to despatch a force to Somaliland, a detachment of two officers (Lieutenants C. V. Edwards and A. L. Godman) and 65 rank and file left Cawnpore in June to join the Mounted Infantry Company concentrating at Fatehgarh for service with the expedition. In addition to the two Yorkshire sections, there was also a section from the Rifle Brigade and another from the Norfolk Regiment. This company took part in all the hard work incidental to the campaign, and the officers and men received the African General Service medal and clasp.[1]

In September, 1904, orders came for a move to Bombay, where the battalion arrived on the 6th October, whence it sent out detachments of two companies to Deesa and one to both Ahmedabad and Deolali.

A draft from England was waiting at Bombay, which brought the battalion up to its greatest strength during its tour abroad, viz., 1,192.

The headquarters remained at Bombay throughout the year 1905, and took part in the various ceremonies in connection

1 An interesting account of the expedition will be found in the "Green Howards' Gazette," Vol. XII., written by Lieutenant Godman.

with the visit of T.R.H. the Prince and Princess of Wales in November.

On the 16th February, 1906, the battalion, 858 strong, after having been in India for just over sixteen years, embarked at Bombay for Cape Town, where it arrived on the 10th March, and on landing proceeded to Wynberg.

In May Colonel Fearon departed for England on leave pending retirement, when the command fell to Major Arthur de Salis Hadow, who was gazetted Lieutenant-Colonel on the 29th September. He had been commissioned to the Green Howards in December, 1878, and with the 1st Battalion had served in the Nile expedition of 1885.

In September a mounted infantry company, under the command of Captain A. L. Godman, with four officers and 146 rank and file, left Wynberg to join the 2nd M.I. Regiment at Harrismith.

The Green Howards remained at Wynberg till December, 1907, when they entrained for Middleburg, in the north of Cape Colony. Here they were joined by the M.I. company in June.

On the 18th February, 1908, the eighth anniversary of the battle of Paardeberg, Lieutenant-Colonel Hadow and a party of eight officers and sixteen men of the battalion proceeded from Kimberley, where they had arrived overnight, to the battlefield of Paardeberg. Two memorials had been erected there, one an obelisk of white marble, on which are recorded the names of Second-Lieutenant A. C. Neave and the forty men who were killed or died of wounds on that eventful day. The other, a marble cross, had been set up on Kitchener's Kopje to mark the graves of Sergeant Richardson and the eight men who lie buried there, in addition to a similar number of Boers.

These monuments were subscribed for by the regiment at large, and it was a happy chance that permitted some of the Green Howards to be present at the unveiling ceremony, fifteen of whom, including Major Christian, D.S.O., had taken part in the battle.

The officers stood in front of the obelisk, the men resting on their arms reversed a few paces away, whilst Lieutenant-Colonel Hadow delivered an impressive and appropriate address, after which three volleys were fired and the "Last Post" sounded.

The party then marched across to Kitchener's Kopje, where Lieutenant-Colonel Hadow performed a similar ceremony, narrating in his address the circumstances in which the men fell, and expressing the hope that Briton and Boer would be allowed to rest there peacefully as they lay, their toil being ended and their warfare o'er.

After nearly two years' service in South Africa the battalion sailed for England in the hired transport "Soudan," where it arrived on the 1st March, 1909.

During its tour abroad 3,379 officers, non-commissioned officers and men served in the battalion, and of these four officers and 219 of other ranks died from various causes. The four officers were:—

Lieutenant and Adjutant W. S. Brooksbank, at Bangalore, on the 18th June, 1891.

Lieutenant D. E. O. Jones, killed in action at Dwatoi, 22nd November, 1897.

Second-Lieutenant H. J. Kirkpatrick, at Benares, 16th April, 1902.

Lieutenant and Quartermaster F. Pettitt, at Wellington, Madras, 7th April, 1894.

Of the 795 who sailed with the battalion for India on the 1st January, 1890, only seven in all returned with it, viz.:—

Lieut.-Colonel Hadow.	Sergeant Brown.
Captain and Qr.-Mr. Greer.	,, Asbrey.
Sergt.-Major Howes.	Corporal Stevens.
Qr.-Mr.-Sergt. Simpson.	Bandsman Colven.

On arrival at Southampton the regiment entrained for York, and was quartered in Fulford Barracks. Owing to limited accommodation there a detachment of two companies, under command of Major Sinclair-MacLagan, was sent to Scarborough.

At the annual celebration of the battle of the Alma this year the first meeting of the Green Howards' Association took place, which was formed by the regiment with the following objects:—

(1) To maintain connection between the warrant officers, non-commissioned officers and men serving in the 1st and 2nd Battalions and old comrades, and to promote friendship and association amongst those now in civilian life, still interested in their old regiment.

(2) To hold an annual dinner and gathering, at which old comrades of all ranks may be able to meet.

(3) To foster *esprit de corps*, and promote recruiting for the regiment, and to make widely known the advantages of service in the regiment; and to create as well as keep up friendship between past and present non-commissioned officers and men of the regiment.

(4) To grant assistance, financially or otherwise, as the funds of the Association may permit, to any member who has fallen into bad health and is unable to earn his own livelihood, or is out of employment and in distressed circumstances

through no fault of his own, also to assist widows and children of members if in need.

(5) To enable members to assist one another in obtaining employment.

The next important event in the history of the battalion was the presentation of new Colours by Her Majesty Queen Alexandra, on the 17th November, at Windsor Castle.

The deputation from the regiment consisted of Major-General W. E. Franklyn, C.B., Colonel A. de S. Hadow, Lieutenant and Adjutant L. Peel, Lieutenant R. S. Ledgard (King's Colour), Lieutenant C. G. Forsyth (Regimental Colour), and Sergeant-Major Howes. Four colour-sergeants formed the escort.

At 3 o'clock their Majesties the King and Queen, accompanied by the Queen of Norway, the Princess Victoria, Princess Louise, the Duke and Duchess of Argyll, and suite, entered the State drawing-room and stood facing the Colours.

The Chaplain then proceeded with the dedication service, after which Lieutenants Ledgard and Forsyth advanced, and taking the Colours off the drums knelt with them in front of Her Majesty.

Queen Alexandra, holding the Colours and crossing the poles, spoke as follows:—

"It gives me the greatest pleasure to commit these Colours to the care of my own regiment, and it gives me still more pleasure to know that your last Colours were given you by my dear old friend, General Sir William Knollys. May they lead you on to honour and glory."

Major-General W. E. Franklyn replied as follows:—

"It was my duty two years ago to convey the humble thanks of the regiment to your Majesty for presenting new

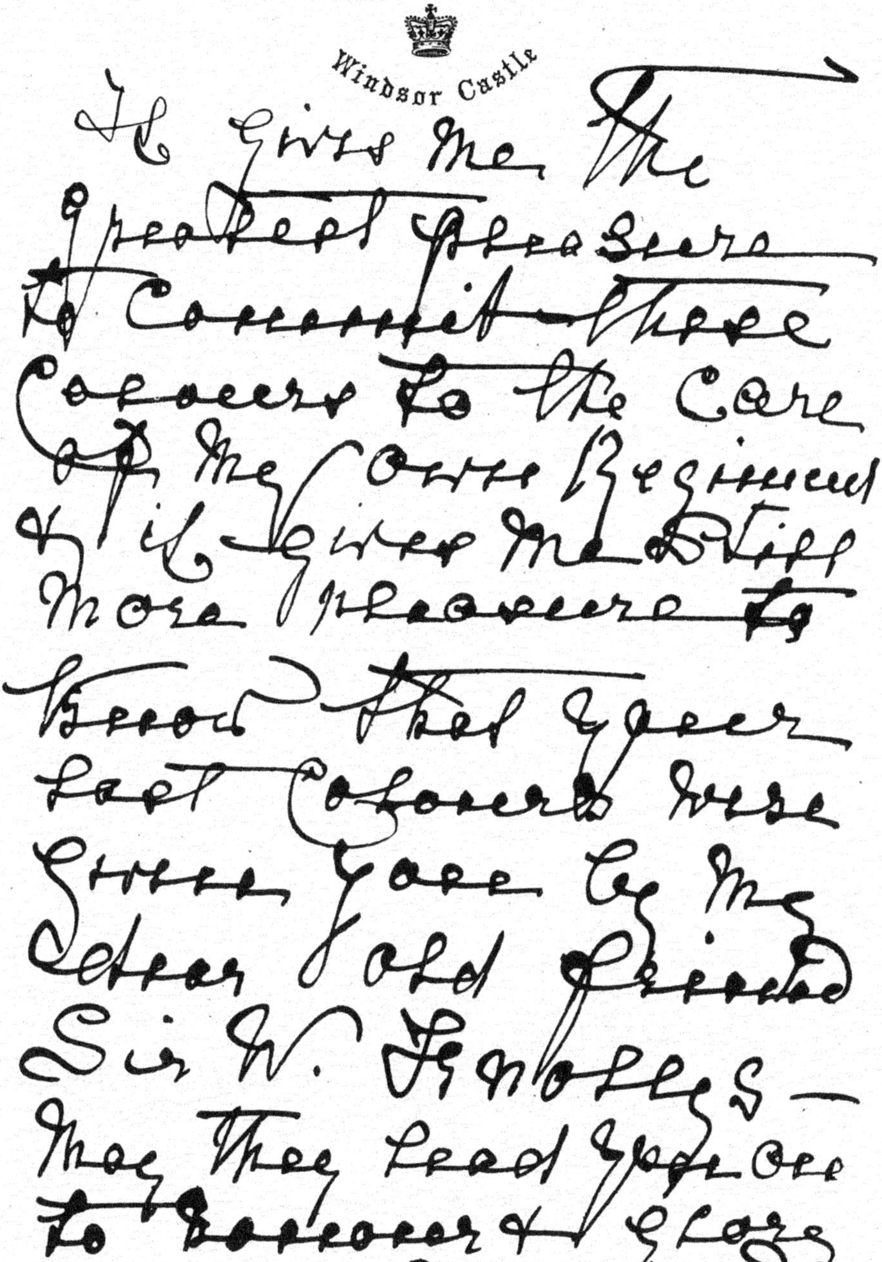

Windsor Castle

It gives me the greatest pleasure to commit these Colours to the care of my own Regiment & it gives me still more pleasure to know that your last Colours were given you by my dear old friend Sir W. Knollys — May they lead you on to honour & glory
Alexandra

FACSIMILE AUTOGRAPH LETTER OF HER MAJESTY QUEEN ALEXANDRA.

Colours to the 1st Battalion. To-day it is once more my privilege to offer equally loyal thanks on behalf of your regiment to your Majesty for presenting to the 2nd Battalion these emblems of duty—emblems which I feel confident will never fail to increase in us a spirit of devotion to our King and country, and this spirit will be enhanced by the presence of our Sovereign here to-day.

"Words fail me to express the feelings which have been engendered in both battalions by these repeated acts of gracious consideration. They will bind the regiment to your Majesty with ever closer ties of loyalty and affection."

After the ceremony the officers were severally presented to their Majesties.

In order to commemorate the first visit of the regiment to the town of Scarborough when the battalion went through its annual training there in August, 1909, the inhabitants presented it with a handsome silver challenge shield. The presentation was made on the 3rd February, 1910, by the Earl of Londesborough, who was supported by the Mayor and many of the leading townspeople.

The two companies at Scarborough rejoined headquarters during April.

On the occasion of the funeral of King Edward VII., the battalion, under command of Colonel Hadow, entrained for London early on the 20th May to take part in lining the streets. After breakfasting at Regent's Park the men marched down the Marylebone Road, where they took up a position in Oxford and Cambridge Road till the procession passed. The battalion returned to York the same day.

On the 3rd July the regiment left York for Ireland to go through brigade training and the annual manœuvres. On arrival at Cork the next day it proceeded to Fermoy, and went into camp at Moore Park, a few miles distant. Here it remained till September the 12th, when the manœuvres commenced, and at their close the battalion returned to York on the 20th September.

On the 28th September the command once more changed hands on the completion of Colonel Hadow's four years' tenure, when Major Charles Arthur Cecil King was promoted to the vacancy. He had been appointed to the Green Howards in September, 1882, and with the 1st Battalion had seen service on the Nile with the Frontier Field Force, and was present at the action of Ginnis. With the 2nd Battalion he had taken part with a column operating in the Katchin Hills, and as Adjutant to the 3rd Battalion accompanied it to South Africa during the Boer War, being twice mentioned in despatches and receiving the brevet of Major.

In order to celebrate the fact that the Green Howards were quartered in the North Riding for the first time since their connection had been established with the county in 1782, the residents determined to present the battalion with a piece of silver plate to mark the occasion. The presentation was made to the battalion at a full parade on the 8th February, 1911, by the Lord Lieutenant of the North Riding, Sir Hugh Bell, Bart., who in an effective speech alluded to its long connection with the county.

The trophy consists of four figures of soldiers in the uniforms of different periods. They are grouped round the Colours and the Alma drums. A silver band round the ebony plinth has the following inscription:—

"Given by the North Riding of Yorkshire to the 2nd Battalion of Alexandra, Princess of Wales's Own (Yorkshire

Photo, Thomson.

LIEUTENANT COLONEL CHARLES ARTHUR CECIL KING

COMMANDING 2ND BATTALION.

Regiment) on the battalion for the first time in its history being quartered in the Riding."

The battalion went to Rhayader, North Wales, on the 31st July, where it was encamped for the purpose of being exercised in field manœuvres with other regiments of its brigade. This course of training was interrupted by the railway strike in August, and it was found necessary to despatch the Green Howards as well as other troops to Liverpool to assist in maintaining order there owing to rioting having broken out in connection with the strike. The battalion afterwards resumed its training at Rhayader, and returned to York on the 11th September.

On the 27th September the battalion moved to Blackdown Camp, near Aldershot.

ALEXANDRA, PRINCESS OF WALES'S OWN (YORKSHIRE REGIMENT).

Regimental District No. 19. (No. 5 District.)

The Cypher of H.R.H. Alexandra, Princess of Wales, in gold (thereon "Alexandra"), interlaced with the Dannebrog inscribed with the date 1875, and the whole surmounted by the Coronet of the Princess.

"Malplaquet," "Alma," "Inkerman," "Sevastopol," "Tirah," "South Africa, 1899-1902," "Relief of Kimberley," "Paardeberg."

Regular and Special Reserve Battalions.

Uniform—Scarlet. *Facings*—Grass Green. *Agents*—Messrs. Holt & Co.

1st Bn. } 19th Foot { *Khartoum.*
2nd " } { *Blackdown, Farnborough.*

Depôt .. ..*Richmond.*

3rd Bn. (5th West York Mil.) .. .. .. *Richmond.*

Record Office .. .. *York.*

Territorial Force Battalions.

4th Bn. .. .. *Northallerton.* | 5th Bn. .. .. *Scarborough.*

Colonel .. .. Franklyn, Lt.-Gen. W. E., *C.B.*, *p.s.c.* .. .. 2 Oct. 06

1st and 2nd Battalions.

Lt.-Colonels.

2 King, C. A. C. 29 Sept. 10
1 Christian, G., *D.S.O.* [*I*] 28 Mar. 11

Majors. (8)

2 Fife, R. D'A. 16 Feb. 06
s. Caffin, E. G. (1) 28 Dec. 07
1 Alexander, W. L., *e.* 5 Feb. 08
2 Hosford, E. A. (*Comdg. Depot*) 4 Mar. 08
Sinclair-MacLagan (2) E. G., *D.S.O.* 28 Oct. 08
1 Noyes, R. E. 29 Sept. 10
2 Chapman, E. H. 11 Feb. 11
1 Edwards, C. V. 28 Mar. 11

Captains. (14)

2 Walker, W. B. 25 June 04
m. *Carey, M.D.* (3) 1 Oct. 04
r. *Stansfeld, T.W.*, D.S.O. (4) 8 Jan. 08 19 Feb. 02
2 Darwin, R. H. 1 Nov. 04
2 Roskell, W. J. 18 Feb. 05
s c. *Swan, N. E.* (5) 18 Feb. 05
c.o. *Humfrey, L. E. H.* (6) 19 July 05 27 Sept. 02
1 Bunbury, C. H. de St. P. 25 Dec. 05
2 Godman, A. L. 16 Jan. 06
Broun, E.S. (7) 5 Apr. 06
McCall, H. W. (8) 29 June 06
1 Leatham, B. H. 21 July 06
Moss-Blundell, B. S. (9) 29 July 06
1 Bastow, H. V. 14 Feb. 08
Marsden, C. H. (10) 10 Oct. 08

Captains—contd.

2 Nash-Wortham, B. C. D. 20 Mar. 09
2 Simonet, K. W. L. (11) 1 Apr. 09
t. *Rollo, W. K.* (12) 12 May 09
1 Gardner, C. J. H. 22 July 09
2 *Jeffery, C. G., Adjt.* 25 Aug. 09
1 Maddison, B. L. 29 Sept. 10
1 Whatford, S. L. 9 Oct. 10
1 Westley, J. H. S. 20 Jan. 11
2 Peel, L. 11 Feb. 11
1 *Godwin, C. C., Adjt.* 11 Feb. 11
1 Ledgard, R. S. 28 Mar. 11
2 Cumberbatch, H.C. 19 July 11

Lieutenants. (19)

c.o. *Oakes, O.* (13) 8 Oct. 06
2 Lanyon, L. F. 8 Oct. 06
1 Levin, H. (14) 8 Oct. 06
2 Burbury, B. T. 27 Apr. 07
c.o. *Jervelund, C. N.* (15) 27 Apr. 07
c.o. *Corser, R. B.* (16) 11 May 07
c.o. *Ramsden, B. V.* (17) 16 Jan. 08
1 Franklyn, H. E. (18) 16 Jan. 08
1 Grant-Dalton, S. 14 Feb. 08
2 Forsyth, C. G. 12 May 09
1 Mintoft, T. C. [*I*] 22 July 09
s. *Hooton, A.C.* (19) 25 Aug. 09
1 Worsdell, G. B. 25 Sept. 09
1 Smith, G. N. N. 17 Nov. 09
c.o. *Magee, D. H.* (20) 29 Sept. 10
Lilley, A. H. (21) 9 Oct. 10
1 Bradford, C. A. 9 Nov. 10
1 Palmer, A. E. G. 28 Nov. 10
2 Nevile, G. L. (22) 11 Jan. 11
1 Blackwood, H. F. 16 Jan. 11
1 Richardson, E. J. 20 Jan. 11
1 Atkinson, R. G. 11 Feb. 11
1 Cuff, B. 11 Dec. 09
2 Phayre, R. H. (23) 11 Dec. 09
1 Manly, M. U. 25 Dec. 09
1 Compton-Smith, G. L. 19 July 11

2nd Lieutenants. (12)

2 Smith, C. R. 20 Apr. 10
1 Bush, G. 18 Feb. 11
1 Bagnall, H. S. 4 Mar. 11
1 Bennedik, S. R. W. 4 Mar. 11
2 Birch, W. C. K. 4 Mar. 11
2 Ledgard, F. C. (24) 25 Mar. 11
2 Kreyer, H. S. 25 Mar. 11
1 Maude, G. W. E. 27 May 11
2 Marriage, L. H. 20 Sept. 11
2 Le Sueur, E. G. C. 20 Sept. 11
2 Chauncy, W. A. A. (25) 20 Sept. 11

Adjutants.

2 Jeffery, C. G., *capt.* 1 Mar. 10
1 Godwin, C. C., *capt.* 16 Jan. 11

Quarter-Masters.

r. *Sheridan, J.* 18 Dec. 99 *hon. capt.* 18 Dec. 09
1 Walker, J., *hon. lt.* 1 Oct. 03
2 Pickard, E., *hon. lt.* 28 July 09

3rd Battalion,

(*Officers serving on* 14 *May* 02 *in the corresponding Militia unit hold honorary Army rank equivalent to the Militia regimental rank they then held. Other officers entitled to honorary Army rank have it shown against their names.*)

Hon. Colonel.

Zetland, Rt. Hon. *L.*, Marquis of, K.T., *late* Lt. R.H.G. 26 May 09 17 Feb. 06

Lt.-Colonel.

p.s. Lodge, J. W., *hon. c.* 14 May 06

Major.

p.s. Turton, R. B. 26 Feb. 02
p.s. Gunter, *Sir* R. B. N., *Bt.*, *hon. l.-c.* (*H.*) 4 July 08
p.s. Aspinall, R. L., *D.S.O.*, *hon. l.-c.*, Capt. ret. pay (Q) (*H*) *p.v.c.*, *P.c.* 14 May 06

Captains. (5)

Somervell, E. (26), *hon. m.*, Capt. ret. pay. (*Res. of Off.*) (*H*) 28 July 96 20 May 05
Stansfeld, H. A. (27), *hon. m.*, Capt. ret. pay (*Res. of Off.*) (*H*) s. (*Brig. Maj.* 1 *W. Rid. Inf. Brig.*) 29 Dec. 00 9 May 06
p.s. Hill, R. 21 Feb. 03
p.s. White, C. R. (*H*) 22 Oct. 04
Stanley, W. B., *late* Lt. W.I.R. (*H*) c.o. 30 June 06
p.s. Gladstone, R.M. (*H*) 15 May 08
Nevile, M. M. H. (28), Capt. ret. pay 29 Jan. 10

Lieutenants. (11)

Raley, W. H. G. (*H*) 30 May 04
Bell, W. F. I. (*H*) 1 Feb. 09
(*p.s.*) Grey, G. L. (*H*) 16 June 09
Hanbury, P. (*H*) 16 June 09
Gladstone, S. P. (*H*) 10 June 10
Wilkie, D. (*H*) c.o. 10 June 10

2nd Lieutenants. (8)

Thorne, A. T. (*on prob.*) 28 July 10
Walmesley, R. (*on prob.*) 18 Aug. 10

Adjutant.

Stansfeld, T. W., *D.S.O.* Capt. York, R. 9 Oct. 10 (*Capt. in Army* 19 *Feb.* 02)

Quarter-Master.

Sheridan, J., *hon. capt.*

---

(1) Recruiting Staff Officer, Dublin
(2) Employed with the Australian Forces
(3) Adjutant 1st Guernsey Light Infantry
(4) Adjutant 3rd Bn., Brother of Major H. A. Stansfeld 3rd Bn.
(5) Student at the Staff College
(6) District Commissioner Southern Nigeria. Son of Lieut.-Col. J. C. T. Humfrey, late 19th Foot
(7) Aide-de-Camp to Sir L. Probyn, Governor of Barbados
(8) Adjutant 5th Bn., Scarborough
(9) Adjutant 4th Bn. Yorkshire Light Infantry
(10) Instructor at School of Musketry, Hythe
(11) Nephew of Lieut.-Col. J. F. Simonet, late 19th Foot and A.P.D.
(12) Adjutant 4th Bn. York and Lancaster Regiment
(13) Employed with Northern Nigeria Regiment
(14) Son of the late Major L. M. Levin, 19th Regiment
(15) Employed with Northern Nigerian Regiment
(16) Employed with the Gold Coast Regiment
(17) Employed with the Gold Coast Regiment
(18) Son of Lieut.-Gen. W. E. Franklyn, C.B., Colonel of the Regiment
(19) Aide de Camp to Sir G. R. Le Hunte, Governor of Trinidad and Tobago.
(20) Employed with the Gold Coast Regiment.
(21) Employed with the King's African Rifles.
(22) Brother of Captain M. H. Nevile, 3rd Bn., and late of 1st Bn.
(23) Son of Lieut.-Col. R. Phayre, late 19th Foot
(24) Brother of Captain R. S. Ledgard, 1st Bn.
(25) Son of Major W. A. Chauncy, late 19th Foot
(26) Late Captain, 1st Bn.
(27) Late Captain. 1st Bn. Brother of Captain T. W. Stansfeld, D.S.O.
(28) Formerly 1st Bn. Brother of Lieut. G. Nevile, 2nd Bn.

The Grand or Slow March.

mp
ff

Regimental "March Past" or Quickstep.

THE BONNIE ENGLISH ROSE.

Arranged by J. MURRELL.

\ England's emblem is the Rose,
There is no other flower
.th half the graces that adorn
This beauty of the bower;
d England's daughters are as fair
.s any bud that blows,
hat son of hers who hath not loved
Some bonnie English Rose?

Who hath not heard of one sweet flower,
The first among the fair,
For whom the best of British hearts
Have breathed a fervent prayer?
Oh, may it never be her lot
To lose that sweet repose,
That peace of mind which blesses now
The bonnie English Rose!

If any bold enough there be
To war 'gainst England's isle,
They soon shall find for English hearts
What charms hath woman's smile;
Thus nerved the thunder of their arms
Would teach aspiring foes,
How vain the power that defies
The bonnie English Rose!

Funeral March.

PIANO.

Lafleur & Son, London, W.

TRIO.

APPENDIX.

SUCCESSION LIST OF COLONELS.

Colonel Francis Luttrell, 20th November, 1688.
General the Right Honourable Thomas Erle, 1st January, 1691.
Br.-General George Freke, 22nd January, 1711/12.
Br.-General Richard Sutton, 3rd April, 1712.
Colonel George Grove, 5th August, 1715.
Lieutenant-General Richard Sutton (reappointed), 24th October, 1729.
General the Hon. Sir Charles Howard, K.B., 1st November, 1738.
Lieutenant-General Lord George Beauclerk, 15th March, 1748.
General David Graeme, 25th May, 1758.
Field-Marshal Sir Samuel Hulse, G.C.H., 24th January, 1797.
General Sir Hew Dalrymple, Bart., 25th June, 1810.
General Sir Tompkins Hilgrove Turner, G.C.H., K.C., 27th April, 1811.
General Sir Warren Marmaduke Peacocke, K.C.H., K.C., 31st May, 1843.
Lieutenant-General Charles Turner, 1st September, 1849.
Field-Marshal Sir William Rowan, G.C.B., 15th June, 1854.
General Sir Abraham Josias Cloeté, K.C.B., K.H., 10th March, 1861.
General Sir Robert Onesiphorus Bright, G.C.B., 27th October, 1886.
Lieutenant-General Edward Chippindall, C.B., 16th November, 1896.
Major-General William Spencer Cooper, 14th September, 1902.
Lieutenant-General William Edmund Franklyn, C.B., 2nd October, 1906.

CASUALTIES.

THE CRIMEAN WAR.

THE ALMA.

OFFICERS.

Killed: Ensign G. Stockwell.

Wounded: Lieutenant-Colonel R. Sanders, Major H. McGee, Captain R. Warden, Lieutenants L. Currie and R. Wardlaw (died subsequently), Ensign and Adjutant A. Cardew.

OTHER RANKS.

Killed: Corporal R. Hines, Drummer J. Calkin, Privates W. Stillwell, F. Giles, T. Pye, P. Gaynor, T. McNicholl, T. Furnival, L. Spencer, G. Bastion, D. Quinn, W. Brown, L. Griffin, R. Jones, J. Leara, W. Thomas, H. Allen, J. Blackburn, J. Doherty, E. Scanlan, T. Baker, C. Conway, T. Downes, A. Eathell, W. Fitzpatrick, J. Hanlon, J. Hitchcock, R. Avery, J. Dobbins, T. Doyle, M. Kelcher, G. Luttrell, P. Lynam, J. Scullen, W. Walsh, W. Ward, A. Young, S. Reeves.

Wounded: Colour-Sergeant W. Rawding, Sergeants J. Rawding, P. Daly, F. Arthur, J. Carvill, G. Hardygrave, J. Smith, F. Lee, C. Price, F. Bergin, Corporals T. Farrar, W. Bromley, W. Pope, J. Fair, M. Young, P. Barlow, J. Annon, J. Alwell, J. Brown, T. Cox, M. Conners, J. Fagg, A. Buchanan, I. Logan, W. Voisey, T. Maddigan, F. Murray, J. Dodds, R. Atkinson, Drummer T. McCarthy, Privates P. Byrne, J. Darsey, W. Eade, W. Jones, S. Farise, J. Williams, T. Lacey, W. Smith, R. Newcom, R. Roberts, J. Rogers, J. Butler, W. Liddle, J. Brennan, J. Nicholson, J. Parker, M. Connolly, M. Cresswell, R. Davis, T. Harriss, F. Lyons, R. Morgan, W. Collett, P. Cowan, G. Grant, D. Lyons, W. Taite, P. Young, J. Sault, M. Donohoe, J. Jolly, P. Peel, F. Armstrong, W. O'Hara, D. Davis, T. Scanlon, J. Carroll, J. Goulding, T. Jones, M. Neale, J. Watson, J. Jarvis, W. Barrett, C. Aistin, J. Alderdice, W. Andrews, W. Bailey, P. Beggins, J. Blythe, M. Burke, P. Campbell, R. Campbell, H. Candling, T. Chadwick, J. Clare, J. Clarke, M. Doolan, T. Donoghoe, W. Doran, W. Gooch, M. Harrigan, P. Kehoe, H. McNamara, P. McNamara, J. McNearney, M. Maroney, T. Bailey, G. Jefferies, W. Battison, W. Harris, L. Hickey, T. McEvoy, W. McGruggan, J. Taylor, W. Walsh, E. Munell, J. Cannon, T. Wheatley, P. Gaffney, E. Fenaker, J. Hanlon, P. Hayes, J. Herrity, J. Hurley, J. Keating, J. Leonard, P. Maloney, A. Murphy, M. Morriss, H. Peacock, E. Rafarley, M. Shannon, T. Williams, J. Bromphy, P. Carse, E. Edwards, R. Hearon, M. Hurst, W. Kelley, P. Logan, M. McCall, T. Martin, R. Pittman, J. Parker, J. Salan, J. Sweeny, T. Turner, J. Vines, J. Wall, T. McGuire, J. Mullins, J. Houraghan, W. Burke, M. Donlan, J. Carroll (2), T. Higgins, W. Smith, P. Quigley, W. Turner, M. Foley, J. Connolly, W. Harrison, M. O'Brien, C. Carroll, A. Smith, T. Downes, B. Raffles, J. Sullivan, J. Murphy, T. Murphy, T. Murdagh, P. Purcell, W. Rawson, G. Richards, J. Sheppard, J. Solomon, W. Twomey, J. Walker, T. White, W. Dubage.

Missing: J. Hennessy, J. Stephens, B. Rourke, J. Boland, P. Lawler, G. Richards.

INKERMAN.

OFFICER.

Wounded: Captain James Ker (died subsequently).

OTHER RANKS.

Killed: Private L. Cooper.

Wounded: Sergeant-Major H. Madden, Privates R. Abbott and John Seal (all of whom died subsequently).

DURING SIEGE OF SEBASTOPOL.

OFFICERS.

Wounded: Lieutenant E. W. Evans, Lieutenant R. D. Barrett, Captain E. R. W. Bayley, Ensign A. Goren.

OTHER RANKS.

Killed: Sergeant-Major William Dunn, Corporal J. Roe, Privates T. Donoughoe, P. M. Gun, J. Donnelly, J. Reynolds, J. Rush, H. Holdsworth, B. McFadyen, G. Moore, F. Osborn, J. Kelly, J. Burke, P. Brown, T. Madding, W. Smith, W. Doyle.

OTHER RANKS.

Wounded: Colour-Sergeant P. Campion, Sergeant H. McClister, Corporals J. Austin, T. Cruikshank, E. Black, T. Keating (twice), W. Costello, T. Robins, S. Reed, Lance-Corporals J. Rudd, T. Toolan, J. Burke, W. Davis, Drummer H. Hanlon, Privates J. Maher, J. Corbett, T. Dixon, M. Laughlin, J. Dunn, J. Renny, J. Beer, G. Haigh, J. Flinn, E. Keating, J. Gomm, W. Crowley, M. Frawley, W. Morey, D. Hawkins, J. Kearney, M. Bennett, C. Aistin, A. Dugan, J. Hales, A. Moore, R. Foyles, J. Behen, E. Tyler, J. Higgings, M. Tolan, M. Hennagan, J. Sheehan, J. Murgatroyd, I. Bridgeman, T. Ash, E. Painting, R. Lynch, J. Keans, 3397 McNamara, M. Carron, W. Hayes, T. Dickenson, R. Newcom, M. Kean, M. Roach, J. Walsh, G. Watson, J. Crowe, J. Feltus, W. Healy, C. Gowing, W. Menzies, T. Plant, T. Collingwood, H. Lavery, J. Regan, J. Thom, P. Lee, P. Ryan, E. Connor, T. Measures, G. Royce, W. Weeden, P. Baldwise, W. Ingram, 2216 W. McHugh, T. Rourke, J. Riley, 2924 W. McHugh, J. Holyoak, D. Cooper, M. Lydon, P. McNamara, P. Burke, G. Russell, W. Lynch, F. Garagan, J. Parker, T. Murphy, J. Kelly, J. Collins, O. Sullivan, J. Armstrong, W. Liddle, W. Hayes, W. O'Hara, W. Crampton, J. Wheelan, P. Minahan, W. Dugan, G. Reed, J. Murphy, R. Bray, J. Crowe, J. Behan, T. Brown, E. Wheatley, J. Clarke, T. Rourke, J. Hennessey, T. Wareing (twice), P. Farrell, J. Carter.

Missing: Corporal J. Dawson.

ASSAULT ON THE REDAN, 8th September, 1855.

OFFICERS.

Wounded: Lieutenant-Colonel T. Unett, C.B. (died subsequently), Captain and Bt.-Major R. Warden, Captain E. Chippindall, Lieutenants P. Godfrey (died subsequently), W. G. D. Massy, A. Goren, R. Molesworth, E. Bayley, Ensigns R. Martin, W. Young.

OTHER RANKS.

Killed: Sergeants P. Taffe, S. Weston, M. Jannon, Corporals W. Bromley, G. Morrish, J. Smith, M. Murphy, Privates B. Baker, T. O'Connor, T. Curtis, P. Dudley, S. Farix, D. Gaiter, T. Gaiter, C. Aistin, P. Hennessy, P. Kennedy, P. Norris, S. Lofthill, G. Roberts, T. Plant, C. Rourke, J. Walsh, G. Wilber, J. Brown, H. Cooke, J. Goulding, G. Bromley, M. Browne, J. Biercliffe.

Missing: Privates T. Walker, W. Sumbling, M. Collins.

Wounded: Acting-Sergeant-Major D. Bell, Colour-Sergeants J. Magner, W. Miller, Sergeants J. Dodds, G. Hogan, W. Murphy, P. Horristall, J. Sherlock, J. Colgan, Lance-Sergeants J. McFaggart, G. Robins, P. Maher, Corporals D. Doyle, E. Doyle, T. Farrer, M. Murphy, A. Newall, H. Strick, C. Ventham, T. Robins, W. Smith, E. Collins, E. Harden, T. Keating, J. Gibson, Drummer H. Hanton, Privates W. Bollen, J. Doorley, J. McCugh, J. Buro, M. Conway, T. Woodman, J. Caldwell, J. Hickey, J. Darkin, J. Hourighaid, J. McDonald, R. Campbell, W. Box, J. Reynolds, W. Ainge, J. Gilgason, P. Bulger, M. McNamara, F. Chivers, R. Lynch, J. Griffin, P. O'Hara, W. Mitchill, C. Collins, F. Didman, P. Clunke, J. Halloran, P. Murphy, J. Murphy, F. Dolan, J. Tillman, J. Gorman, J. Anderson, F. Riggs, T. Wardle, J. Harley, A. McRoberts, W. Ingram, D. Cooper, G. Gilpin, F. Nelson, H. Anscombe, J. Ralph, J. Bernan, H. Neil, R. Buck, J. Robinson, S. Frost, J. Rogers, F. Stillman, P. Donelly, J. Clarke, J. Hutchings, T. Jones, J. Beer, W. Cormick, G. McDonald, J. Brodrick, J. Walsh, J. Herrity, S. Evans, M. Daly, J. Brown, D. Shea, W. Clarke, G. Rixson, J. Grill, J. Murphy, H. Simms, J. Duffy, H. Dasson, W. Lynch, T. Brown, M. Geary, J. Hazel, S. Shea, A. Handly, J. Healy, J. Lawrence, P. Brophy, G. Nihill, J. McHugh, A. Levi, M. Mack, W. Shaw, J. Finsley, J. Carsen, J. O'Connor, J. Dacey, W. Chadwick, D. Moriarty, F. Barnard, A. Hanley, J. Mexal, W. Holland, T. Behan, T. Norwell, M. Faulkner, P. Daley, J. Burke, G. Thaxted, D. Lyons, M. Stafford, T. Toolair, J. Connelly, J. Dobbin, W. Fogarty, E. Murray, P. McNamara, H. Stevens, T. Williams, T. Ash, J. Madden, R. Abbott, T. Simpson, J. Sherridan, J. Mooney, A. Saittings, J. Griffin, J. Green.

Wounded by the explosion of a magazine in camp, 15th November, 1855: Corporal J. Doorly, Privates J. O'Brien, J. Caldwell, R. Sommerville, J. Gearedd, J. Stillham, E. Herwood, W. Beech.

DIED ON SERVICE IN EGYPT AND THE SOUDAN—1884-1887.

OFFICERS:

Captain and Bt.-Major F. B. Briggs	Cairo	8th September, 1885.
Lieutenant W. M. Douglas	Korosko	22nd February, 1886.

OTHER RANKS:

Private Thomas Jennison	Ramleh	25th August, 1884.
Private Henry Taylor	,,	29th September, 1884.
*Private Thomas McDonnell	,,	16th October, 1884.
*Private Emmerson Clarke	,,	29th October, 1884.
*Private Henry Bell	Port Said	6th December, 1884.
*Private George Exley	Ramleh	8th January, 1885.
†Lance-Corporal Thomas Bushby	Sohag, Nile	13th March, 1885.
†Private John Kelly	,,	13th March, 1885.
*Private Henry Marshall	Assouan, Nile	12th April, 1885.
*Private Samuel Wright	Suakin	29th April, 1885.
*Quartermaster-Sergeant James Park	Assouan, Nile	12th May, 1885.
Private Thomas Connor	,,	15th May, 1885.
*Private David Wood	,,	22nd May, 1885.
*Private Thomas Kay	,,	30th June, 1885.
Private Enoch Hirst	,,	18th July, 1885.
Private John Shaw	,,	23rd July, 1885.
*Private George Chapman	,,	29th July, 1885.

Private Thomas Lowe	Assouan, Nile...	1st August, 1885.
Private John Hanlon	,, ...	1st August, 1885.
*Private John Robinson...	,, ...	10th August, 1885.
Private Harry Baker	,, ...	12th August, 1885.
Private Phillip North	,, ...	18th August, 1885.
*Sergeant William Ronksley	,, ...	24th August, 1885.
*Private James O'Neil	Ramleh ...	30th August, 1885.
*Private Aquilla Mansley	Assouan, Nile...	19th October, 1885.
*Private George Gardner	Korosko, Nile...	31st October, 1885.
*Private John Lane	Ramleh ...	4th November, 1885.
†Private Thomas Dougherty	Nile	9th November, 1885.
Private Peter Fox	Korosko, Nile...	25th November, 1885.
*Corporal J. J. O'Connor	,, ...	2nd December, 1885.
Sergeant George Hargreaves	,, ...	3rd December, 1885.
*Private Joseph Edwards	,, ...	5th December, 1885.
Private A. Taylor	,, ...	8th December, 1885.
*Private John Banks	Wady Halfa, Nile	17th December, 1885.
Private J. Taylor	Korosko, Nile...	18th January, 1886.
†Private James Evans	Nile	18th January, 1886.
*Private Charles Applin...	Korosko, Nile...	12th February, 1886.
†Corporal William Harley	Nile	10th February, 1886.
Private Peter McHugh...	Korosko, Nile...	22nd February, 1886.
*Private Arthur Barton	,, ...	13th March, 1886.
*Private John Johnson	,, ...	14th March, 1886.
*Private William Wedgwood	,, ...	18th March, 1886.
*Private James Franklin	,, ...	21st March, 1886.
Private George Hunter	,, ...	22nd March, 1886.
*Private Lewis Cook	,, ...	27th March, 1886.
*Private Isaac Collins	,, ...	5th April, 1886.
*Boy Arthur Seymour	,, ...	14th April, 1886.
*Corporal Percy Shaw	,, ...	14th April, 1886.
*Private James Thompson	,, ...	16th April, 1886.
*Private Charles Godfrey	,, ...	16th April, 1886.
*Lance-Corporal John Chapman ...	,, ...	18th April, 1886.
*Private Edward Pearce	,, ...	22nd April, 1886.
Private John Hird	,, ...	25th April, 1886.
*Private William Clayton	,, ...	29th April, 1886.
*Private Thomas Downes	,, ...	30th April, 1886.
†Private William O'Rorke	Kalabshi, Nile	2nd May, 1886.
†Private John Smith	,, ...	2nd May, 1886.
*Private William Hammond	Assouan, Nile...	3rd May, 1886.
*Private George Barkway	,, ...	4th May, 1886.
*Private Arthur Smith	Korosko, Nile...	5th May, 1886.
Private George Knight	Assouan, Nile...	15th May, 1886.
*Private Frederick Black	,, ...	22nd May, 1886.
*Private William Gallagher	,, ...	23rd May, 1886.
*Lance-Corporal Robert Newbegin ...	,, ...	26th May, 1886.
*Private William Knowles	,, ...	27th May, 1886.
*Private Henry Sharp	,, ...	30th May, 1886.
*Private Albert Roberts	,, ...	30th May, 1886.
*Private George Saville	,, ...	31st May, 1886.
*Private Frederick Miller	,, ...	1st June, 1886.
*Private Frederick Judson	,, ...	5th June, 1886.
*Private George Dobson	,, ...	11th June, 1886.
Private James Higham	,, ...	14th June, 1886.
Private Joseph Jordan	,, ...	16th June, 1886.

Private Thomas Lee	Cairo	16th June, 1886.
Private George Turner	Assouan, Nile	18th June, 1886.
Private Thomas McGowen	,,	18th June, 1886.
Private Lawrence Doolan	,,	21st June, 1886.
†Private Patrick Feeley	,,	24th June, 1886.
Lance-Corporal John Jones	Cairo	1st July, 1886.
*Private Charles Longden	Assouan, Nile	3rd July, 1886.
Private Henry Deeks	Assiut, Nile	16th June, 1886.
Private Walter Hallett	Assouan, Nile	23rd July, 1886.
†Private Thomas Harris	,,	30th July, 1886.
Private John Beadle	,,	18th August, 1886.
Private George Siddle	,,	8th September, 1886.
*Corporal Ernest Garner	Ramleh	13th September, 1886.
Corporal Charles Oliver	,,	8th November, 1886.
Private Harry Heckingbottom	Assouan, Nile	25th December, 1886.
Private J. Brown	,,	7th January, 1887.
Private E. Fawcett	,,	31st January, 1887.
Private J. Birkenshaw	,,	20th February, 1887.
Private A. Jenkins	Alexandria	23rd March, 1887.
Lance-Corporal Frederick Povey	,,	15th July, 1887.
Private John Foster	Ramleh	27th October, 1887.
Corporal William Purvis	Alexandria	31st December, 1887.

Of the above, those marked with * died of enteric fever, and those marked with † were drowned in the River Nile.

THE TIRAH CAMPAIGN.

CASUALTIES IN ACTION.

OFFICERS.

Killed:

22nd November, 1897, Lieutenant D. E. O. Jones, at Dwatoi.

Wounded:

2nd November, 1897,	Lieutenant E. G. Caffin, at Bagh.
22nd ,, ,,	Second-Lieutenant O. C. S. Watson, at Dwatoi.
29th ,, ,,	Lieutenant B. C. Williams, in the Khanki Valley.

NON-COMMISSIONED OFFICERS AND MEN.

Killed:

23rd October, 1897,	Private J. Mulroy, at Kharuppa.
23rd ,, ,,	Private E. Phillips, at Kharuppa.
18th November, 1897,	Sergeant W. House, at Bagh.
18th ,, ,,	Private H. Smithson, at Bagh.
26th ,, ,,	Private A. Hine, in the Bazar Valley.

Died of Wounds:

30th October, 1897,	Sergeant T. Bromwich, at Gundaki.
27th November, 1897,	Private C. Matthews, in the Khanki Valley.
27th ,, ,,	Sergeant W. Calvert, in the Khanki Valley.
26th December, 1897,	Private J. Carling, in the Bazar Valley.

Wounded:

23rd October, 1897,		Private W. Watson, at Karuppa.
23rd	,, ,,	Private G. Wheatley, at Karuppa.
23rd	,, ,,	Private C. Ferguson, at Karuppa.
23rd	,, ,,	Private J. Tobin, at Karuppa.
29th	,, ,,	Private G. Cullen, at Sampagha Pass.
29th	,, ,,	Private W. Kent, at Sampagha Pass.
29th	,, ,,	Private H. Scarborough, at Sampagha Pass.
29th	,, ,,	Private S. Rix, at Sampagha Pass.
16th November, 1897,		Private R. Kirke, at Maidan.
16th	,, ,,	Sergeant J. Hornsby, at Maidan.
16th	,, ,,	Private D. Reardon, at Maidan.
18th	,, ,,	Colour-Sergeant B. Wyatt, at Bagh.
18th	,, ,,	Private H. Pawson, at Bagh.
18th	,, ,,	Private J. Wright, at Bagh.
18th	,, ,,	Private A. Poole, at Bagh.
18th	,, ,,	Private F. Johnson, at Bagh.
18th	,, ,,	Private E. Donovan, at Bagh.
19th	,, ,,	Private J. Elliot, at Bagh.
22nd	,, ,,	Lance-Corporal F. Brunton, at Dwatoi.
28th	,, ,,	Private G. Williams, in the Khanki Valley.
28th	,, ,,	Private D. Simpson, in the Khanki Valley.
29th	,, ,,	Sergeant B. Richardson, in the Khanki Valley.
29th	,, ,,	Lance-Corporal W. King, in the Khanki Valley.
29th	,, ,,	Lance-Corporal D. Perry, in the Khanki Valley.
29th	,, ,,	Private J. Turner, in the Khanki Valley.
29th	,, ,,	Private W. Connell, in the Khanki Valley.
26th December, 1897,		Sergeant W. Roche, Bazar Valley.
26th	,, ,,	Private P. Feeney, Bazar Valley.

Died from other causes:

NON-COMMISSIONED OFFICERS AND MEN.

Sergeant H. Jackson, Lance-Corporal J. Clarke, Privates J. Addison, E. Beaumont, T. Mortemore, H. Sykes, J. Coates, J. Bentley, J. Wilson, T. E. Benson, W. Brophy, G. Jackson, I. Senior, R. Ricketts, H. Ward, J. Morton, J. Pickles, G. Brown, H. Gibson, T. Cummings, W. Mansell, A. Richinson, S. A. Wilson, S. Holliday, A. Thwaites, J. King, W. Heppinstall, A. Wilson, G. Bridges, G. Smith, E. Knight, W. Carr, E. Howbrigg, W. Emms.

THE BOER WAR, 1899-1902.

AT SLINGERSFONTEIN, CAPE COLONY, 15th January, 1900.

Wounded: Major M. H. Orr.

Killed: Colour-Sergeant F. Roberts, Sergeant D. Jamieson, Privates A. Smith, B. Duffy, E. Ward.

Wounded: Privates J. Biggs, S. Clough, F. Evans, C. Flood, J. Hayes.

AT KLIP DRIFT, O.F.S., 16th February, 1900.

Wounded: Second-Lieutenant W. G. Tarbet, 4th M.I., Private W. Hamley, 4th M.I.

AT PAARDEBERG, O.F.S., 18th February, 1900.

Killed: Second-Lieutenant A. C. Neave, Colour-Sergeant P. Hughes, Sergeant R. Tate, Corporals J. H. Kearns, T. Frankland, Lance-Corporal D. Horton, Bandsman T. W. Davis, Privates J. Brown, F. Carley, C. Chaplin, W. Diplock, S. Greetham, C. Harrison, E. Horton, H. Johnson, F. Jones, T. Langan, J. McHale, W. Mulligan, H. Newton, O'. O'Shaughnessy, A. Robinson, H. Royston, G. Smith, G. Sowerby, R. W. Tinkler, T. Trinder, F. Turvey, W. Ward, J. Watson, J. Webb.

Wounded: Lieutenant-Colonel H. Bowles, Major T. D. Kirkpatrick, Captain A. C. Buckle (South Staffordshire Regiment, attached), Lieutenant C. V. Edwards, Sergeant A. Atkinson,* Corporals J. Collins, J. Henderson, G. H. King, W. Stalker, W. Wiles, Lance-Corporals J. Barrett, J. Dullingham,* A. Hatton, W. Holdsworth, H. Venison, Drummer A. Coombs,* Privates J. Ainsley, E. Albany, W. Allen,* H. Baggaley,* J. Barker, P. Bartley, J. Benson,* J. Best, F. Bingley, W. Booth, E. Brown, A. Bruce,* J. Carson, T. Cave, G. Challis, F. Cole, C. Coldwell,* N. Cooling, R. Donald, J. Elliott, W. Fowler, S. Francis, R. Hand, H. Harrison, J. Hellings, A. Hopwood, W. Ingleson, J. Jefferson, T. Kielty J. Kennedy, J. Lilly, G. Lonsdale, T. Lown, J. Lupton, J. Marchant, J. McGann, J. Mee, W. Meehan, F. Moore, J. Mullan, W. Murray, R. Nixon, G. Nolan, J. Parker, H. Parker, H. Plenty, H. Powell, C. Raymond, L. Ricci, C. Riley, J. Robinson, W. Rodgers,* E. Root, W. Rose, W. Rushworth,* J. Sharpe, J. Simmonds, W. Skipsey, J. Smith, G. Spence, H. Satters, T. W. Swift, T. Taylor, J. Prest, S. Thompson, J. Williams, T. Wimbleton, J. Windsor, J. Winzar, J. Woodman, J. Parker, J. Harrison, J. Lacy, J. Handley, C. Rawlings, W. Bradbury, H. Garside, J. Burroughs, M. Gargon, J. Wood, H. Rolfe, G. London, D. Jones.

Those marked * died of their wounds.

Missing: Privates Barrett, Bowler, and Poole.

These men were taken prisoners and were released at Pretoria.

AT KITCHENER'S KOP, PAARDEBERG, 23rd February, 1900.

Killed: Sergeant B. Richardson, Privates J. Clarke, J. Cram, A. Grigg, R. Hardy, R. Raw, G. Scruton, J. Wilkinson.

Wounded: Captain G. Pearson, Lieutenant M. E. T. Gunthorpe, Second-Lieutenant E. V. L. Wardle, Corporals A. Dutfield, W. Taylor, Privates C. Alton, P. Bartley, J. Bailey, H. Buckle, J. Burke, G. Burrows, M. Gibson, E. Wilson, T. Jackson, G. Jessop, J. Holmes, A. Hunter, R. Ruddy, R. Rusted, J. Christie, J. Williams.

AT DRIEFONTEIN, O.F.S., 10th March, 1900.

Killed: Private G. Birks.

Died of Wounds: Privates J. Bolton, D. Smith.

Wounded: Sergeants F. C. Hatton, H. Mackay, Privates J. Castle, W. Jones, J. Rimmington, J. Wootton, P. Carroll, B. Keeley, F. Brown, F. Townsend, J. Bennett, J. Lynch, J. Watson, J. Davis, F. C. Burgess, J. Lister, M. Hughes, E. Bennett, J. Lockwood, G. Newton, G. Woodhead, J. Clancy, J. Duffy, W. Wilson.

AT SANNAS' POST, O.F.S., 31st March, 1900.

Wounded: Lance-Corporal A. Carter, Privates E. Holland, J. Gamble (all of the 4th M.I.).

NEAR THABA N'CHU, O.F.S., 27th April, 1900.

Killed: Private C. Humphries (2nd M.I.).

AT KAREE KLOOF, O.F.S., 30th April, 1900.

Wounded: Private E. Simpson (4th M.I.).

AT PRETORIA, 4th June, 1900.

Wounded: Private T. Hicken.

AT BALMORAL, TRANSVAAL, 12th August, 1900.

Killed: Private W. H. Varlow (4th M.I.).

AT BELFAST, TRANSVAAL, 24th August, 1900.

Wounded: Lieutenant W. G. Tarbet, Privates W. Hawkins, F. Axup, P. Broker, W. Keighley, J. Ryan (all of the 4th M.I.).

AT VILJOEN'S DRIFT, O.R.C., 6th November, 1900.

Wounded: Private J. Daley (2nd M.I.).

AT KLIPPLAAT'S DRIFT, TRANSVAAL, 16th July, 1901.

Wounded: Private C. F. White (4th M.I.).

AT HOLFONTEIN, O.R.C., 29th July, 1901.

Died of Wounds: Private D. Hawthorne (21st M.I.).

AT HONING'S KLOOF, TRANSVAAL, 30th July, 1901.

Wounded: Privates F. Hainstock, H. Dearman (both of the 4th M.I.).

AT KRUGER'S POST, TRANSVAAL, 3rd October, 1901.

Wounded: Drummer J. Tiller, Privates R. Winkle, J. Green (all of the 4th M.I.).

IN THE ORIGSTAD VALLEY, TRANSVAAL, 4th October, 1901.

Wounded: Lieutenant M. R. Liddon (4th M.I.)

AT JORDAN SIDING, O.R.C., 14th December, 1901.

Wounded: Sergeant-Instructor of Musketry W. Moy (P.S. 3rd Battalion).

NEAR SPRINGS, TRANSVAAL, 25th April, 1902.

Wounded: Private J. Wardle.

AT PAARDEPLAATS, TRANSVAAL, 10th May, 1902.

Wounded: Private J. Greenwood (4th M.I.).

Died from various causes during the campaign: Colour-Sergeants H. Battye,* G. Stephenson,* Band-Sergeant W. Hall, Corporals J. Nagginton, A. Stephenson (vol.), J. Bailey, G. Collings, D. Duffy, B. Smithers,* Lance-Corporals W. Stead, H. Hobson, J. McCardle, R. Stanton, R. Johnson, J. McHale,* Drummer H. Haines,* Privates J. Ackroyd,† G. Appleton, H. Adamson,† J. Butterworth,† W. Baines, C. Bryant, O. Birmingham, A. Benson, J. Frost (vol.), H. Fitton, A. Gosling, R. Goodwill, A. Giles, W. Gibbon, J. Gains, J. Newton, J. Nixon, E. Outhwaite (vol.), J. Pinnock, H. Portbury, T. Burton,† J. Cavanagh, G. Carlile, J. Conroy, J. Cameron, J. Carpenter, P. Carroll, J. Dibble,* A. Woollett, R. Rolfe, F. Rouse, W. Rose, W. Simmons, J. Smith, J. Stephenson, G. Sullivan, A. Sollars,* W. Sykes, W. Tighe, J. Trickett,* C. Turton, A. Smith,† G. Wilson, A. Wilkinson, G. Webster, H. White, R. Walker,* J. Pullan, E. Perkins,† J. Purvis, J. Hagen, F. Horton, A. Dawson, J. Kenney, H. Lane, G. Lofthouse, R. Leeming, H. Marrow, J. Milner (vol.), C. Matthews, J. Pease,* J. Pattison, G. Pye,* M. Dunn (vol.), P. Foley,* J. Stubbs (vol.), H. Webb, E. G. Worne,* J. Dunford,† J. Dixon.

Those marked with * were attached for duty to the 3rd Battalion. The men marked † belonged to the 3rd or 4th Battalion and were attached to the 1st.

Missing, presumed to be dead: Privates T. Dunn, J. Swales.

Died on passage home from South Africa: Corporals J. Buxton, A. Tribe.

Died on passage to South Africa: Private Sewell (Vol. Co.).

"In Nature's hallowed keeping safe; no need there be to mourn,
The Guard dismounts at sunrise and the Sentries watch the dawn;
Yet, England, through the watches of the long calm night serene,
For him, in thine own honour bound, oh, keep thy memory green."

REGIMENTAL PLATE.

As far as can be ascertained most of the ordinary plate of the 1st Battalion was purchased on the return of the regiment to England in 1820. It is improbable that any plate existed before this time, as the battalion was practically on active service during all its tour abroad from 1796 to the above date. From the various hallmarks it would appear that all the beautiful candelabras, wine coolers, soup tureens and ordinary plate in daily use belong to the latter part of the George III. period.

The plate was added to considerably in 1832, probably on an increase in the establishment of officers. The most valuable piece is a solid silver epergne candelabra.

The earliest presentation plate now in the mess is the cigar lighter given by Lieutenant-Colonel Mundy in 1857, but it is well known that the officers were in possession of several cups and valuable snuff-boxes prior to the Crimean War, and that these were all stolen from the depôt at Walmer before the battalion returned home.

After the presentation of new Colours by H.R.H. the Princess of Wales in 1875, a centre piece was bought by the officers to mark the occasion. It represents the Princess in the act of presenting one of the Colours, and is considered a very fine piece of modelling. On the ebony stand are silver tablets, one showing the actual ceremony and another the Nineteenth crossing the River Alma, whilst the remaining two sides are devoted to the inscription and crest of the regiment. The centre piece rests on a mirrored plaque edged with Sheffield plate, made up of six pieces, in all about nine feet in length and over sixteen inches in width.

The following is a list of the presentation plate:—

DATE.	ARTICLE.	PRESENTED BY.
1857	Cigar Lighter	Lieutenant-Colonel Mundy, C.B., on joining.
1860	Tankard	Lieutenant-Colonel Mundy, on promotion.
1869	Tankard	Colonel R. O. Bright, C.B., on leaving.
1871	Goblet	Captain P. D. Vigors.
1871	Goblet	Captain W. Bennett.
1872	Cup	Lieutenant-Colonel H. de R. Pigott.
1872	Goblet	Colonel W. S. Cooper, on appointment.
1874	Goblet	Captain G. A. Skipton.
1874	Goblet	Captain F. O. Openshaw.
1876–1877	Tankard	Lieutenant C. J. Spottiswoode and Lieutenant H. Bowles, on promotion.
1877–1878	Tankard	Lieutenant G. H. Mathison and Lieutenant J. A. Fearon, on promotion.
1878	Tankard	Captain W. G. McClintock, on promotion.
1878	Cup	Officers U.S.A. Training Frigate "Saratoga," Bermuda, 19th July, 1878.
1878–1879	Tankard	Lieutenants C. S. Molony and G. C. S. Handcock, on promotion.
1879	Tankard	Major A. M. Handley, on promotion.
1879	Tankard	Lieutenants W. O. Chauncy and A. de S. Hadow, on promotion.
1879	Tankard	Captain J. O. Moller, on exchange.
1880	Tankard	Lieutenants A. G. Cartwright and J. T. Cotesworth, on promotion.
1880	Tankard	Captain F. B. Briggs, on promotion.
1880	Tankard	Captain R. Phayre, on promotion.
1881	Tankard	Lieutenants M. H. Orr and M. L. Ferrar, on promotion.
1881	Tankard	Lieutenant-Colonel C. Hereford, on promotion.
1881	Tankard	Lieutenants W. L. Mercer, and A. T. Newnham, on promotion.
1881	Tankard	Major W. Bennett, on promotion.
1881	Tankard	Lieutenants H. O. Hickman and G. Pearson, on promotion.
1882	Tankard	Lieutenants G. N. Prendergast and C. A. King, on promotion.
1882	Tankard	Lieutenant-Colonel J. Jameson, on retirement.
1883	Tankard	Captain J. H. Eden, on promotion.
1886	Milk Jug & Sugar Bowl	Lieutenant C. W. Gale, on appointment.
1887	Cream Jug	Lieutenant E. Somervell, on appointment, 1885.
1889-1892	Five Menu Holders ...	Lieutenants H. A. Stansfeld (1889), C. M. Kemble, A. F. Robinson and W. A. F. Williamson (1890), H. G. L. Corbett (1892), all on appointment.
1889	Cigar Lighter	Lieutenant G. Pearson.
1890	Dinner-table Bell ...	Lieutenant P. Lees, on joining.
1890	Two Fruit Dishes ...	Major J. W. Parker, on promotion.
1890	One Fruit Dish ...	Major C. J. Spottiswoode, on promotion.
1890	One Fruit Dish ...	Captain G. Pearson, on promotion.
1891	Coffee Pot	Lieutenant M. H. Tomlin, on promotion.
1891	Two Fruit Dishes ...	Lieutenant A. F. Robinson, on joining.
1891	Two Fruit Dishes ...	Captain W. L. Mercer, on promotion and marriage.

DATE.	ARTICLE.	PRESENTED BY.
1892	Two Fruit Dishes ...	Captain E. J. Buckle, on promotion.
1893	Claret Jug	Lieutenant-Colonel E. A. Bruce, on promotion.
1895	Table Dinner Bell ...	Captain the Hon. J. G. Byng, 10th Hussars.
1895	Cup	Captain D. L. Hartley, on joining.
1895	Statuette of Officer, 1830	Lieutenant F. H. Hodge, on retirement.
1896	Cup won at Light-weight Point-to-Point, Portsmouth, 1890, by Lieutenant F. Connop.	Mrs. N. Connop, in memory of Captain F. Connop, 1884-1896.
1898	Cup	Major J. T. Cotesworth and Captain B. C. Williams, on promotion.
1896	Four small Dessert Bowls	Lieutenant M. E. Gunthorpe, on joining.
1897	Statuette of Drummer, 1747	Lieutenant H. A. Stansfeld, on his marriage, 22nd September, 1897.
1898	Statuette of Officer, 1779	Captain W. L. Mercer, on retirement.
1902	Cup	Captain M. L. Bell, Special Service Company, In remembrance of South Africa, April to November, 1900.
1903	Gold Snuff Box ...	Captain B. C. Williams, on retirement.
1903	Goblet	Captain H. J. Charrington.
1903	Statuette of Private, representing 1900-1902, S.A. War period.	Second-Lieutenants A. C. Pearson, B. Nash-Wortham, S. C. Whatford, J. H. Westley, B. L. Maddison, C. E. Moss-Blundell, and G. P. Stevens, on appointment.
1904	Three Cream Jugs ...	Captain W. G. Tarbet, D.S.O., on promotion to Worcestershire Regiment.
1905	Milk Jug	Captain H. F. Lea, on appointment.
1907	Coffee Pot	Major M. H. Tomlin, on promotion.

The following cups and challenge shields are also in possession of the battalion:—

Silver Musketry Challenge Shield, purchased in 1881.

Lord Roberts' Musketry Cup, won at the Curragh, 1897.

The Morrison Cup, won at Gibraltar, 1898-1899.

Bayonet Fighting Cup (replica), Royal Military Tournament, for dismounted officers, 1904. Team: Lieutenant Morgan, Second-Lieutenants Godwin, Cumberbatch, Gilbert and Gore.

The Battalion Athletic Shield, purchased in 1905.

The Marlborough Cup, Aldershot Command Rifle Meeting, 1906, "B" Company. Captain of team: Lieutenant Leatham.

Officers' Challenge Cup, 3rd Prize, Aldershot Command Rifle Meeting, 1906. Team: Lieutenants Marsden and Leatham, Second-Lieutenants Mallinson and Grant-Dalton.

Officers' Challenge Cup, 2nd Prize, Aldershot Command Rifle Meeting, 1907. Team: Captain Leatham, Lieutenant Marsden, Second-Lieutenants Grant-Dalton and Smith.

Evelyn Wood Cup (miniature), for shooting and marching. Open to the Army. Company teams. 1907. "B" (Captain Leatham's) Company.

The Gunthorpe Challenge Cup (for musketry), presented by Colonel E. J. Gunthorpe in memory of his son, the late Captain M. E. Gunthorpe, 1896-1907.

Officers' Cup, Cairo Command Rifle Meeting, 1910, 1st Prize. Team: Captain C. V. Edwards, Lieutenant S. Grant-Dalton, Second-Lieutenants Smith and Bradford.

In the Officers' Mess there is a collection of war medals, which have been either bought or presented. They represent the campaigns of late years in which the battalion has taken part. These comprise:—

The Victoria Cross and Legion of Honour of Private John Lyons, together with his photograph (a daguerrotype) in an antique gilt frame, the latter presented by Captain M. L. Ferrar.
Crimean Medal and 3 clasps (Alma, Inkerman, and Sebastopol).
The French War Medal (awarded to ten men per regiment).
The Sardinian Medal (Sergeant John Halloran).
The Turkish Medal.
North-West Frontier of India, 1868.
The Egyptian Medal, 1885-1886.
The Khedive's Bronze Star.
The Queen's South African Medal and 6 clasps.
The King's South African Medal and 2 clasps.
Medal for distinguished conduct in the field.
The Medals of the late Private S. Evans, V.C., left by him to the regiment. These include:—
- The Victoria Cross.
- The China Medal, 1842.
- The French War Medal.
- The Crimean and 3 clasps.
- The Turkish Medal.

Other Medals of interest are the following commemorative ones in silver and copper presented by Captain M. L. Ferrar:—

Battle of Aughrim, 1691.
Sieges of Galway, Limerick, &c., 1691
Battle of Steenkirke, 1692.
Battle of Landen, 1693.
Siege of Namur, 1695.
Battle of Malplaquet, 1709.
Sieges of Douay, Bethune, &c., 1710.
Siege of Bouchain, 1711.
Battle of Fontenoy, 1745.
Battle of Roucoux, 1746.
Battle of Lauffeldt, 1747.
Siege of Belle Isle, 1761.

The battalion also possesses five Russian Drums taken at the Alma, one of which was presented by Mrs. Robert Warden. The Colour Belts used in the Crimea have been preserved, and the one which Ensign Stockwell wore when killed at the Alma has been suitably mounted. The latter, as well as the drums, are used on the annual Trooping of the Colour on the anniversary of the Alma.

In the same case is the Sword Belt with its handsome badge worn by the late Major Lidwill all through the Crimea, and presented by him to his old regiment.

PICTURES.

SUBJECT.	PRESENTED BY.
Two signed Photographs of H.R.H. The Princess of Wales	H.R.H. The Princess of Wales, 1875 and 1890.
Colonel Francis Luttrell (photograph) ...	G. F. Fownes-Luttrell, Esq., of Dunster Castle.
General Thomas Erle (photograph) ...	Captain M. L. Ferrar.
Major-General the Hon. Sir C. Howard, K.B. (oil painting)	Lieutenant A. F. Owen Lewis.
Richmond Castle (artist proof engraving) ...	Lieut.-Col. E. S. N. Dickenson.
T.R.H. Prince Albert Victor and Prince George of Wales (signed photographs)	The Princes at Bermuda, 1879.
Wellington Crossing the Pyrenees	——
The Dinner Round Light Co., 19th Foot	S. M. Milne, Esq.
Field-Marshal Sir Samuel Hulse, G.C.B. (engraving)	Captain M. L. Ferrar.
General Sir Hew Dalrymple (engraving)...	Captain R. H. Darwin.
Field-Marshal Sir Charles Rowan, G.C.B. (pencil drawing)	Colonel Rowan, of Mount Davy's, co. Antrim.
General Sir R. O. Bright, G.C.B. (photograph)	General Bright.
Major-General C. Wills (engraving) ...	Lieutenant W. J. Roskell.
Captain G. Lidwill (coloured photograph)...	Major G. Lidwill.
Lieutenant W. Mercer, 19th Foot (painting)	His Grandson, Captain W. L. Mercer, 19th Foot.
The Battle of the Boyne (engraving) ...	Captain M. L. Ferrar.
An Infantry Charge	Lieutenant F. Connop.
Three artist proof engravings by de Neuville: Incidents in the Franco-German War	Second-Lieut. A. F. Robinson, 1890.
Lieutenant-General E. Chippindall, C.B. (photograph)	Lieutenant-General Chippindall.
The Duke of Wellington and Lord Nelson (coloured prints with autographs)	Lieut.-Colonel G. Egerton, C.B.
H.R.H. The Duke of Connaught (signed photograph)	The Duke of Connaught, 1892.
Framed relics of the Colours carried in the Crimea	——

2ND BATTALION—PRESENTATION PLATE.

DATE.	ARTICLE.	PRESENTED BY.
1870	Silver Snuff-Box ...	Major P. D. Vigors.
1876	Two Burmese Bowls...	Colonel Chippindall, C.B.
1877	Gold Snuff-Box and Case	Bequeathed by Lieutenant Abney Hastings Cameron, 2nd Battalion XIX. Regiment, in which he served 15 years. This relic fell into the hands of General Cameron (father of the above) who was present and wounded at Waterloo, and was a gift from the Emperor Napoleon to Marshal Ney, from whose carriage it was taken on the field.
1879	Silver Goblet	Lieutenant Villiers Sankey.
1880	Silver Goblet	Lieutenant A. B. Elton.
1880	Silver Goblet	Colonel H. Cook.
1881	Claret Jug	Captains A. C. Fryer and A. Money Kyrle.
1881	Hand Bell	Major C. Barton.
1881	Silver Goblet	Lieutenant Percy Lees.
1881	Silver Goblet	Lieutenant John Parker.
1881	Silver Goblet	Captain J. F. Simonet.
1881	Silver Goblet	Captain W. E. Franklyn.
1881	Silver Goblet	Captain Crosbie Barton.
1881	Silver Goblet	Lieutenant W. J. Todd.
1881	Silver Goblet	Lieutenant E. L. Herapath.
1881	Silver Goblet	Lieutenant T. D. Kirkpatrick.
1881	Silver Tankard ...	Major A. J. Paterson and Major G. H. Reynolds.
1881	Cigar Lighter	Lieutenant Villiers Sankey.
1882	Cigarette Box	Detachment 16th Lancers.
1882	Silver Goblet	Lieutenant E. W. Mills.
1882	Challenge Cup ...	For musketry. By the Officers.
1882	Menu Holder	Lieutenant C. L. Nicholson.
1882	Cigar Box	Colonel H. E. Davidson.
1884	Menu Holder	Lieutenant W. S. Brooksbank.
1885	Silver Goblet	Captain E. W. Mills.
1885	Cigar Lighter	Captain C. T. Hennah.
1885	Cigar Lighter	Lieutenants B. C. Hood and D. L. Hartley.
1885	Menu Holder	Lieutenant E. S. Bulfin.
1885	Cigar Cutter	Lieutenant Hooper, 33rd Punjabis.
1886	Inkstand	Lieutenant Spankie, Royal Irish Rifles.
1886	Salad Bowl	Colonel H. E. Davidson.
1886	Sugar Caster	Lieutenant-Colonel G. Oakes.
1886	Silver Goblet	Major A. W. Money Kyrle.
1887	Menu Holder	Lieutenant G. Christian.
1887	Cigar Cutter	Signalling Class, Aldershot.
1887	Two Cruet Stands ...	Major H. Bowles.
1891	Fruit Dish	Captain W. J. Todd and Lieutenant E. L. Vans-Agnew.
1891	Fruit Dish	Second-Lieutenants Jones, Caffin, and Bond.
1891	Fruit Dish	Lieutenant R. D'A. Fife.
1892	Madras Bowl	Lieutenant A. L. Napier.
1894	Two Burmese Bowls...	Captain H. F. Lea.
1896	Challenge Shield for musketry	Lieut-Colonel W. G. McClintock.
1897	Milk Jug	Lieutenant W. L. Alexander.

DATE.	ARTICLE.	PRESENTED BY.
1898	Sugar Bowl	Captain B. C. W. Williams.
1898	Punch Bowl	Lieutenant-Colonel W. E. Franklyn.
1899	Claret Jug	H.H. The Rajah of Sirmoor.
1899	Cigarette Box ...	Captain Buchanan, Gordon Highlanders.
1900	Two Silver Salvers ...	Lieutenant W. J. Roskell.
1900	25 Finger Bowls ...	Lieutenant-Colonel C. J. Spottiswoode.
1901	Inkstand	Major Hore, R.A.M.C.
1904	Cruet Stand	Captain A. L. Napier.
1905	Horn Snuff Box ...	Lieutenant-Colonel Newcomen, Cawnpore Light Horse.
1906	Four Menu Holders ...	Lieutenants Ledgard, Peel and Cumberbatch.
1907	Silver Calendar ...	Major Brooksmith, R.G.A.
1909	Musketry Challenge Shield	Residents of Scarborough.
1910	Silver Barometer ...	Second-Lieutenant B. Cuff.
1911	Trophy	Residents of the North Riding of Yorkshire.

A handsome Silver Tankard commemorates the winning of the Inter-regimental Rifle Match in 1883.

PICTURES.

DESCRIPTION.	PRESENTED BY.
Colonel The Hon. Charles Howard (oil painting)	Captain A. L. Godman.
General Sir A. J. Cloeté, K.C.B.	H. Graham Cloeté, Esq.
General The Right Hon. Thomas Erle ...	Major G. Christian.
Polo at Hurlingham (engraving)	Lieutenant-Colonel G. Oakes.
Officers 2nd 19th at Aldershot (oil painting by Cecil Cutler)	Purchased.
Rorke's Drift (engraving)	Captain L. M. Levin.
General Sir Hew Dalrymple (engraving) ...	——
The Dinner Round, 19th Foot (coloured print)	Captain W. J. Roskell.
Frances I'Anson, "The Lass of Richmond Hill"	——
General Sir R. O. Bright, G.C.B.	——
Richmond Castle (engraving by David Law)	Purchased.
Pigsticking (four engravings)	Lieutenants C. J. Jeffery and G. C. Denton.
H.M. Queen Victoria (engraving)	Purchased.
General Sir A. Gaselee, K.C.B.	General Sir A. Gaselee.
Major-General W. S. Cooper	——
The Meeting of the Allied Generals before Sebastopol (engraving)	Purchased.
H.R.H. George Duke of Cambridge ... (engraving)	Purchased.
H.R.H. The Princess of Wales (Queen Alexandra)	——
"Scotland for Ever" (Lady Butler)... ...	——
Hounds in Full Cry (engraving)	Captain G. O'N. Segrave.
Lieutenant-General E. Chippindall, C.B. ...	——
The Bramham Moor Hunt (five coloured prints)	Lieutenant W. S. Brooksbank.

INDEX.

Printed and bound by CPI Group (UK) Ltd, Croydon, CR0 4YY
08/10/2026
02230073-0001